Lecture Notes in Electrical Engineering 1153

The book series *Lecture Notes in Electrical Engineering* (LNEE) publishes the latest developments in Electrical Engineering—quickly, informally and in high quality. While original research reported in proceedings and monographs has traditionally formed the core of LNEE, we also encourage authors to submit books devoted to supporting student education and professional training in the various fields and applications areas of electrical engineering. The series cover classical and emerging topics concerning:

- Communication Engineering, Information Theory and Networks
- Electronics Engineering and Microelectronics
- Signal, Image and Speech Processing
- Wireless and Mobile Communication
- Circuits and Systems
- Energy Systems, Power Electronics and Electrical Machines
- Electro-optical Engineering
- Instrumentation Engineering
- Avionics Engineering
- Control Systems
- Internet-of-Things and Cybersecurity
- Biomedical Devices, MEMS and NEMS

For general information about this book series, comments or suggestions, please contact leontina.dicecco@springer.com.

To submit a proposal or request further information, please contact the Publishing Editor in your country:

China

Jasmine Dou, Editor (jasmine.dou@springer.com)

India, Japan, Rest of Asia

Swati Meherishi, Editorial Director (Swati.Meherishi@springer.com)

Southeast Asia, Australia, New Zealand

Ramesh Nath Premnath, Editor (ramesh.premnath@springernature.com)

USA, Canada

Michael Luby, Senior Editor (michael.luby@springer.com)

All other Countries

Leontina Di Cecco, Senior Editor (leontina.dicecco@springer.com)

**** This series is indexed by EI Compendex and Scopus databases. ****

Zhongliang Jing · Xingqun Zhan
Editors

Proceedings of the International Conference on Aerospace System Science and Engineering 2023

Springer

Editors
Zhongliang Jing
School of Aeronautics and Astronautics
Shanghai Jiao Tong University
Shanghai, China

Xingqun Zhan
School of Aeronautics and Astronautics
Shanghai Jiao Tong University
Shanghai, China

ISSN 1876-1100 ISSN 1876-1119 (electronic)
Lecture Notes in Electrical Engineering
ISBN 978-981-97-0549-8 ISBN 978-981-97-0550-4 (eBook)
https://doi.org/10.1007/978-981-97-0550-4

This Springer imprint is published by the registered company Springer Nature Singapore Pte Ltd.
The registered company address is: 152 Beach Road, #21-01/04 Gateway East, Singapore 189721, Singapore

Paper in this product is recyclable.

Contents

Investigation into Provisions and Validation for Cockpit Smoke Evacuation on Civil Aircraft 1
Dan Wu, Zhi Yang, Xudong Yan, and Zhong Shi

SysML-Based Approach for Functional Quantitative Modeling of Civil Aircraft Systems 13
Shuijin Ye, Yong Chen, and Tianlong Wang

Research on MagicGrid-Based Requirements Development Process of Flight Control System 32
Hangqi Liu, Yong Chen, Meng Zhao, and Zhuo Yin

Airport Collaborative Decision-Making in Single Pilot Operations of Commercial Aircraft 52
Yue Luo, Miao Wang, Yong Chen, Kelin Zhong, and Guoqing Wang

Research on Capability Catalog Generation of UAV Intelligent System Based on DoDAF 61
Yuexing Wang, Xiaoxu Dong, Jiazhi Yan, and Miao Wang

Research on the Contribution Rate of Shipboard Manned/Unmanned Aerial Vehicle Cooperative Operation Based on Wargame Data Mining 75
Ling Peng, Xin Wang, Gechen Wang, Shitong Zhao, Miao Wang, and Wenjun Xu

Cooperative Organization and Application Mechanism Based on Intention Environment Target for Maritime Ship-Aircraft Cooperation 89
Jingshi Wang, Zehua Zou, Miao Wang, and Guoqing Wang

LPV Robust Filter Based Fault Diagnosis Method for Aeroengine Control System 104
Huihui Li, Linfeng Gou, and Zongting Jiang

A Research on the Intelligent Flight Deck Development Trend for the Civil Aircraft 111
Hua Meng, Qinpeng Dong, Chunling Zhao, Xianchao Ma, and Dayong Dong

Angle of Attack of the Aircraft Calibration ... 125
Hui Yang, Chao Xi, Zhengdi Zhu, and Yang Fang

Influence of Radius of Microphone Array on Virtual Rotating Array Beamforming for Rotating Noise Source Localization ... 134
Rachael Amoah, Cheng Wei Lee, and Wei Ma

A Finite Element Method of Research on Transverse Mechanical Properties of Fiber-Reinforced Composites with Random Matrix Void Defects ... 145
Wei Zhang, Yin Yu, Yile Hu, Haomin Li, and Zekai Zhao

A Review of Some Key Issues in CFD-Based Throughflow Simulation ... 159
Zhihao Hu, Qitian Tao, and Xiaohua Liu

Influence of Multiple Parameters on the Efficiency of a Single Nozzle in a Heavy Gas Turbine Combustor ... 178
Lei Hu, Fang Chen, Yu Meng, Zhenkun Sang, and Yuhuang Chen

Analysis of the Influence of Total Pressure Error on Air Data Calibration ... 192
Chao Xi

Analysis of Impact Properties of Nanocomposites at Micron Scale ... 200
Biao Yan and Zhongcheng Mu

Simulation Research on Air Distribution Optimization of Civil Aircraft Cabin Ventilation System ... 211
Rongjia Jin

Investigation of the 2-D Distribution Form of Bearing Stress in a Single-Bolt Single-Shear Metal-Composite Hybrid Joint ... 220
Yuxiang Lin and Xiaojing Zhang

The Impact of Bleeding Slot Angles on the Performance of a Compressor ... 231
Jiang Junjian, Qiang Xiaoqing, and Zhao Ziheng

Numerical Investigation of Diffuser Curvilinear Meridional Shape on Centrifugal Compressor Stage Performance ... 240
M. O. Goryukhin, V. V. Eremenko, A. E. Mikhailov, A. B. Mikhailova, and D. G. Krasnoperov

Modeling Method of Specimen Repair Techniques from Polymer Composite Material ... 250
Artem Shubin and Konstantin Shramko

Process Approach as a Tool of the Knowledge-Intensive Industry Organization Management System 261
Alena Zastrovskaya, Natalia Khudoliy, and Ekaterina Rybakova

Author Index 269

Investigation into Provisions and Validation for Cockpit Smoke Evacuation on Civil Aircraft

Dan Wu(✉), Zhi Yang, Xudong Yan, and Zhong Shi

Shanghai Aircraft Design and Research Institute, Shanghai, China
Wudan@comac.cc

Abstract. In case of smoke occurrence in the cockpit on civil aircraft, whether the cockpit can readily evacuate smoke directly influence safety of the pilots and passengers. The airworthiness regulation CCAR25.831 (d) puts forward requirements for cockpit smoke evacuation. However, the provision does not provide explicit indexes or instructions and cannot be directly used for aircraft design or validation. This paper mainly investigates CCAR25.831 (d) and related advisory circular, based on which the validation method is proposed in detail and put into practice. Through theoretical research and flight test, it provides ideas and methods for design and airworthiness validation for cockpit smoke evacuation on civil aircraft.

Keywords: Smoke Evacuation · CCAR25.831(d)

1 Introduction

During flight of civil aircraft, smoke may appear in cockpit or cabin in case of fire, explosion and system failure, etc. Smoke in cockpit or cabin will influence respiration of the crew and passengers, affect eye-sight of the pilots and weaken their ability to operate in emergency cases, endangering safety of the aircraft and passengers. Therefore, cockpit smoke evacuation capability is very important and it is one of the main concerns in airworthiness regulations. The civil aviation authorities such as FAA, EASA and CAAC are paying high attention to the smoke handling capacity of civil aircraft, and the major aircraft manufacturers also take smoke evacuation and smoke protection capacity as an important indicator for design of civil aircraft.

In airworthiness provisions, there is 25.831(d), 25.795(b)(1), 25.795(b)(2) regulations for smoke evacuation and protection of cabin and cockpit. This paper mainly focuses on 25.831 (d) [1], which is a requirement for ventilation system. A well-designed ventilation system can guarantee cockpit smoke evacuation capacity. We investigate into main terms in 25.831(d) and its related advisory notices, analyzes the precautions and provides compliance ideas and validation methods for 25.831(d).

Z. Jing and X. Zhan (Eds.): ICASSE 2023, LNEE 1153, pp. 1–12, 2024.
https://doi.org/10.1007/978-981-97-0550-4_1

2 CCAR25.831(d) Regulation

CCAR25.831(d) is quoted as follows: If accumulation of hazardous quantities of smoke in the cockpit area is reasonably probable, smoke evacuation must be readily accomplished, starting with full pressurization and without depressurizing beyond safe limits.

3 25.831(d) Regulation Interpretation

We can see from 25.831(d) that there are no specific indexes in the regulation, which brings uncertainty to aircraft design and validation. We need to investigate into the key terms first.

1) 'Reasonably possible'. In aircraft safety assessment, function failure cases are divided into five categories based on severity of consequences: catastrophic, hazardous, major, minor and no safety effect. For each category, there is a probability defined for the failure event. The more severe the consequence is, the less probable it is allowed to happen. For smoke evacuation regulation 25.831(d), reasonably possible means we only consider smoke evacuation in the cases with a probability greater than10^{-5} [2]. Therefore, during smoke evacuation, the ventilation system, the anti-fire system and other systems all functions normally. We do not consider combined failure cases which happen at a probability much less than 10^{-5}.
2) 'Hazardous quantities of smoke'. In the test, we use smoke generator to simulate smoke in the cockpit. According to advisory notice AC25-9A [3] chapter 12 e (2): 'When the cockpit instruments are obscured (dial/panel indicator numbers or letters become indiscernible), smoke generation should be terminated, and the appropriate AFM and operations manual (if applicable) fire and smoke procedures should be initiated.' However, if the applicant wants to demonstrate protection for a continuous source in the cockpit, the smoke should be generated continuously. It's not mandatory.
3) 'Starting with full pressurization'. According to the provisions of AC25-9A chapter 8.b:

Section 25.831 allows the use of depressurization within safety limits to evacuate smoke from the cockpit and passenger compartment. Aircraft manufacturers have developed smoke evacuation procedure and tested that. Those techniques have been found to be effective for the specific airplane models that utilize them.

Reducing cabin pressure can speed up smoke evacuation. Therefore, 'full pressurization' defines the initial status of cabin pressure before smoke evacuation starts and should be understood as 'starting from maximum pressurization'.

4) 'Without depressurizing beyond safe limits'. This restriction includes two aspects: cabin pressure altitude and cabin pressure change rate.

Cabin pressure altitude is mainly based on the constraints of human physiological endurance, aircraft structure and mass. According to CCAR25.841a, "Pressurized cabin and compartments to be occupied must be equipped to provide a cabin pressure altitude of not more than 2,400 m (8000 ft) at the maximum operating altitude of the airplane under normal operating conditions. (1) If certification for operating above 7600 m (25000 ft) is requested, the airplane must be designed so that occupants will not be exposed to cabin pressure altitude in excess of 15000 ft after any probable failure in the pressurization system."

In addition, the rate of pressure change needs to be considered to prevent negative impact on the eardrum. Normally, the rate of pressure change is related to flying height, the rate of aircraft ascending or descending, which is specified in pressurization system design specification.

5) 'Smoke evacuation must be readily accomplished'. According to AC25-9A chapter 12 a (1): Cockpit smoke evacuation test verify that smoke from sources within or contiguous with the cockpit, can be readily evacuated as required by 25.831(d). Typical commercial large transport airplanes are capable of evacuating dense cockpit smoke within approximately 90 s after the AFM fire and smoke emergency procedures are initiated. Three minutes is an acceptable maximum time to evacuate smoke from any transport category airplane cockpit.

Based on this, we define three minutes as the maximum acceptable time for cockpit smoke evacuation. The faster the smoke evacuation is, the better the pilot can recover from ability degradation, and the less the impact will be caused on flight safety.

4 Validation for CCAR25.831(d)

For validation of CCAR25.831(d), we conduct ground test and flight test to show cockpit smoke evacuation capacity on the ground and in the air. Smoke in flight has a greater impact on aircraft safety and it is more difficult to carry out, therefore this paper mainly focuses on flight test.

The objective of in-flight smoke evacuation test is to demonstrate that the AFM emergency fire and smoke procedures provide means to clear the cockpit of dense smoke at an acceptable rate. This test should also demonstrate that the flight crew can use the procedures without introducing any additional hazard. Refer to Table 1 for the test cases.

Table 1. Test Conditions for Cockpit Smoke Evacuation

No	Height(ft)	Bleed Configuration	AMS configuration
1	39000	L&R Engine Bleed	L&R Pack
2	31000	L&R Engine Bleed	Single Pack

Note: In both test cases, trim air is set to be AUTO, cargo ventilation OFF, cabin pressurization mode AUTO

The flight test procedure for cockpit smoke evacuation is as follows:

a) Adjust the flying height, air bleed mode and air conditioning system settings according to Table 1, turn on all lights in the cockpit (non-emergency lights), turn off cargo ventilation.
b) Crew on board shall wear protective breathing apparatus (or oxygen mask) as required, and confirm that the interphone communication functions normally.
c) Turn on the smoke generator. Stop generating smoke when the cockpit instrument becomes obscure. Meanwhile the aircraft starts to descend from the initial altitude at normal descent rate. Record the time when the smoke generator shuts down and smoke removal completes (the standard for smoke evacuation completion is that remained smoke will not affect crew operation).
d) Emergency pressure relief can be conducted as required during the test. If emergency pressure relief is carried out, related status message and alarms will appear in the cockpit, which can be ignored. This information should be transferred to the pilots before the test to avoid any confusion.

In addition, the aircraft should get out of cabin depressurization mode before it reaches the value when emergency oxygen module will automatically drop down due to low cabin pressure.

e) Level off the airplane when smoke evacuation test finishes as required.

Here are more explanations for the test procedures.

According to AC25-9A chapter 12 section e, when the cockpit instruments are obscured (dial/panel indicator numbers or letters become indiscernible), smoke generation should be terminated, and the appropriate AFM and operations manual (if applicable) fire and smoke procedures should be initiated.

In the test, we follow above instructions. There were two pilots in the cockpit and both of them judge whether the instruments become obscured. When the pilots indicate the instruments are clear again, the test stops. It's a common practice and highly reliable. Therefore, the pilots' subjective judgements are the criteria for the test to start and stop.

When the test starts, smoke begins to accumulate with smoke generators on. When the pilot indicates panels are obscured, smoke generation stops. During the whole process, the air conditioning system functions normally and therefore smoke will be extracted outboard through ventilation system. However, if the applicant wants to demonstrate protection for a continuous source in the cockpit, the smoke should be generated continuously. It's not mandatory.

5 Test Configuration and Devices

In smoke evacuation test, the smoke is generated by a smoke generator. In AC25-9A chapter 9 section b [3], it provides acceptable smoke generator types for smoke evacuation test.

Before the cockpit smoke evacuation test, the cockpit is equipped with test devices.

a) A smoke generator is installed in the cockpit, as shown in Fig. 1;

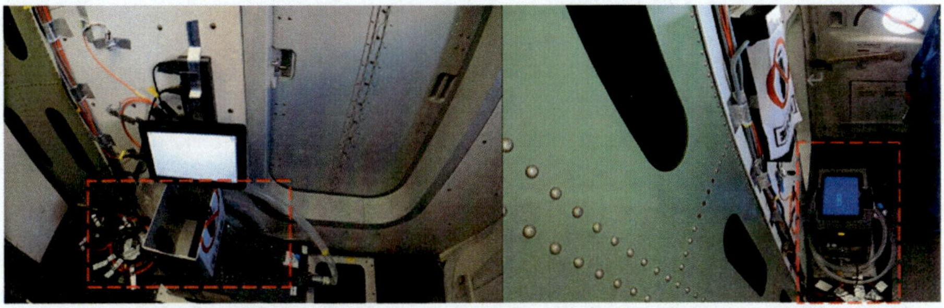

Fig. 1. Smoke Generator Installation

b) A total of three portable cameras are prepared in the cockpit: one is fixed on the right rear panel of the cockpit to take an overall view of the cockpit, the second one behind the observer seat, and the third one is hand-held by the crew.

The test equipment is shown in Table 2.

c) We use stopwatch to keep record of time.

Table 2. Test Devices

No.	Instrument	Type	Quantity
1	Smoke generator	Imbps27er	1
2	Portable camera	GoPro	3
3	stopwatch	SE7–211	1

6 Test Results and Analysis

In 2022, flight test for cockpit smoke evacuation was carried out at the height of 39000 ft and 31000 ft.

a) 39000 ft

At 12:15, the aircraft maintained a level flight at 39000 ft. The aircraft configuration was set in accordance with test condition 1 in Table 1.

At 12:15:32, the crew gave instructions and smoke generator was turned on. The overall view of the cockpit right before smoke generation is shown in Fig. 2.

20 s later (at 12:15:52), the smoke filled in the cockpit and smoke generation was stopped (Fig. 3).

28 s later (at 12:16:20), both pilots reported that they could see the instrument panel clearly without affecting the flight operation (Fig. 4).

The cockpit smoke evacuation took 28 s, satisfying the test criteria of 3 min.

Fig. 2. At 12:15:32 An overall View of the Cockpit Before Smoke Generation

Fig. 3. At 12:15:52 An overall View of the Cockpit When Smoke Generation Stops

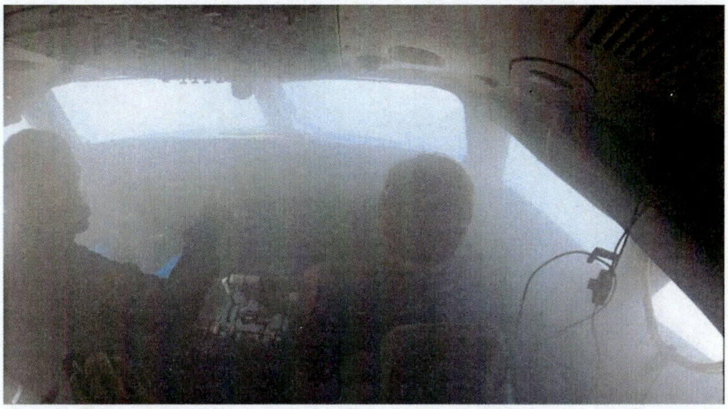

Fig. 4. At 12:16:20 An overall View of the Cockpit when Smoke Evacuation Test Stops

During smoke evacuation, cabin altitude and fly altitude status are shown in Fig. 5 Status of engine bleed, packs, trim air, cargo ventilation, pressure control mode are shown in Figs. 6, 7, 8 and 9. As we can see, during smoke evacuation, both engine bleed were on, L &R packs on, cargo ventilation off, trim air automatically on, cabin pressure manual mode off (means automatic mode on) and emergency pressure relief uninitiated.

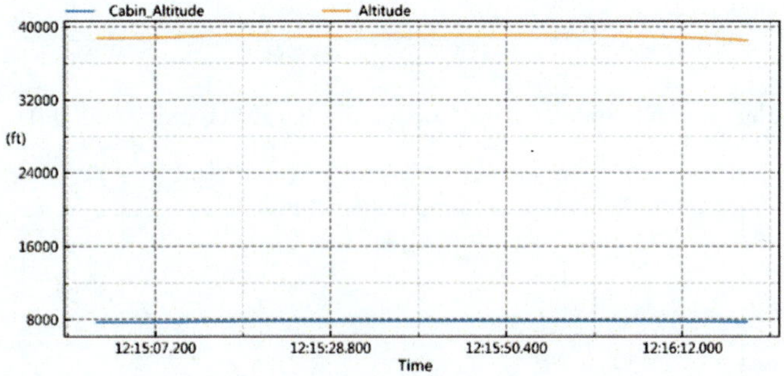

Fig. 5. Cabin Altitude and Fly Altitude during Cockpit Smoke Evacuation (39000 ft)

b) 31000 ft

The test procedure at the height of 31000 ft is basically the same as 39000 ft. The only difference is that we use only one pack for ventilation.

At 12:28, the aircraft maintained a level flight at 31000 ft. The aircraft configuration was set in accordance with test condition 2 in Table 1. The right pack was closed at 12:27:40.

At 12:28:23, the crew gave instructions and smoke generator was turned on (Fig. 10).

27 s later (at 12:28:50), the smoke filled in the cockpit and smoke generation was stopped (Fig. 11).

34 s later (at 12:29:24), both pilots reported that they could see the instrument panel clearly without affecting the flight operation (Fig. 12).

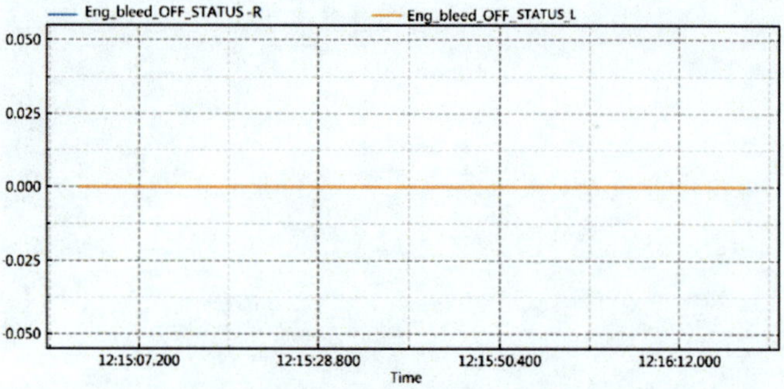

Fig. 6. Engine Bleed Status during Cockpit Smoke Evacuation (39000 ft)

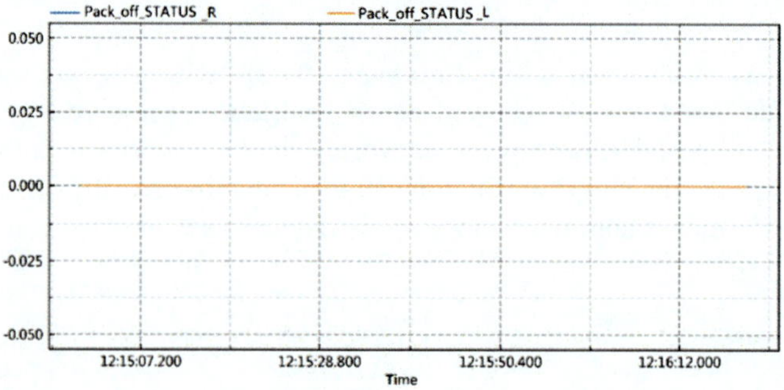

Fig. 7. Pack Status during Cockpit Smoke Evacuation (39000 ft)

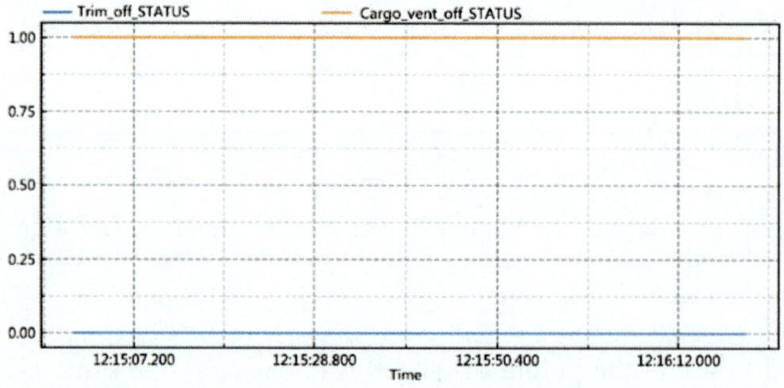

Fig. 8. Trim Air and Cargo Ventilation Status during Cockpit Smoke Evacuation (39000ft)

In this test condition, the cockpit smoke evacuation took 34 s, satisfying the test criterion of 3 min. Figures 13, 14, 15, 16 and 17 show status of the main systems during the test.

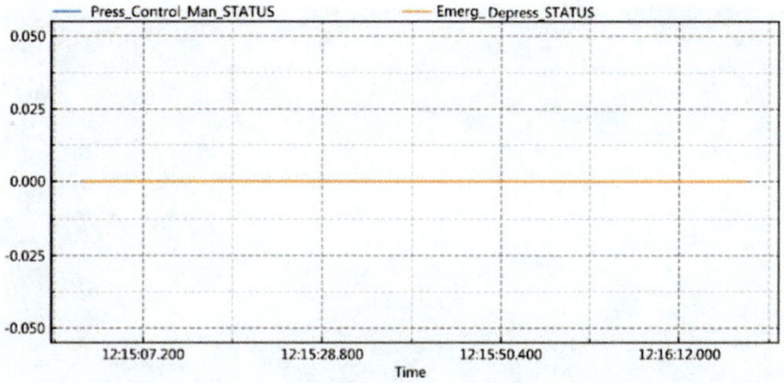

Fig. 9. Pressure Control and Emergent Depression Status (39000 ft)

Fig. 10. At 12:28:23 Overall View of the Cockpit before Smoke Generation

Fig. 11. At 12:28:50 Overall View of the Cockpit when Smoke Generation Stops

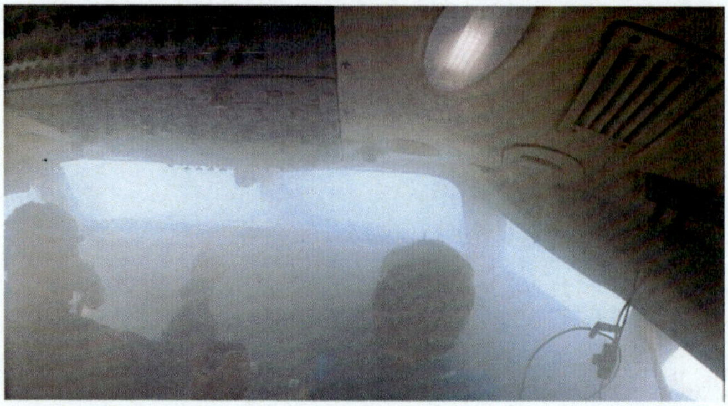

Fig. 12. At 12:29:24 Overall View of the Cockpit when Smoke Evacuation Test Stops

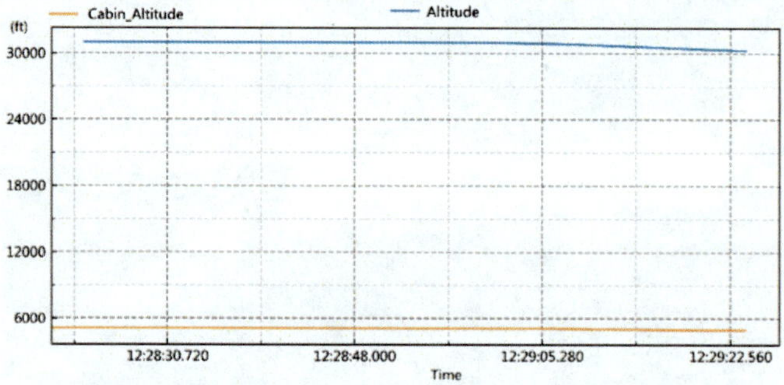

Fig. 13. Cabin Altitude and Fly Altitude during Cockpit Smoke Evacuation (31000 ft)

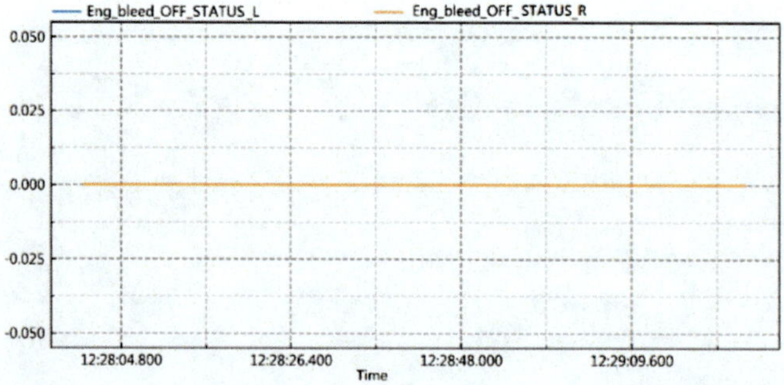

Fig. 14. Engine Bleed Status during Cockpit Smoke Evacuation (31000 ft)

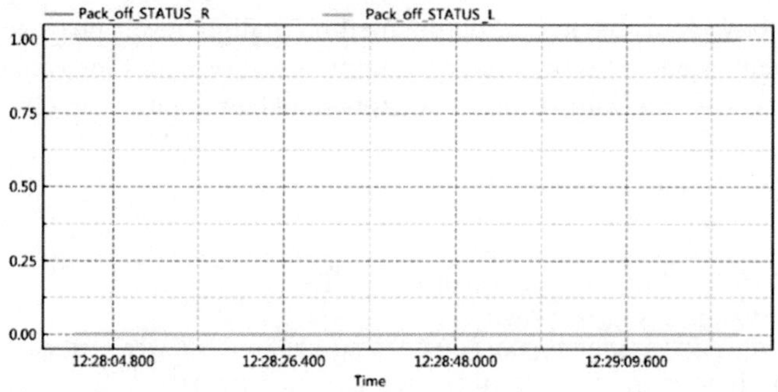

Fig. 15. Pack Status during Cockpit Smoke Evacuation (31000 ft)

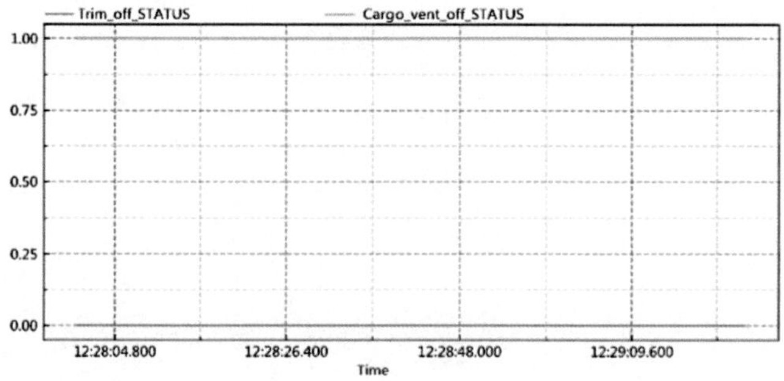

Fig. 16. Trim Air and Cargo Ventilation Status during Cockpit Smoke Evacuation (31000 ft)

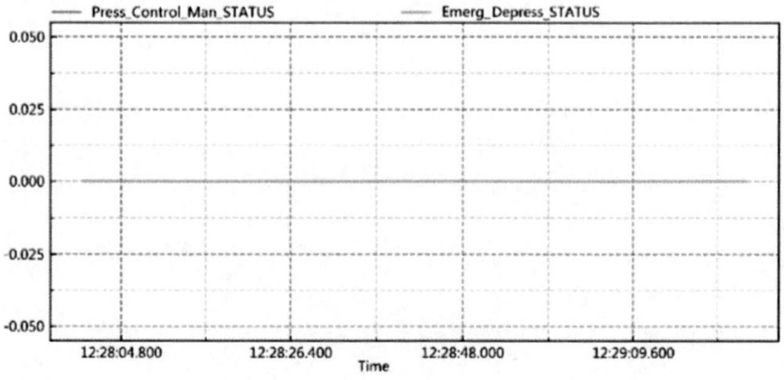

Fig. 17. Pressure Control and Emergent Depression Status (31000 ft)

7 Conclusion

This paper mainly focuses on CCAR 25.831 (d) and related advisory circular, investigating into the main terms of probability of smoke occurrence, amount of hazardous smoke, safety cabin pressure limits and acceptable smoke evacuation time limit.

Based on above analysis, the validation method of flight test is proposed and put into practice. Through theoretical research and flight test, it provides ideas and methods for design and airworthiness validation for cockpit smoke evacuation on civil aircraft.

References

1. CCAR-25-R3 China Civil Aircraft Regulation 25 Rev.3
2. AC25.1309 System Design and Analysis
3. AC25–9A, SMOKE DETECTION, PENETRATION, EVACUATION TESTS AND RELATED FLIGHT MANUAL EMERGENCY PROCEDURE

SysML-Based Approach for Functional Quantitative Modeling of Civil Aircraft Systems

Shuijin Ye, Yong Chen(✉), and Tianlong Wang

Shanghai Jiao Tong University, Shanghai, China
{zszysj9899,aerocy,389434860}@sjtu.edu.cn

Abstract. During the conceptual design phase of civil aircraft systems, it is imperative that engineers accurately establish the functional representation of the system and ensure its correctness to provide a solid foundation for subsequent detailed design. Traditional document-based methods for functional representation frequently result in ambiguity, making it difficult to precisely communicate design requirements. Therefore, the academic community is committed to using model-based system engineering to achieve functional representation. However, existing model-based functional representation techniques primarily focus on describing functional input/output, rather than explicitly expressing functional input/output conversions, which can impede system engineers from verifying functional completeness. In response to this issue, this paper proposes a functional quantitative modeling approach for civil aircraft systems based on SysML. Firstly, a quantitatively-analyzable flow model is established, and a general functional model of integrated structure, state, and behavior is subsequently developed based on this flow model. Next, a hierarchical functional model of the system is introduced. Finally, to verify the proposed approach, the elevator surface control system is employed as an example, and a quantitative analysis of the functional model is presented.

Keywords: Model-Based Systems Engineering · SysML · Functional Modeling · Quantitative Analysis · Civil Aircraft System

1 Introduction

In recent years, complex products have become increasingly complex. Compared with traditional mass-manufactured products, their complexity is reflected in the multi-field coupling and correlation, the complex composition of systems and subsystems, the high degree of coupling, complex functions and functional levels, and complex R&D organizations [1]. During the conceptual design process of civil aircraft systems, engineers need to comprehensively design the aircraft system, clearly describe the relevant functional requirements, and conduct reasonable verification of the system and internal functions, so as to convey clear design intentions and technical requirements to project participants.

Document-based systems engineering has been difficult to meet the requirements of functional accuracy, completeness, and process description, while model-based system

Z. Jing and X. Zhan (Eds.): ICASSE 2023, LNEE 1153, pp. 13–31, 2024.
https://doi.org/10.1007/978-981-97-0550-4_2

engineering (MBSE) is gradually applied to the design of complex product systems, which can identify possible faults and design defects in advance [2]. MBSE can carry out unified modeling management on various design elements in the design process to ensure the consistency between system design data [3]. MBSE can realize the structured expression of functions and provide support for the structured expression of system functions.

According to the study of Chen et al. [4], function refers to the abstraction of the system's effect on its external environment (object) formed in the brain of engineers, so as to expect to change the unsatisfactory external environment (object) to achieve a satisfactory state. Functional representation is the basis of functional modeling, and the representative functional representation methods include: 1) Gerund pair representation, where the verb represents the action performed by the functional object, and the noun represents the functional object; 2) Input-output stream representation, according to the design theory of Pahl et al. [5], the objects of input and output stream can be matter, energy, information and their sub-types; 3) Integrate the representation of verbs and input-output streams, that is, verbs are used to express the purpose of functions, and input-output streams are used to express the interaction between environment and functional object.

With the development of MBSE, there are more and more model-based functional representations, each with its own characteristics. For example, the Function–Behavior–Structure (FBStr) model proposed by Gero et al. [6] focuses on the design process of transition from intent to structure. The Function–Environment–Behavior–Structure (FEBS) model proposed by Deng et al. [7] focuses on the impact of environment on function. The Function–Behavior–State (FBS) model proposed by Umeda et al. [8] focuses on the connection between function and behavior. The Structure–Behavior–Function (SBF) model proposed by Goel et al. [9] focuses on the relationship between system structure and function. The ontology-based product function description model proposed by Guo et al. [10], focusing on functional semantics and reasoning. The State-Behavior-Function (SBF) model proposed by Chen et al. [11] focuses on the relationship between the state change of system structure and the transformation of input and output streams.

Although existing model-based functional representation methods can effectively describe the qualitative aspects of functions, they lack the capability to enable quantified representation and verification of input/output flows in functions. Consequently, they are unable to support the comprehensive representation and verification of system functions.

To address the aforementioned challenges, it is necessary to express the flow and its transformation and conversion involved in the civil aircraft system function, establishes a functional model that can be qualitatively or quantitatively analyzed, which in turn provides the foundation for quantitative analysis and verification of the system function. Specifically, this paper proposes the establishment method of flow models for quantitative analysis based on SysML, in addition to general functional models that integrate structure, state, and behavior based on the established flow model. Moreover, the paper proposes a representation method for the hierarchical functional model. Finally, using the elevator surface control system as an example, this paper illustrates how to establish the general functional models, hierarchical functional model, and how to carry out quantitative model simulation. This example provides a basis for system functional verification.

2 Flow and General Functional Modeling

Existing functional modeling methods often ignore the structural expression of functional flow, and use a noun to represent the input/output flow, which can satisfy the simple representation of functions, but cannot express the relationship between output flow and input flow. For example, it is difficult to determine whether the value of the flow attribute and the type of the flow have changed after the input flows through the functional principal. For this reason, this section provides a modeled representation of the input/output flow, so that the relationship between the input/output flows and the function can be more accurately described.

This section also expresses the function in a structured way. The general functional model refers to a model that integrates structure, state, and behavior without considering the internal composition of the functional object during the modeling process. It can be used to describe the input and output flow, internal state, behavior, etc. In the conceptual design stage of civil aircraft systems, functional modeling is procedural. When a system does not consider internal components, a general functional model of the system can be established; when a function cannot be separated, a general functional model can be established. Based on the flow model, a qualitative/quantitative functional model can be established to provide a basis for quantitative analysis of the general functional model.

2.1 Flow Model Establishment

Flow is the medium through which functional objects interact with the environment, including material flow, energy flow, and information flow. Flow modeling is the process of modeling and expressing flows, that is, to describe flow types and attributes in the form of "type name + attribute set", where "attribute set" is optional. The attributes referred to in this article include physical attributes, chemical attributes, etc., and flow attributes refer to the attributes related to participation in the flow and functional utility (the interaction between the functional object and the environment). For example, electrical energy flow can be used as a flow to participate in some functional utilities. The attributes of electrical energy objects generally include current, voltage, power, frequency, etc. When functional utility involves energy conversion (such as transforming electrical energy into light energy or mechanical energy), the attributes involved in the functional utility may be current, voltage, and power, so current, voltage, and power can be regarded as the flow attributes of electric energy, while frequency is not regarded as its flow attributes.

The basis of functional utility quantitative analysis is that the input/output flow has attributes, and the corresponding attributes have certain constraints. When quantitative analysis is required, the flow must have attributes with definite constraints, but when some constraints are clearly understood in the upstream and downstream of the design, the corresponding constraints do not be displayed. It should be noted that the quantitative analysis does not mean that the results of functional modeling include the principle understanding. The quantitative analysis mentioned here is only related to the attributes of the input/output flow, and has nothing to do with the functional principle understanding. The Block Definition Diagram (BDD) of SysML is a structural diagram that can be used to describe a complex flow model. The block elements in it can describe a single flow model, and are generally identified by a partition box with <<block>>. For example, when performing a qualitative analysis on an electric lamp, the input flow (*electricity*) may not contain any attributes, as shown in Fig. 1; when performing a quantitative analysis on it, since the electric lamp has a voltage requirement for the electricity flow, the *electricity* flow model must include the attribute of voltage at this time, and the voltage needs to be defined clear constraints, as shown in Fig. 2.

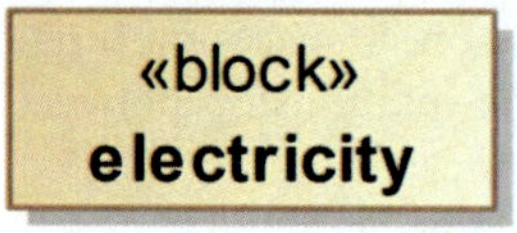

Fig. 1. Electricity flow for qualitative analysis

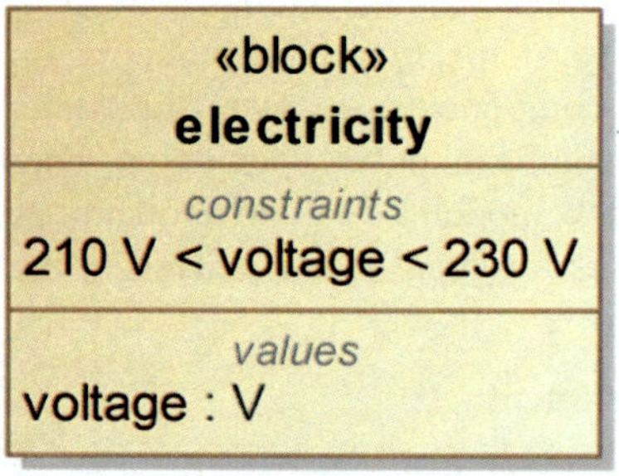

Fig. 2. Electricity flow for quantitative analysis

2.2 The Structure of General Functional Model

The structure of the general functional model consists of internal flows, input/output flow ports, flags, etc. The internal flow is an intermediate flow in the transformation and conversion process of the input and output flow; the input/output flow port is the channel through which the functional object interacts with the environment, that is, the transmission channel of the flow; the flag is an object used to describe the internal state or characteristics of the model. A block in BDD can describe a single model structure: the internal flow and input/output flow ports of a functional model can be described by the component attributes of the block; flags can be expressed by the component attributes or value attributes of the block. The Internal Block Diagram (IBD) of SysML is a structural diagram that can display the specific structure inside the block; the Parameter Diagram (PD) is a specific internal block diagram that can describe the binding connection relationship of related value attributes in the internal composition of the block. Therefore, the internal composition and connection of the general functional model structure can be accurately described by PD or IBD.

Taking the simplified elevator surface power control unit (PCU) as an example, the PCU receives the electrical signal output by the primary actuator control electronics (P-ACE), controls the output of the hydraulic components, and manipulates the position of elevator surface through a sector disk, cable, or connecting rod assembly. The structure of the PCU functional model can be described by using block and IBD, as shown in Fig. 3 and Fig. 4. For example, "*inSOVH*" and "*inSOVL*" are the internal flows of the PCU functional model, and the input flow "*SOVH*" passes through the input port "*pSigSOVH*", stored by the internal flow of "*inSOVH*", provides an intermediate medium for flow transformation and conversion.

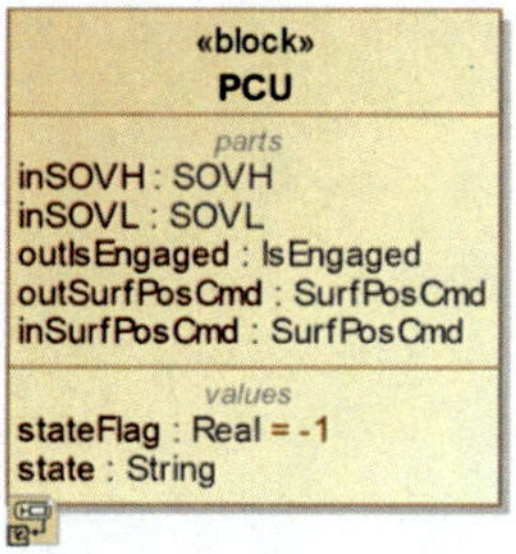

Fig. 3. The structure of PCU functional model

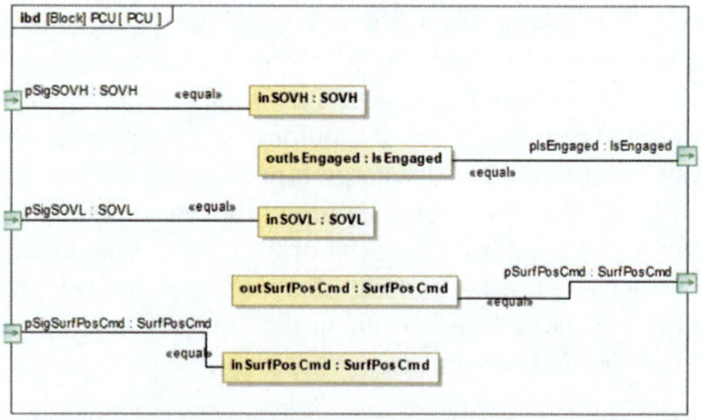

Fig. 4. Internal connections of PCU functional model

2.3 The State of General Functional Model

The state of general functional model is an abstraction of a set of specific attribute values, where attributes include physical properties (such as physical connections inside functional object) and chemical properties. Engineers can use the form of "state name + description" to express the general functional model state, where "description" is optional. For example, when an operational amplifier power supply circuit has no effective voltage (that is, state = "*off*"), the function of the operational amplifier cannot work properly. The State Machine Diagram (SMD) of SysML is a kind of behavior diagram, which can be used to describe the state of the model and the transition between states [12]. Model states can be described by state elements in a state machine diagram.

The simplified PCU has two states: "*NotEngaged*" and "*Engaged*"; in order to facilitate model expression, engineers can define the auxiliary state "*off*", as shown in Fig. 5.

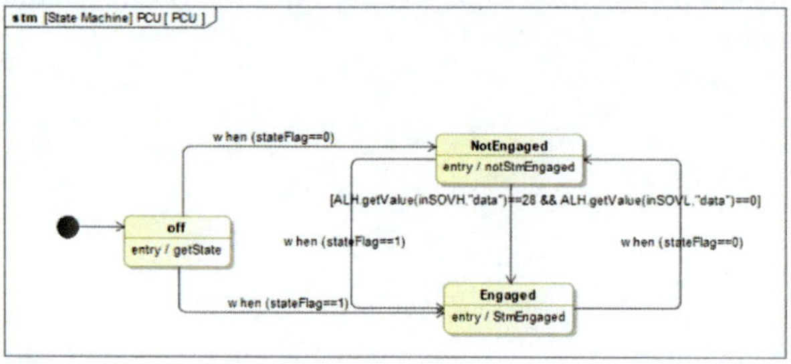

Fig. 5. The states and state transitions of PCU functional model

2.4 The Behavior of General Functional Model

The types of interaction between functional objects and the environment mainly include: 1) There is input but no output: environmental input causes the state of the object to change, such as the water storage function of a pool, that is, the water flow input causes the water level of pool to change; 2) There is output but no input: the object directly acts on the environment, such as battery discharge function, that is, battery outputs electric energy to the outside; 3) There is input and output: the environmental input is output to the environment through the internal function of the object. Functional model behavior can be defined as an activity in which a functional object interacts with the environment to change one or both of them. General functional model behavior can be divided into state transition behavior and flow-processing behavior.

(1) **State Transition Behavior**
There are state switching activities in the multi-state general functional model, and such activities can be called state transition behaviors. The trigger event is the event that causes the model state to change. The trigger event includes input/output flow interaction events, time events, and model attribute change events. The state transition behavior of the model can be represented by SMD: a solid line with an open arrow can be used to mark the transition of the model from the source state to the target state, and the trigger event can be described by the conditions on the solid line.

The state transition behavior of the PCU is shown in Fig. 5. For example, when the "$stateFlag = 1$" event occurs, the PCU state will switch from "*off*" to "*Engaged*".

(2) **Flow-Processing Behavior**
In a specific model state, the model can complete the corresponding flow transformation or conversion activities, which can be called flow-processing behavior. The "input flow + output flow" binary expression is difficult to express the flow-processing behavior of complex models in different states, that is, it is impossible to determine the type of flow-processing behavior: a) the flow attribute value changes, b) the flow type changes. But based on the flow model and flow-processing behavior, the function can be expressed qualitatively or quantitatively. When the flow attributes and attribute constraints are clear, the flow-processing behavior of the functional model can be quantitatively analyzed. It should be noted that the flow-processing behavior should follow general physical rules, such as the law of energy conservation. For example, in the qualitative analysis of lamp function, the input flow is electrical energy (*flow_electricity*), the output flow is light energy (*flow_opticalEnergy*), the flow-processing behavior can be described as: when the input is *flow_electricity*, *flow_electricity* transforms to *flow_opticalEnergy*, and outputs *flow_opticalEnergy*. In the quantitative analysis of lamp function, the input flow is electrical energy with clear attributes and attribute constraints (*flow_electricity*: *voltage* = 220 V, *power* >= 0 W), and the output flow is light energy with clear attributes and attribute constraints (*flow_opticalEnergy*: *power* >= 0 W), the flow-processing behavior can be described as: inputs *flow_electricity*, *flow_electricity* is converted into *flow_opticalEnergy*, and *flow_opticalEnergy* is output externally, and the *power* of *flow_opticalEnergy* is equal to the *power* of *flow_electricity*.

Flow-processing behavior can be represented by Activity Diagram (AD) of SysML, and the trigger of flow-processing behavior can be defined by SMD: if the state in the SMD has executable events (entry, do, exit, etc.), then the functional model will execute the corresponding popularity behavior after entering the state according to the corresponding event. AD can be used to express the transition and conversion from input flows to output flows, and the specific processing process of flow-processing behavior can be described by opaque expressions in SysML. According to the flow model and opaque expressions, qualitative or quantitative analysis of the flow-processing behavior of the functional model can be realized. It should be noted that the flow-processing behavior processing of different states of the model can be expressed by the same/different opaque expressions.

The flow-processing behavior of the simplified PCU is shown in Fig. 6. The opaque expression can be represented by the pseudocode in Fig. 7, which uses the same opaque expression to implement the detailed process of the flow-processing behavior in the various states of the PCU. And it can be known that the output flow of the PCU (such as *aOutIsEngaged*, *aOutSurfPosCmd*) is generated internally by the PCU, the attribute value of the output flow may be related to the input flow (such as *aOutSurfPosCmd.data* = *aInSurfPosCmd.data*).

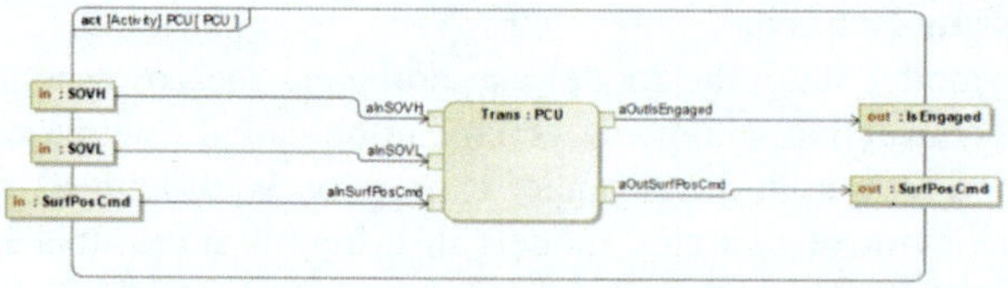

Fig. 6. The flow-processing behavior of PCU

```
if(insovh==28 && insovl==0){stateFlag=1}else if(insovh<28 || insovl>0){stateFlag=0}else{stateFlag=-1}
createObject aOutIsEngaged
createObject aOutSurfPosCmd
if("NotEngaged"){
    aOutIsEngaged.data = -1
    aOutSurfPosCmd = null //outputFlow invalid
    state="NotEngaged"
}else if("Engaged"){
    aOutIsEngaged.data = 1
    aOutSurfPosCmd.data = aInSurfPosCmd.data
    state="Engaged"
}else{
    state="stateError"
}
```

Fig. 7. The pseudocode of PCU flow-processing behavior

3 Hierarchical Functional Models

In the modeling process, the sub-level functional models (including general functional models and hierarchical functional models) can be integrated to establish the parent-level functional model, namely the hierarchical functional model. Different from the general functional model, the hierarchical functional model focuses on describing its internal composition and connection. In the conceptual design stage of civil aircraft system, when a system considers the internal composition, the hierarchical functional model of the system should be established; when multiple functional models are integrated, the corresponding hierarchical functional model should be established.

3.1 The Structure of Hierarchical Functional Model

The structure of the hierarchical functional model should contain all sub-level functional models and the connection between functional models. When the parent-level functional model does not have independent ports, the ports of the sub-level functional models can be used as the ports of the parent-level functional model; when the parent-level functional model has independent ports, it is necessary to clarify the relationship between the parent-level ports and the sub-level functional models.

Based on the method in Sect. 2.2, the structure of the hierarchical functional model can be represented by BDD, IBD and PD. When the parent-level model has independent ports, IBD can clearly describe the boundary of the parent-level model and the connection relationship between the ports on the boundary and the ports of the sub-level functional models. As shown in Fig. 8, the outer frame of IBD is the boundary of the parent-level model. The input port of the parent-level model will be connected to the input ports of the sub-level functional model, and the output port of the sub-level functional model will be connected to the output port of the parent-level model.

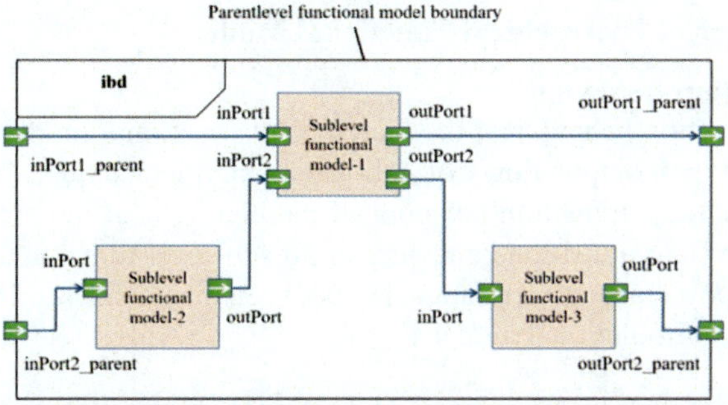

Fig. 8. Illustration of the internal connections hierarchical functional model structure

3.2 The State of Hierarchical Functional Model

The state of the hierarchical functional model is jointly determined by the states of all sub-level functional models. The determinacy of the hierarchical functional model's state is predicated upon the collective states of its subordinate sub-level functional models. In other words, the hierarchical functional model's state can be precisely inferred solely upon the attainment of unambiguous states of all sub-level functional models.

Based on the method in Sect. 2.3, the state of the hierarchical functional model can be defined by a block in SysML, and it is imperative that the state of the hierarchical functional model aligns with the state of its constituent sub-level functional models in a one-to-one correspondence, as shown in Fig. 9. The process of establishing the state of the hierarchical functional model is as follows: establish the state of the hierarchical functional model using block in SysML; then correspond to the state of sub-level functional model, which can be represented by a PD.

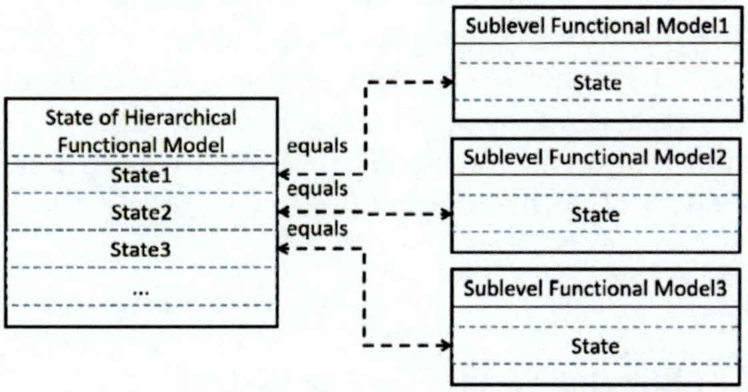

Fig. 9. Illustration of the state of hierarchical functional model

3.3 The Behavior of Hierarchical Functional Model

(1) **State Transition Behavior**
The state transition behavior of the hierarchical functional model can be triggered by external flows, output flows of sub-level functional models, time events, etc. Therefore, the state transition behavior of the hierarchical functional model needs to include the state transition behaviors of all sub-level functional models, and the input-output port connection relationship between the parent-level functional model and the sub-level functional models.

Based on the method in Sect. 2.4, IBD and SMD can be used to describe the state transition behavior of the hierarchical functional model. Connect the defined parent-level model's ports with the input ports of the corresponding sub-level functional models, expressing that the flow outside the system will trigger the state transition behavior of the system functional model; connect the sub-level functional model output ports with the corresponding sub-level functional model input ports, indicating that the output of a certain sub-level functional model will trigger the state transition behavior of another sub-level functional model, which will trigger the state transition behavior of the hierarchical functional model.

(2) **Flow-Processing Behavior**

The flow-processing behavior of the hierarchical functional model should include the flow-processing behavior of the sub-level functional models, and include the input-output port connections between the parent-level functional model and the sub-level functional model.

Based on the method in Sect. 2.4, IBD and AD can be used to describe the flow-processing behavior of the hierarchical functional model. Connect the defined parent-level model boundary ports with the ports of the corresponding sub-level functional models to express the flow transfer relationship of the hierarchical functional model; connect the output ports of the sub-level functional model with the input ports of the corresponding sub-level functional model to represent a certain sub-level functional model, which indicates that the output flows of the sub-level functional model will become the input flows of another sub-level functional model, which can trigger the flow-processing behavior of the hierarchical functional model.

4 Case Study

4.1 Introduction of Elevator Surface Control System

The elevator surface control system is mainly composed of P-ACE, PCU, elevator surface and structural support components. The main functions of the elevator system are: 1) Locate the elevator surface according to the pilot's instruction; 2) Locate the elevator surface according to the steering rod force required to maintain the normal acceleration of the aircraft; 3) Perform feedback control of the elevator surface position according to the pilot's instruction and the position of the elevator surface to realize the stability control of the elevator surface. This paper takes elevator surface control system as an example to illustrate the above functional modeling methods.

The purpose of this study is to support engineers to express functional schemes clearly, and provides a basis for quantitative analysis of functional models, so as to complete functional completeness verification. In this section, Cameo Systems Modeler (a modeling software for SysML) can be used to establish the main sub-functional models and the hierarchical functional model based on the function scheme of the elevator surface control system of one type of civil aircraft.

4.2 Establishment of Flow Models and Sub-level Functional Models of the Elevator Surface Control System

(1) **Flow Modeling**
There are many interactive flows in the modeling process of the elevator surface control system. For example, P-ACE determines its connection priority through the connection priority signal ("*EngagePrior*"); the related signal of the surface position control command ("*SurfPosCmd*"), and the P-ACE validity signal ("*FCMValid*") output by the flight control module. Based on the method in Sect. 2.1, the required input/output flow models in the system can be established to support the qualitative and quantitative analysis of the hierarchical functional model, as shown in Fig. 10.

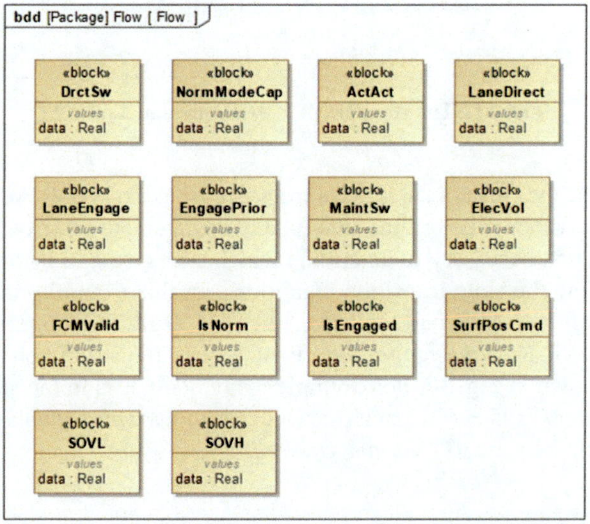

Fig. 10. Input/output flow modeling for elevator surface control system

(2) **P-ACE and PCU Functional Modeling**
P-ACE receives instructions from the flight control module, adjacent P-ACE, etc., and processes these instructions and then sends the results to the PCU. The general functional model structure of P-ACE is shown in Fig. 11. The process of establishing the PCU functional model has been introduced in Sect. 2, so it will not be repeated here.

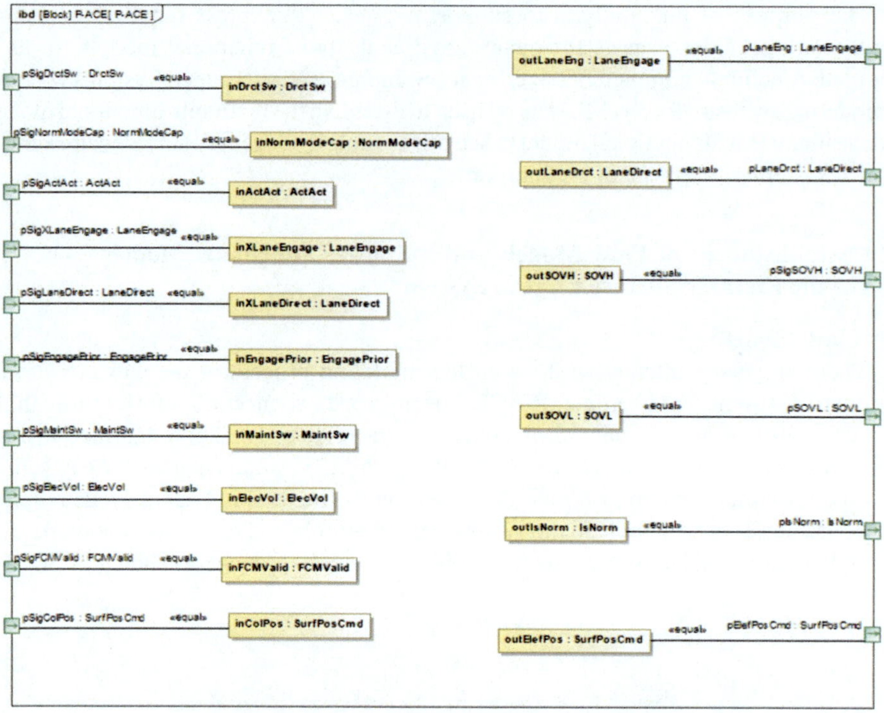

Fig. 11. The structure of P-ACE functional model

In this example the elevator surface control system has a pair of P-ACE, and the value of signal "*inEngagePrior*" determines the P-ACE engagement priority. When modeling any P-ACE, it is first necessary to determine which states exist in its normal operation, and then determine the state transition conditions. In this example, in addition to the auxiliary state "*off*", P-ACE has six states, such as "*Direct-UnCapEngage*", "*Direct-CapEngage*", "*DNR-UnCapEngage*", etc. P-ACE state transition behavior and trigger conditions are shown in Fig. 12. For example, when P-ACE is in the state of "*Normal-CapEngage*", "*stateFlag5* = *0*" (that is, the value of *inFCMValid* equals to 0) will trigger an event to switch the state to "*Normal-UnCapEngage*".

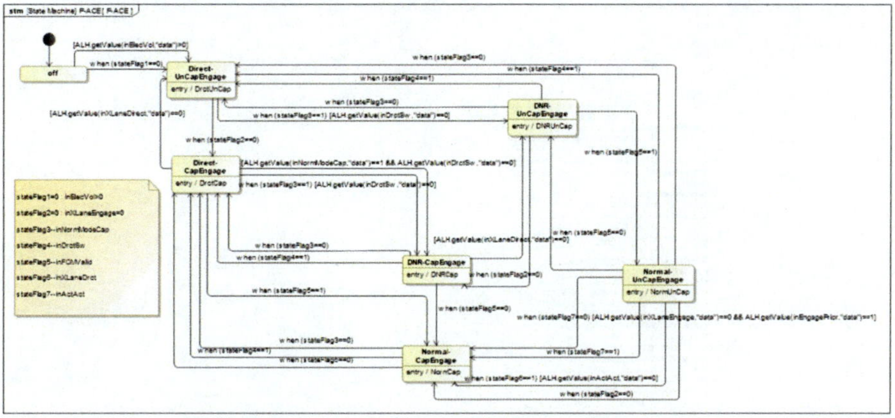

Fig. 12. State and state transition behavior of the P-ACE functional model

P-ACE has different flow-processing behaviors in different states, and the specific processing flow of its flow-processing behavior is described by the opaque expression of "*PACE*" referred to by "*Trans*", as shown in Fig. 13. When the opaque expression of "*PACE*" is processed only for the flow model type, the flow qualitative analysis of P-ACE functional model can be carried out. When the properties and values in the flow model are processed by the opaque expression "*PACE*", flow quantitative analysis can be carried out on the P-ACE functional model, as shown in Fig. 14.

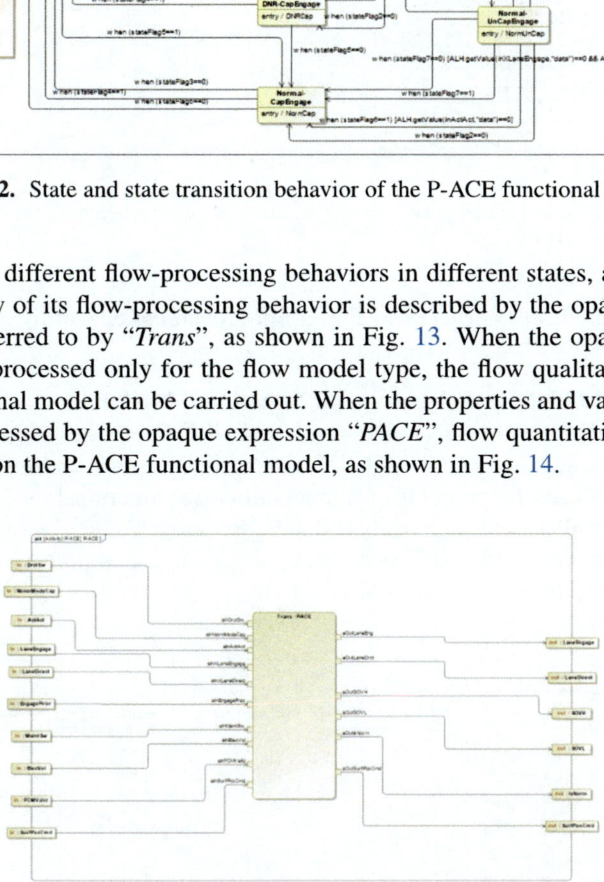

Fig. 13. The flow-processing behavior of the P-ACE functional model

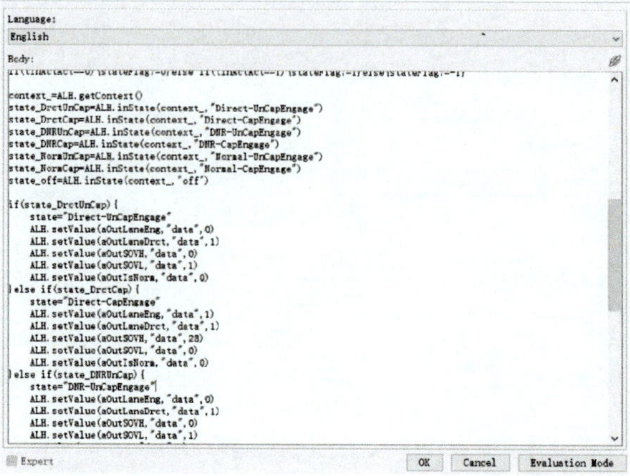

Fig. 14. Enlarged section of opaque expression of flow-processing behavior for P-ACE functional model

4.3 Establishment of Hierarchical Functional Model of Elevator Surface Control System

(1) **Structural Modeling of Hierarchical Functional Model**
Based on the method in Sect. 3.1, combined with the sub-level functional model, the hierarchical functional model structure of the elevator surface control system is established. As shown in Fig. 15, the hierarchical functional model includes two P-ACE general functional models ("*PACEOB*" and "*PACEIB*"), two PCU general functional models ("*PCUOB*" and "*PCUIB*"), PCU signal processing ("*PCUHL*") and simplified elevator surface ("*EleSurf*").

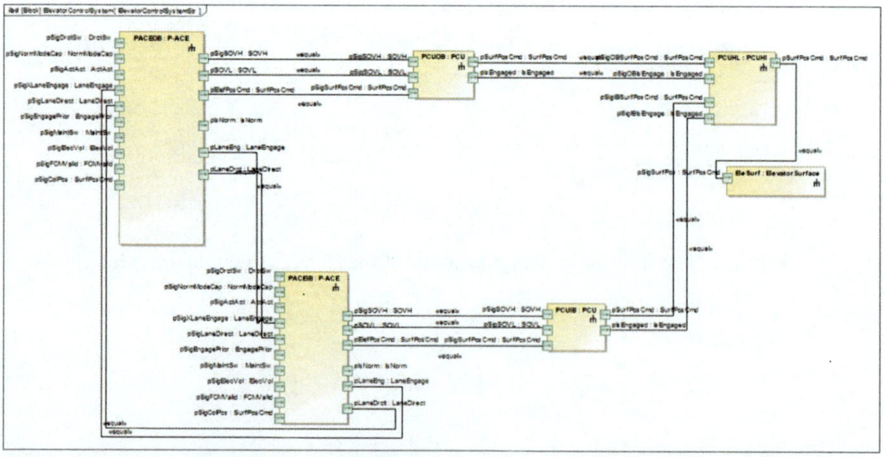

Fig. 15. The structure of hierarchical functional model

(2) **State and State Transition Behavior Modeling of Hierarchical Functional Model**
Based on the methods of Sects. 3.2 and 3.3, the states of the sub-level functional models are integrated to establish the relationship between them and the state of the elevator surface control system, as shown in Fig. 16. Based on the state transition behavior of the sub-level functional models and the structure of the hierarchical functional model, the state and state transition behavior of the hierarchical functional model are modeled. As shown in Fig. 17, describing the state and state transition behavior of the hierarchical functional model of the elevator surface control system, the input flows to P-ACE will trigger the state transition behavior of P-ACE and the corresponding sub-level functional models, and then trigger the state transition behavior of the hierarchical functional model.

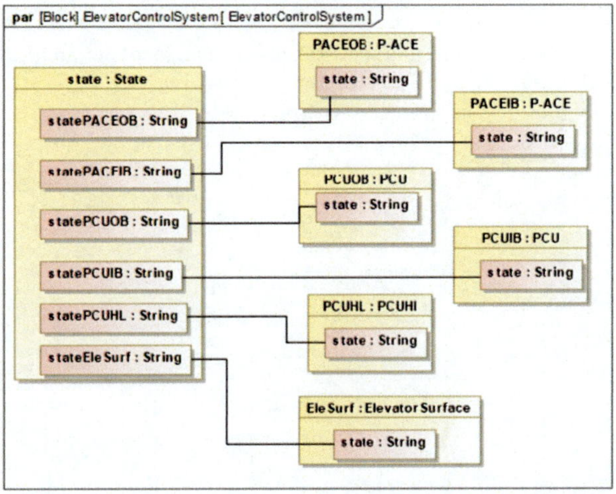

Fig. 16. State representing for hierarchical functional model

(3) **Flow-Processing Behavior Modeling for Hierarchical Functional Model**
Based on the method in Sect. 3.3, the flow-processing behavior of the elevator surface control system is modeled. As shown in Fig. 18, the input flow of P-ACE is output to the corresponding functional model under the action of the specific model state and flow-processing behavior, and causes their flow-processing behavior to be executed. Then, the flow-processing behavior of the hierarchical functional model is triggered to change the output flow of the hierarchical functional model.

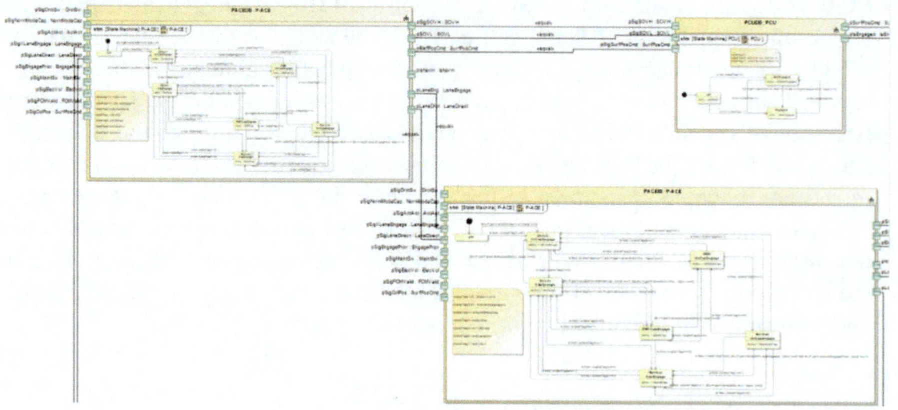

Fig. 17. Enlarged section of the state and state transition behavior of hierarchical functional model

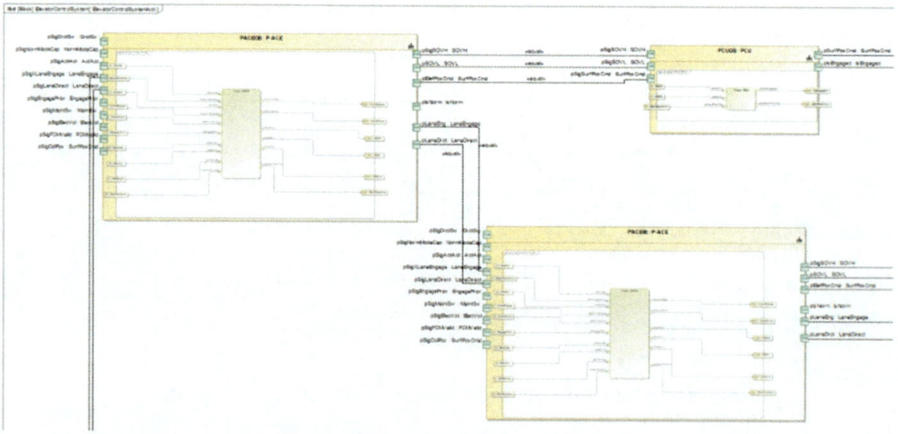

Fig. 18. Enlarged section of flow-processing behavior of hierarchical functional model

4.4 Simulation of Hierarchical Functional Model of Elevator Surface Control System

In the conceptual design of complex systems, it is necessary to simulate the functional model of the system to eliminate errors related to functional logic. Based on the above hierarchical functional model, the functional simulation of the elevator system, including quantitative analysis of the flow-processing behavior of the functional model, is helpful to verify the accuracy of the system functional model. This section gives the simulation cases of the general functional model and the hierarchical functional model respectively, and verifies that the functional model can complete the corresponding traffic behavior in a specific state.

(1) **Simulation of General Functional Model**
Based on the sub-level functional model and given the expected input, the flow-processing behavior and state transition behavior of the functional model are verified. For example, simulate the PCU functional model: the state of PCU switches from "*off*" to "*NotEngaged*" when the input is valid. When the data value of the input flow "*inSOVL*" is 28, the data value of "*inSOVH*" is 0, and the data value of "*inSurfPosCmd*" equals to 10, the PCU state switches to "*Engaged*". At this time, the data value of the output flow "*outIsEngaged*" of the PCU equals to 1, and the data value of "*outSurfPosCmd*" is 10 (as expected), as shown in Fig. 19 and Fig. 20.

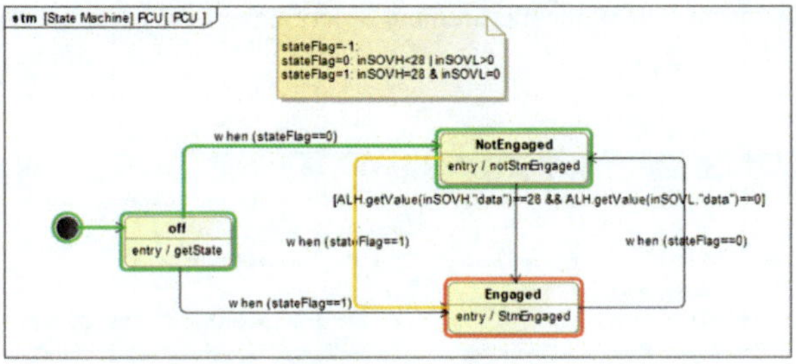

Fig. 19. PCU in "Engaged" state

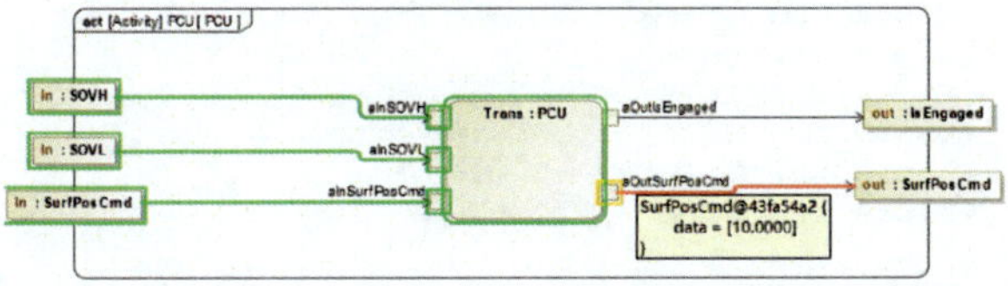

Fig. 20. The flow-processing behavior of PCU in "Engaged" state

(2) **Simulation of Hierarchical Functional Model**
Initialize the inputs and state of the hierarchical functional model, change the inputs of the system functional model based on the state transition behavior and flow-processing behavior of the hierarchical functional model, track the state transition and flow transition of the corresponding functional model in the system, and then complete the simulation of the functional model. For example, the hierarchical functional model of the elevator surface control system above is simulated: after the inputs of the hierarchical functional model are initialized, the state of the hierarchical functional model is shown in Fig. 21; when the input flows are adjusted to switch the state to a normal working state, verify whether the surface position output instruction is compound expectation at this time. It can be seen from Fig. 22 that when the model is in this state and the given input is 11.5°, the output of the elevator surface position is 11.5°, which is in line with the expected output, and the verification of this state flow-processing behavior is completed. Other state transition behaviors and flow-processing behaviors can be simulated according to the same step, and then the simulation verification of system functions can be completed.

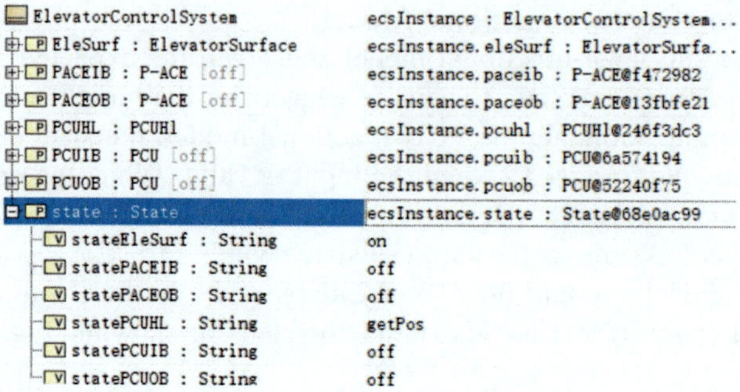

Fig. 21. The initialization state of the hierarchical functional model of the elevator surface control system

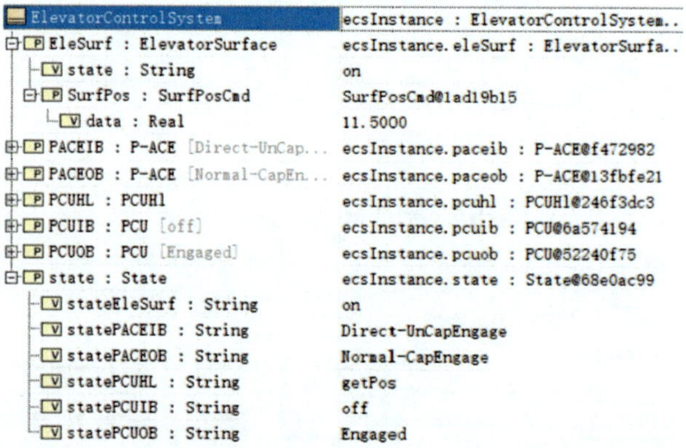

Fig. 22. Surface position control signal output in a certain state

It can be seen from Fig. 19, 20, 21 and 22 that the hierarchical functional model of the elevator surface control system can verify state transition behaviors and flow-processing behaviors. In addition, it can be determined whether the state transition behaviors and flow-processing behaviors of the hierarchical functional model meet expectations, so as to establish an accurate and reliable system functional model.

5 Conclusion

In the conceptual design process of civil aircraft system, accurate establishment of the functional representation of the system serves as the cornerstone for subsequent detailed design. To this end, this paper presents a functional quantitative modeling methodology for civil aircraft systems based on SysML. This approach enables the quantitative expression of the functional model and provides the foundation for its quantitative verification. Specifically, the methodology leverages flow modeling to facilitate quantitative analysis of flow transformation and conversion, while also formally expressing the structure, state, and behavior of the general functional model. Additionally, the approach proposes a method for establishing the hierarchical functional model. To demonstrate the feasibility of the proposed method, the functional modeling of an elevator surface control system is conducted, and the establishment of the functional model and quantitative simulation are succinctly described and demonstrated. The results indicate that the functional modeling methodology described in this paper facilitates the quantitative analysis of the input/output flow of the functional model, thereby providing the necessary foundation for functional completeness verification of the system.

One limitation of the current study is that the proposed method does not address the automated provision of input flows required to complete the state switching of the functional model. As a result, it cannot facilitate automated simulation verification of the state transition and flow-processing behavior of the functional model. To address this shortcoming, future work will involve encapsulating the functional model in the proposed method and developing a system functional simulation model. This will enable the realization of automated simulation verification of complex system function integrity.

References

1. Kwapień, J., Drożdż, S.: Physical approach to complex systems. Phys. Rep. **515**(3–4), 115–226 (2012)
2. Wang, W., Bi, W., Zhang, A., Fan, Q.: Functional modeling of civil aircraft systems based on MBSE. In: Systems Engineering and Electronics (2021)
3. Madni, A.M., Sievers, M.: Model-based systems engineering: motivation, current status, and research opportunities. Syst. Eng. **21**, 172–190 (2018)
4. Chen, Y., Tian, B., Li, Z., Xie, Y.: Philosophical thoughts on some basic concepts of civil aircraft development system engineering. Civil Aircraft Des. Res. **000**(003), 35–41 (2017)
5. Pahl, G., Beitz, W., Feldhusen, J.A., Grote, K.H.: Engineering design: a systematic approach (2007)
6. Gero, J.S., Kannengiesser, U.: The situated function-behaviour-structure framework. Des. Stud. **25**(4), 373–391 (2004)
7. Deng, Y.M., Shu, B.T., Britton, G.A.: A computerized design environment for functional modeling of mechanical products (1999)
8. Umeda, Y., Ishii, M., Yoshioka, M., Shimomura, Y., Tomiyama, T.: Supporting conceptual design based on the function-behavior-state modeler. AI Edam **10**(4), 275–288 (1996)
9. Goel, A.K., Rugaber, S., Vattam, S.: Structure, behavior, and function of complex systems: the structure, behavior, and function modeling language. In: Artificial Intelligence for Engineering design ANALYSIS and Manufacturing, vol. 23 (2008)
10. Guo, G., Tang, H., Luo, Y.: Formalized modeling of product functions based on semantics. Comput. Integr. Manuf. Syst. **17**(6), 7 (2011)
11. Chen, Y., Zhao, M., Huang, J.: A state-behavior-function based approach for functional modeling of multi-state systems and its application. In: ASME 2016 International Design Engineering Technical Conferences and Computers and Information in Engineering Conference (2016)
12. INCOSE SE Handbook Working Group. INCOSE systems engineering handbook v. 3.2. 2. INCOSE-TP-2003-002-03.2.2 (2011)

Check for updates

Research on MagicGrid-Based Requirements Development Process of Flight Control System

Hangqi Liu[1], Yong Chen[1](✉), Meng Zhao[1], and Zhuo Yin[2]

[1] School of Aeronautics and Astronautics, Shanghai Jiao Tong University, Shanghai, China
{liuhangqi,aerocy,magicmeng}@sjtu.edu.cn
[2] Shanghai Institute of Satellite Engineering, Shanghai, China

Abstract. With the development of aviation technology, the complexity of aircraft systems has been increasing. The development of requirements for aircraft systems has become increasingly important. However, traditional requirements development methods often overly rely on text and manual operations, resulting in unclear requirements, difficulties in changes, and low efficiency. They are no longer able to meet the needs of designers and developers. As an emerging requirements development method, the MagicGrid method fully considers the complexity and scalability of the system, and through the decomposition and organization of system-level requirements, designers can understand the system requirements more clearly, and is being continuously applied to aircraft system design.However, in current engineering practice, the MagicGrid method has not yet been fully implemented due to its high complexity and insufficient tool support. Therefore, this paper describes the process of requirements development using the MagicGrid method in detail, and demonstrates the specific process through a practical case of requirements development for the flight control system.Finally, the advantages and disadvantages of the method were further analyzed, and some feasible solutions were proposed, providing new practice and experience for the application of the MagicGrid method in aircraft system design.

Keywords: MagicGrid · Civil Aircraft System · Requirements development

1 Introduction

As a typical complex product system, aircraft have the characteristics of high research and development costs, high degree of technological intensity, long development cycle, and strong interdisciplinary nature [1]. It requires comprehensive multi-disciplinary technologies to optimize the overall performance of the aircraft. Requirements development is the beginning of civil aircraft development, the basis of civil aircraft development, and runs through the entire life cycle of aircraft development [2]. It transmits the customer's requirements layer by layer, so that the final produced aircraft meets the customer's requirements and the requirements of operational safety.

Z. Jing and X. Zhan (Eds.): ICASSE 2023, LNEE 1153, pp. 32–51, 2024.
https://doi.org/10.1007/978-981-97-0550-4_3

The traditional requirements development methods can often be termed as document-based systems engineering, with documents as the core in the requirements development process. However, in the face of highly dynamic and interactive activities, the information expression with the document-based systems engineering approach was often incomplete, making it difficult to ensure the integrity and consistency of engineering requirements, and ambiguity is easy to occur [3]. Additionally, the lack of testing and verification methods makes the requirements development process difficult, and it is easy to bring early requirements problems into the later design process, resulting in substantial changes to the system design process. Therefore, the traditonal document-based systems engineering approach cannot effectively support the development of complex systems of today [4].

Therefore, there should be some new methods for developing the requirements for complex engineering systems. This research examines the MagicGrid approach to see whether it can be applied to the requirements development of complex systems,especially for the development of aircraft systems. MagicGrid is an approach based on system engineering thought, which aim to solve the shortcomings of traditional requirements development methods. Compared with the traditional requirements development method, MagicGrid is expected to provide several advantages. First, it is a model-based systems engineering method [5], which has the advantages of visualization and automation. This allows systems engineers and other stakeholders to better understand the system, reducing manual errors, and improving the efficiency and quality of system development. Second, the MagicGrid method emphasizes the traceability of requirements, which can help designers better manage the requirements development process. Finally, the MagicGrid method has better scalability and can be flexibly adjusted and optimized according to actual requirements. Therefore, the MagicGrid method can well solve the existing problems in the field of requirements development and improve the quality and efficiency of requirements development.

However, there are still some challenges in the practical application of the MagicGrid method. Firstly, the MagicGrid method is relatively complex and requires developers to have high capabilities in order to understand the required technology. In addition, the MagicGrid method requires a lot of experience accumulation to identify and handle problems that arise. Secondly, the types and quantities of aircraft system requirements are enormous, requiring an efficient method for requirement collection and management. Finally, the MagicGrid method needs the support from corresponding tools, and the current tools still have some shortcomings. Therefore, in order to better apply it to aircraft system design, further research and exploration are needed.

The purpose of this article is to investigate the applicability of the MagicGrid method in aircraft system requirements development, and to exploit how to use this method to solve the existing problems of traditional requirements development methods. Through the in-depth analysis of the MagicGrid method and the actual case studies of the development of flight control system requirements, the feasibility of the MagicGrid method in aircraft system requirement development is further verified, providing reference for developers and designers. Finally, some existing deficiencies are identified, and suggestions for improvement are proposed to better apply the MagicGrid method to solve problems in the requirement development process.

2 Overview of the MagicGrid Method

The MagicGrid modeling method, initially developed by No Magic and later integrated into the Dassault Systems 3DE platform, is a relatively recent addition. The MagicGrid approach is based on a framework, which can be represented as a Zachman-style matrix [6]. Its main process is shown in the Fig. 1.

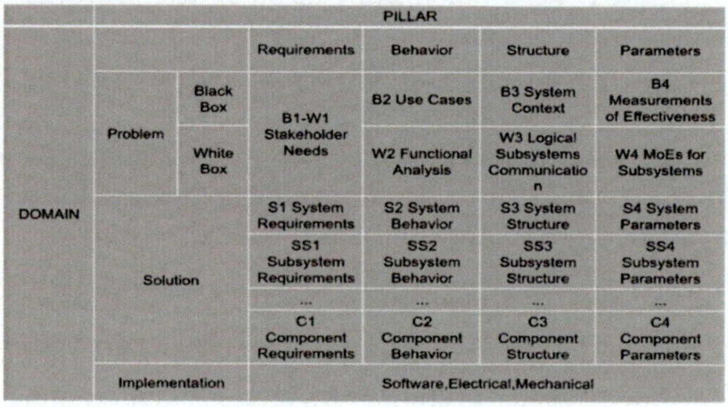

Fig. 1. The Structure of the MagicGrid

From Fig. 1, we can know that the MagicGrid method mainly includes three domains: the problem domain, the solution domain and the implementation domain [7]. Each domain considers and captures four aspects of the system, and these four aspects correspond to the SysML pillars as requirements, behavior, structure, and parameters [8].

The problem domain modeling is the starting point of the MagicGrid method. Its main task is to analyze the requirements of stakeholders [9], and refine them using SysML model elements. The problem domain is divided into two levels, black box and white box [10]. As shown in Fig. 2, the black box mainly focuses on how it is related to the outside world. After completing the black-box modeling of the problem domain, it enters the white-box modeling stage, which involves expanding the black-box model and analyzing it from the inside. In essence, white-box modeling relies on the black-box model, and the output at this stage will be more detailed and complete in describing the system [11].

The main modeling process of the solution domain is shown in Fig. 3. Through modeling in the solution domain, it can help developers understand the problem domain, better discover and understand user requirements [12], reduce the risk and cost of system design.

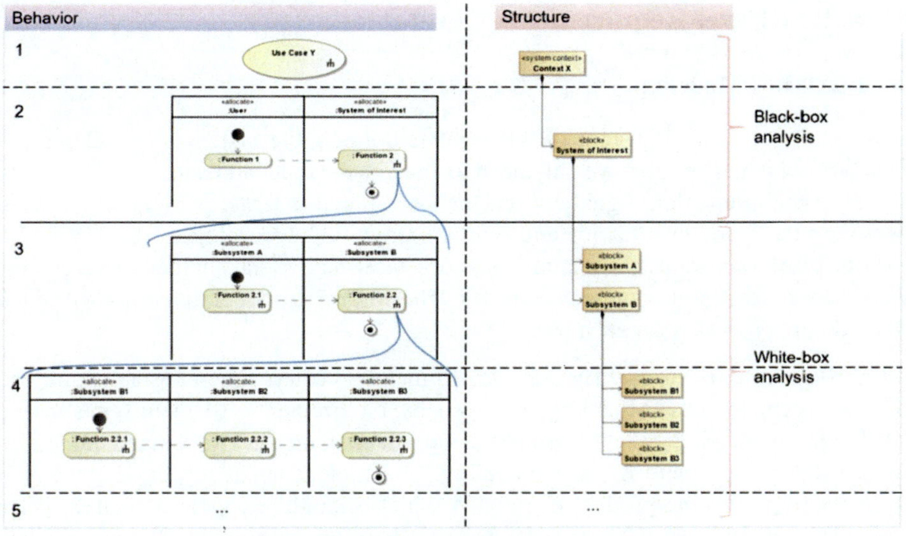

Fig. 2. The Process of the Problem Domain

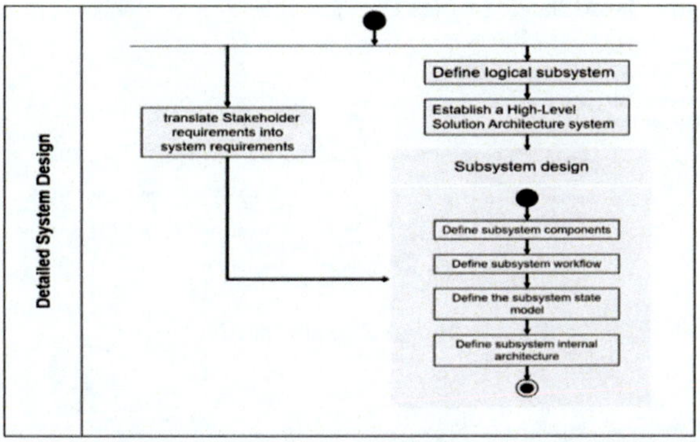

Fig. 3. The process of the Solution Domain

So far, the two core levels of the MagicGrid method have been fully implemented. Although the MagicGrid method defines the physical requirement specification of the implemented system, its detailed design is not part of MBSE, so it will not be analyzed. The following chapters 3 and 4 will describe the development process of the problem domain and solution domain through examples, providing a clearer understanding of the process of requirements development based on the MagicGrid method.

3 System-Level Requirements Development

3.1 Black Box Modeling

In the black box modeling stage of the problem domain, the system is considered as a black box, which means that we only need to analyze the input and output of the system, as well as the interaction logic between the system and external participants, without specifying its dynamic behavior and internal structure. According to the MagicGrid method, black-box modeling mainly includes three parts: stakeholder requirements, system use cases, and system environment. The following model diagrams will detail the modeling processes for each part.

Initial Stakeholder Needs. The MagicGrid method centers around establishing executable system models. Modeling is driven by requirements, so identifying system stakeholders, capturing stakeholder needs, and establishing SysML demand models are prerequisites for problem domain modeling. Stakeholder needs are those directly or indirectly related to the product or project, which should be captured, analyzed. But these needs obtained are often very rough, and, therefore, require detailed analysis and refinement [13]. Because this article mainly focuses on the development of functional requirements, as shown in the Fig. 4, three pieces of stakeholder needs related to flight control functions are listed as examples.

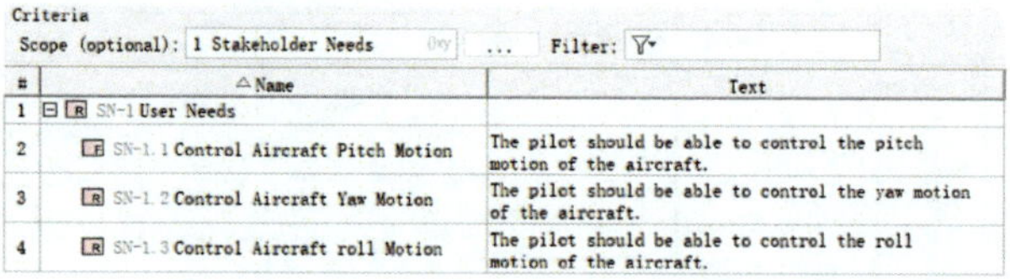

Fig. 4. Initial Stakeholder Needs

Definition and Analysis of Use Case. Use case is a model element derived from the abstraction of the functional requirements in the requirements model, which represents those functional activities that can be directly triggered or participated by external actors. The use case is the core element of the use case diagram and the basis for subsequent modeling and analysis. It will run through the entire project life cycle. In the MagicGrid method, the use cases mainly exist in the problem domain modeling stage, and the use cases are extracted by the top-level functions of the attitude control module. The use case diagram shown in Fig. 5 is established.

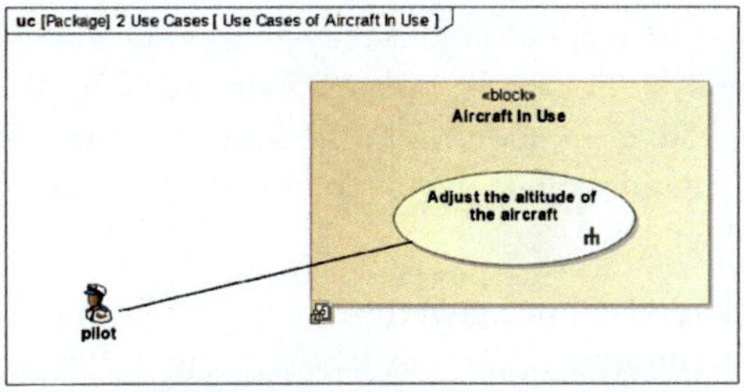

Fig. 5. The Use Case Digram

The use case analysis is mainly to describe the use cases defined in the above process from the perspective of behavior, that is, to establish the dynamic behavior model of the aircraft system. Because we are currently in the black box stage, the established SysML behavior model mainly focuses on the interaction logic with external participants. The Fig. 6 shows a black-box activity diagram of the use case, displaying the system workflow from an external perspective. There are mainly two swim lanes, representing the pilot and the aircraft respectively, where the pilot represents the external participant in the context of the use case.

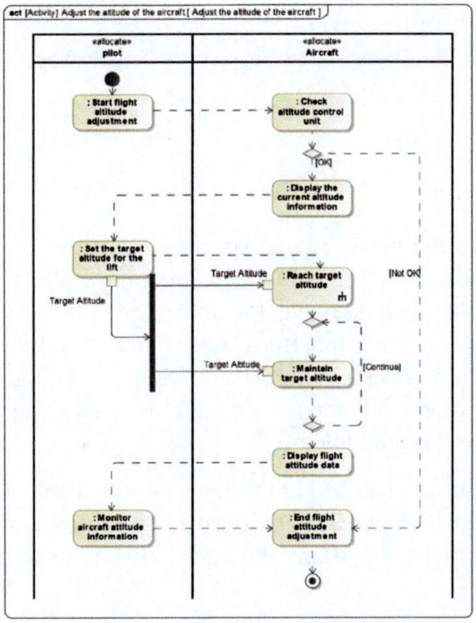

Fig. 6. The activity diagram of adjusting the altitude of the aircraft

System Context. The system environment or system context mainly describes the external environment of the system, that is, creates a structural view of different use cases from an external perspective. In addition to the system itself, elements in a particular system's environment should include the external systems or users that interact with the system in that environment.

Through the analysis and definition of the system environment, developers can better understand the system requirements and ensure that system entities can interact well with the outside world. As shown in Fig. 7, the system environment of the aircraft, the pilot is an external participant, providing control to it.

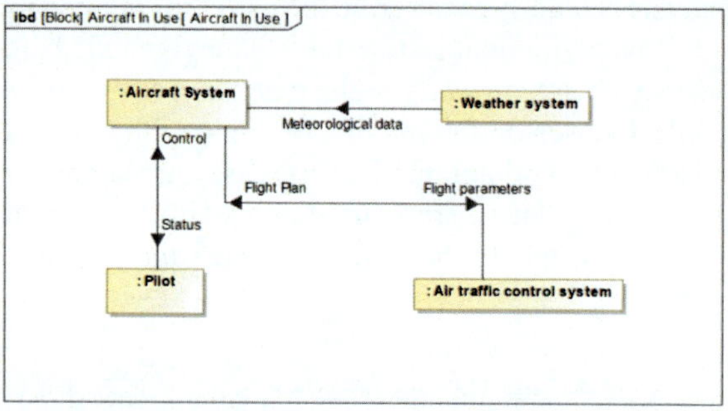

Fig. 7. The System Context for the aircraft system

3.2 White Box Modeling

After the completion of the black-box perspective modeling, we can enter the second stage of problem domain modeling: white-box modeling. White-box means the breakdown of black-box to gain deeper understanding of how the system operates. Through in-depth analysis of the system's functions, functional modules and associated logical subsystems are identified. In essence, the white box modeling stage is to refine and decompose the black box model, and assign more detailed requirements, activities and structures to system function modules.

System-Level Function Analysis. At this stage, establish a more detailed activity diagram model, and assigns behavioral activities to specific subsystems based on certain control logic, that is, which subsystems should satisfy which system behaviors.

Because this paper mainly studies the flight control system, it mainly analyzes the systems related to flight control.In the specific workflow, after receiving the target height command, it is necessary to carry out comprehensive calculation and processing of the height command, and then output the operation commands to other activities through the object flow.As shown in the Fig. 8, the thrust command, pitch command and trim Command are respectively output to other activities that change the thrust, control aircraft pitch, and perform trim operations. The four swim lanes represent the defined logical subsystems, and the corresponding swim lane implement corresponding activity.

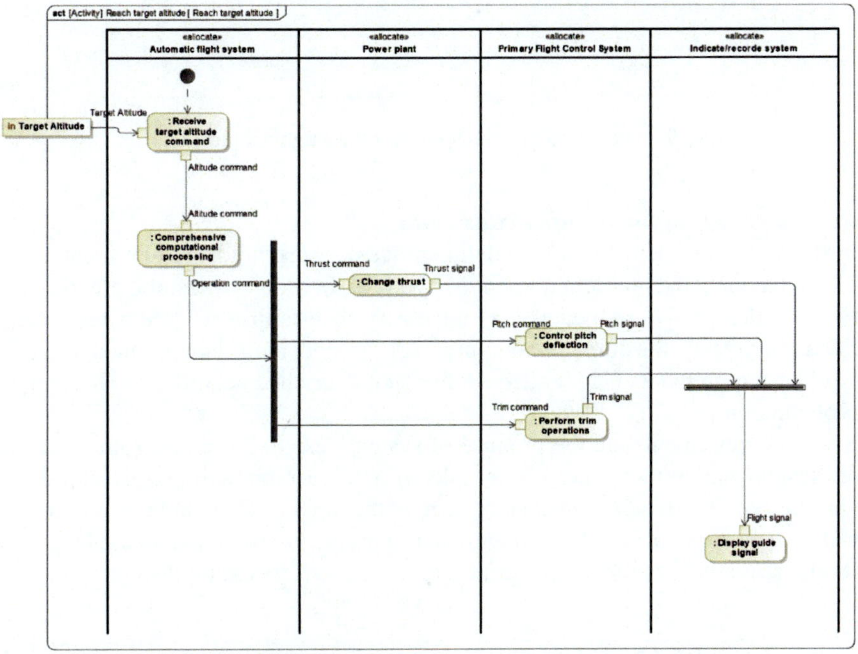

Fig. 8. The SwimLane diagram of reaching target altitude

System-Level Architecture. With the completion of dynamic behavior modeling in the problem domain,the model describes the external and internal behavior activities of the aircraft system in detail from two perspectives of system use cases and functional analysis. However, we still need to further analyze the internal composition structure according to the results of functional analysis.

As shown in the Fig. 9, after further analysis, the related systems of the aircraft are decomposed, the internal IBD diagram of the system is established, the internal functional architecture model is described, and the port transmission of the internal system is established. And according to the events or data information transmitted by the port, the internal systems are connected to each other or to external ports.

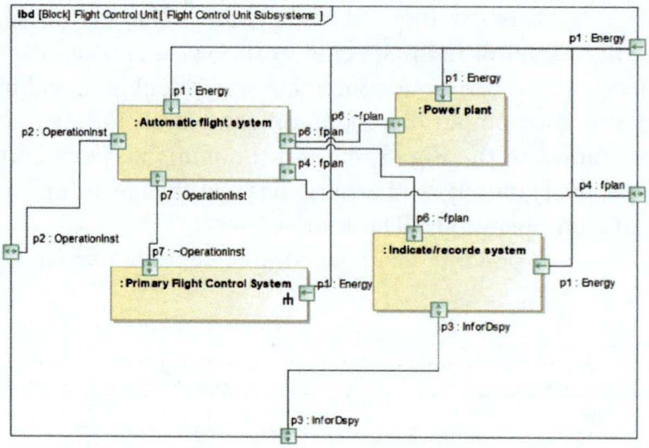

Fig. 9. The internal architecture for the Flight Control Unit

System-Level Requirements and Traceability

During the above problem domain modeling process, a stakeholder requirement model is established to meet the problem domain design task. In order to meet the precise system architecture design requirements, it is necessary to establish a system requirements specification to guide the domain modeling work. System requirements are derived from stakeholder requirements [12], and are refined and enriched according to the model of the problem domain.

System requirements are the premise of system design. Different from stakeholder needs, system requirements are observable, system-specific, and precise expressions. This article mainly discusses the development of the functional requirements of the flight control system, so we only list the functional requirements. As shown in the Fig. 10, five functional requirements for the aircraft system are listed for examples.

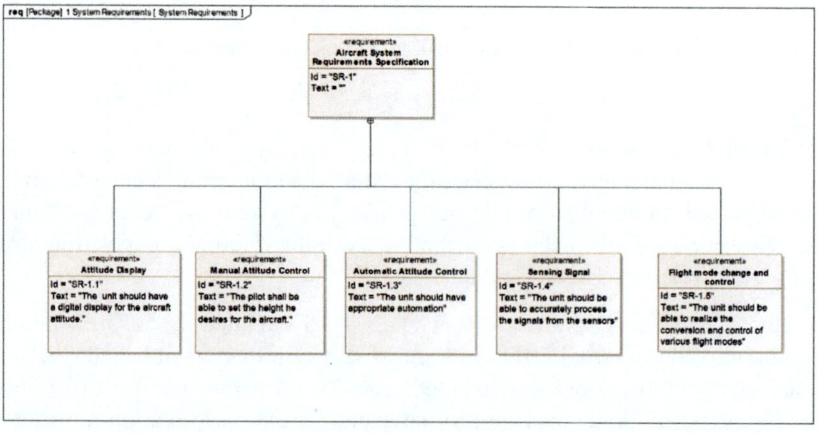

Fig. 10. The functional requirements for the aircraft system

In addition, it is necessary to establish traceability from system requirements to stakeholder requirements, that is, to determine which system requirements are derived from which stakeholder needs.At the same time,traceability relationships between system requirements and problem domain model elements need to be established to determine which system requirements are refined from which elements.

As shown in the Fig. 11, the left part of the picture is the traceability relationship between system requirements and stakeholder requirements. The horizontal row of the correlation matrix represents the system requirements, and the vertical row represents the stakeholder requirements. The arrow indicates that the system requirements of this row point to the stakeholder requirements of this column. The right part of the picture is the traceability relationship between system requirement and problem domain model elements. Similarly, the horizontal rows of the matrix represent the system requirements, and the vertical rows represent the model elements of the problem domain.

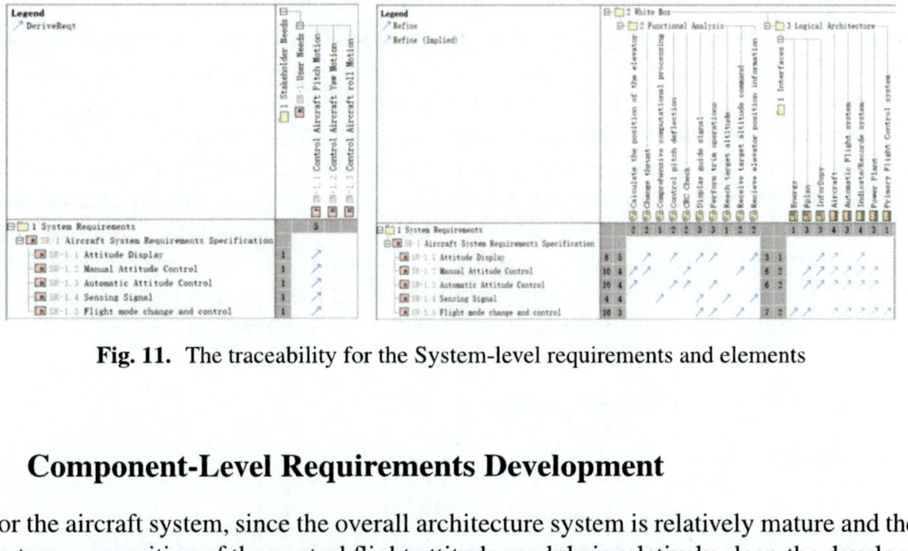

Fig. 11. The traceability for the System-level requirements and elements

4 Component-Level Requirements Development

For the aircraft system, since the overall architecture system is relatively mature and the system composition of the control flight attitude module is relatively clear, the development of system-level requirements can be realized in the field of white-box modeling without re-analysis in the solution domain. Next, the component-level requirements will be further analyzed based on MagicGrid's solution-domain method.

4.1 System-Level Modeling

Defining Subsystems. After the completion of system requirements, the high-level solution structure of the aircraft needs to be established. The HLSA model specifies all logical subsystems in a single hierarchy.Since this article only focuses on the development of the flight control system, only the relevant systems are analyzed. Furthermore, as the aircraft system is a relatively well-structured product system, its system composition is relatively clear, so its HLSA model composition is the same as the problem domain. As shown in Fig. 12, not only the logical subsystem is specified, but also the subsystem ports and the data flow transmitted by the ports are specified.

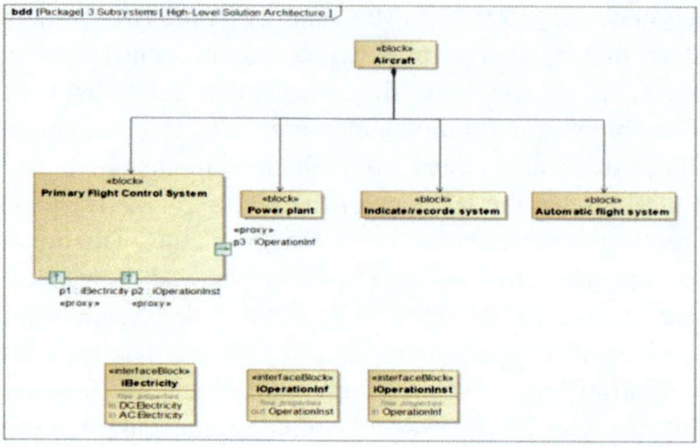

Fig. 12. The High-level Solution Architecture for the aircraft system

Structural Modeling of Primary Flight Control System. This paper selects the primary flight control system as the research object, and based on the required functions,six components of the main flight control system are mainly defined, including the elevator position sensor, remote electronic control unit, elevator aileron computer, actuator,main flight control computer and the actuator control electronic device. The connection relationships between each component and their respective port information are also defined, as shown in the Fig. 13. Three interaction ports between the main flight control system and the external system, the first is the pitch command transmitted by the automatic flight system; the second is the energy supply provided by the power unit; and the third is the pitch signal that needs to be sent to the indication/recording system.

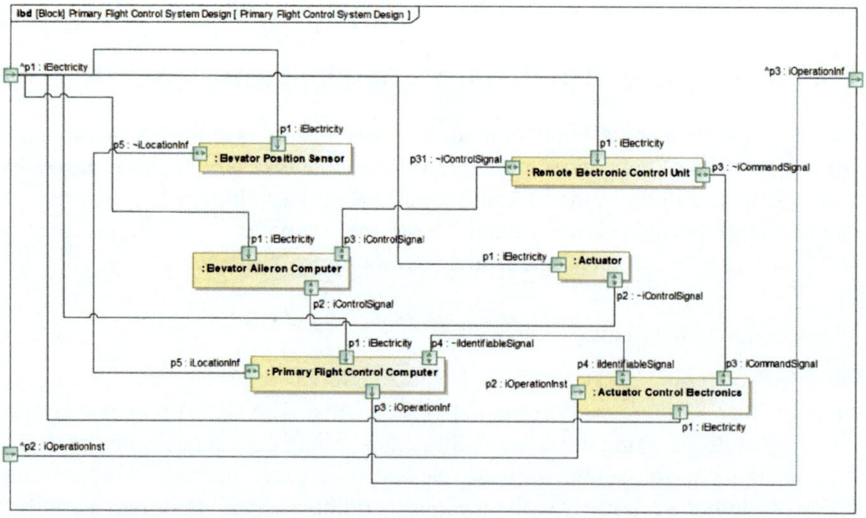

Fig. 13. The internal architecture of the Primary Flight Control System

Behavioral Modeling of Primary Flight Control System. After clarifying the subsystem structure model, it is also necessary to implement the subsystem behavior modeling, to ensure the integrity of the requirements-structure-behavior system design. At this stage, it is necessary to establish an independent behavior model for each subsystem. These models can be relatively abstract and not too detailed, but must completely describe the behavior characteristics and functions of each subsystem. After the behavior models of each subsystem are established, they need to be integrated to ensure the coordination between the various subsystems and the consistency of the system as a whole.

The following Fig. 14 is the activity diagram for the subsystem design, which assigns the activity elements of the main flight control system to their respective components, clarifying which system behaviors should be performed by which components. First, the actuator control electronic device receives the pitch command and converts it into a recognizable signal, which is sent to the main flight control computer for preliminary calculation. Finally, the elevator aileron computer drives the actuator to work according to the control signal. When the pitch adjustment is completed, data is transmitted back to the main flight control computer through the elevator position sensor, and the completed pitch signal is transmitted to the external system.

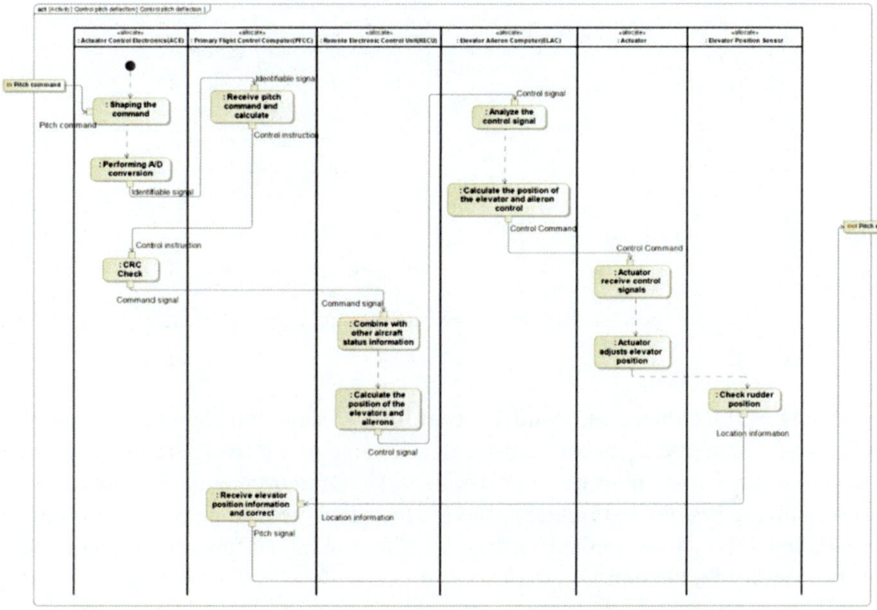

Fig. 14. The swimlane diagram of controlling pitch deflection

Based on the above subsystem activity diagram, it is possible to capture the state of the subsystem and the transition relationships between these states, in order to establish a SysML state machine that describes the changes in the subsystem's operational states. As shown in Fig. 15, the state model of the primary flight control system. Initially, when the system receives the pitch command, it will enter the conversion state. Through the conversion, the pitch command becomes an identifiable signal. It then proceeds through the calculation state and execution state. When the execution action is completed, the system enters the check state to determine whether the altitude requirement has been met. If the required altitude is not reached, the system will continue executing until the it is satisfied. Finally, when the pitch command is completed, the system enters the maintenance state to maintain the current flight attitude.

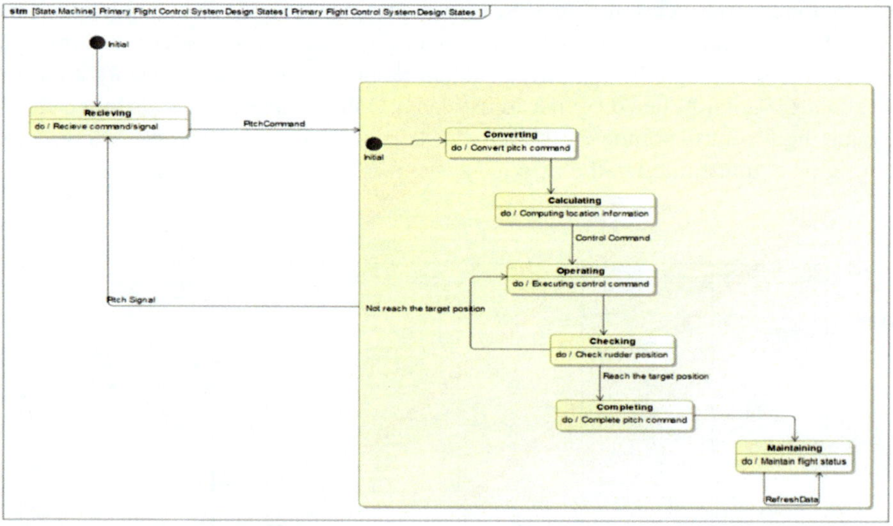

Fig. 15. The state diagram of thePrimary Flight Control System

Subsystem-Level Requirements and Traceability. As the refinement and concretization of system requirements, subsystem-level requirements defines the requirements that the subsystem must meet in order to satisfy the system requirements. They serve as the basis for realizing system requirements and the foundation for component requirements at the next level. As shown in the Fig. 16, there are 5 obtained functional requirements about the primary flight control system.

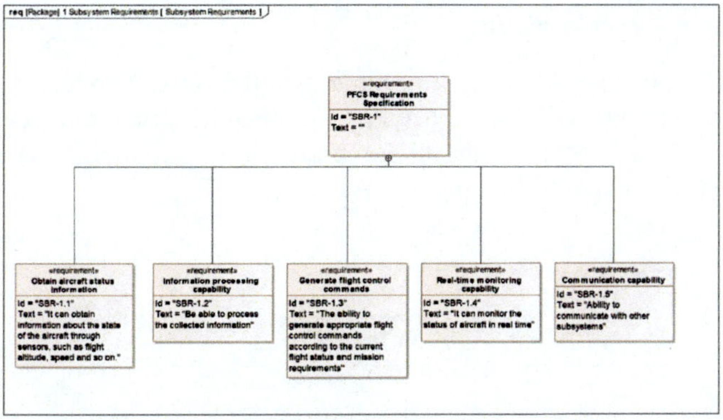

Fig. 16. The funtional requirements for the Primary Flight Control System

In addition, the traceability from subsystem requirements to system requirements can be established, that is, to determine which subsystem requirements are derived from which system requirements, as well as the traceability between subsystem requirements and solution prototype elements. As shown in Fig. 17, the left part of the picture shows the traceability relationship between subsystem requirements and system requirements where the horizontal rows represents subsystem requirements and the vertical columns represents system requirements. Similarly, the right part of the picture shows the traceability relationship between system requirements and subsystem elements. In this case, the horizontal rows represent the model elements of the subsystem, and the vertical columns represent the system requirements.

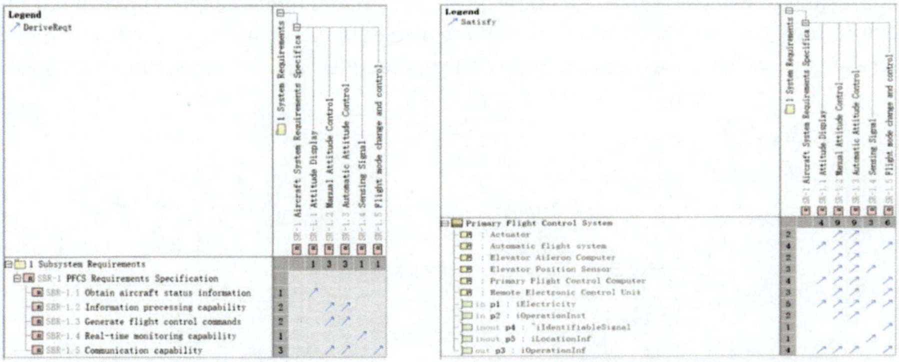

Fig. 17. The traceability for the Subsystem-level requirements and elements

4.2 Component-Level Modeling

Definition of components. After the requirements at the subsystem level are completed, enter the component layer. First, we need to establish the structure model of the main flight control system. Similar to the flight control system model, all components are specified in a single hierarchical structure. As shown in Fig. 18.

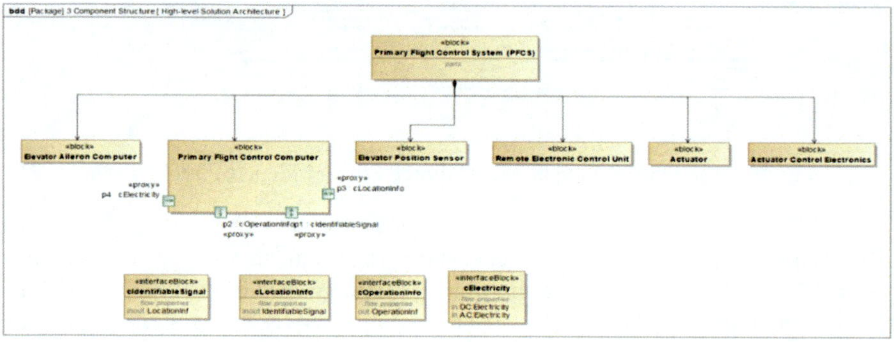

Fig. 18. The High-level Solution Architecture of the Primary Flight Control System

Component structure modeling. When the function allocation is completed within the primary flight control system, its internal solution structure can be studied. This paper selects the main flight control computer of the main flight control system as the research object. According to the required functions, five components were mainly defined, including the input/output interface, central processing unit (CPU), controller memory and sensor data filter. The connection relationships between the components and the port information of each component were also defined, as shown in the Fig. 19. Four interaction ports between the main flight control computer and external components were

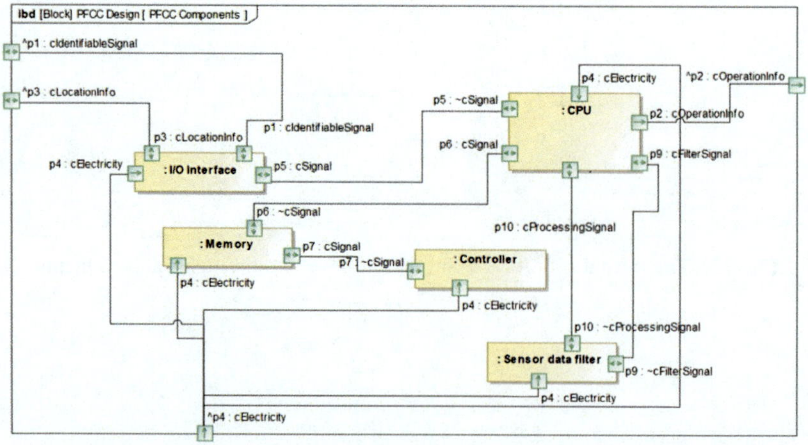

Fig. 19. The internal architecture for the Primary Flight Control Computer

defined, including the identifiable signal received by the main flight control computer; the position information received by the main flight control computer; the power supply provided by the power unit,and the control signal that needs to be sent to the remote electronic control device when the main flight control computer completes execution.

Component Behavior Modeling

After the structure model of the main flight control computer is clarified, behavioral modeling should be carried out. Figure 20 shows the activity diagram. Firstly, the input/output interface receives the identifiable signal and passes it to the central processing unit (CPU). The CPU reads the signal and stores it in the Memory. The Controller monitors the data in the Memory, and when new pitch signal data is detected in Memory, the Controller will send a signal to notify the CPU. The CPU reads the latest pitch signal data from Memory according to the signal from Controller, and inputs it into the Sensor Data Filter for filtering. The signal data processed by the Sensor Data Filter is then passed back to the CPU for calculation, and finally generate corresponding control command.

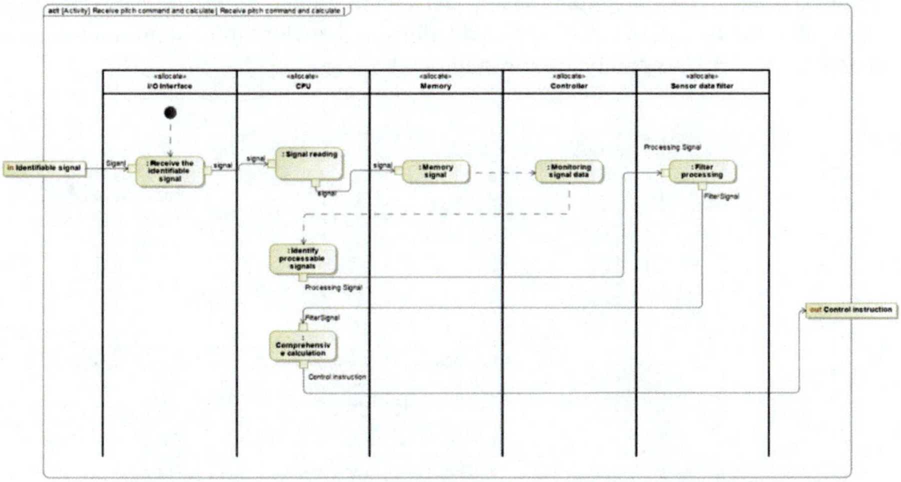

Fig. 20. The Swimlane diagram of recieving pitch command

Component-Level Requirements and Traceability. Component requirements, as the specificization of subsystem requirements at the component level, can enable developers to better understand and implement the system. According to the analysis of the main flight control computer structure and behavior model, component-level requirements can be obtained, as shown in Fig. 21.

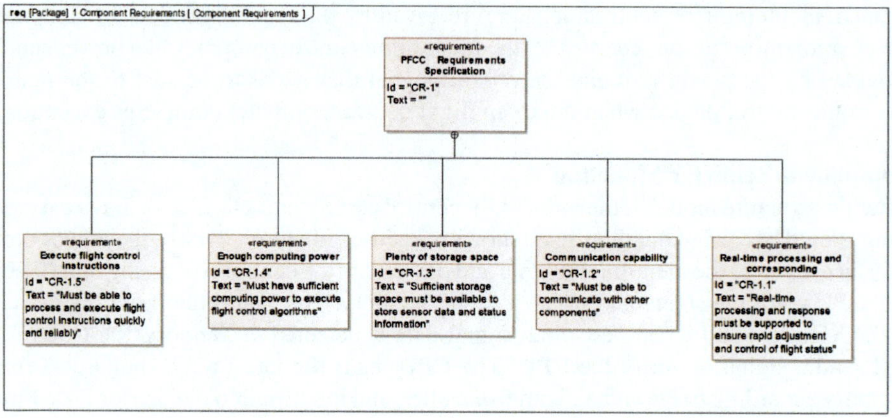

Fig. 21. The functional requirements for the Primary Flight Control Computer

Similarly, the traceability relationship between component requirements and subsystem requirements and the traceability relationship between subsystem requirements and component elements can be established, as shown in Fig. 22.

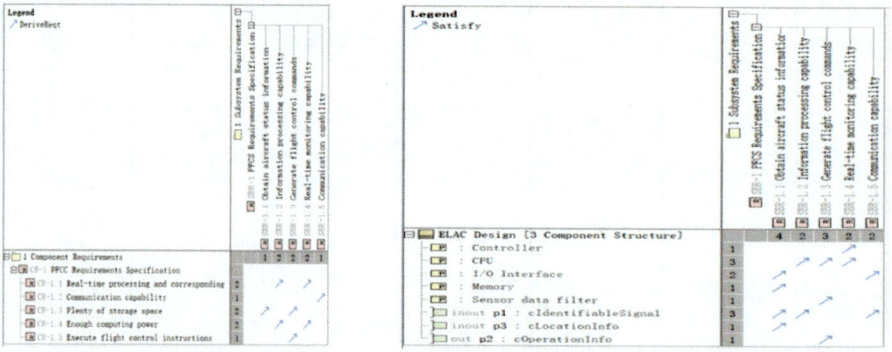

Fig. 22. The traceability for the Component-level requirements and elements

Through the solution domain process of MagicGrid, the high-level system requirements have been successfully transformed into implementable component-level requirements, and traceability relationship has been established between each layer of requirements to ensure that the source of each requirement can be traced, ensuring the accuracy of requirements, and providing clear guidance and assurance for the subsequent developers.

5 Discussion

Through the description of the MagicGrid method and its application in the actual requirements development process, it can be found that the MagicGrid method can produce good results in the requirements development process of complex systems; it can deeply understand and analyze the needs of stakeholders, and it is clear Requirements representation can also help team members better understand and collaborate, track and manage requirements more accurately; they can also have a clearer understanding of the development system for the layered thinking of the problem domain and solution domain raised by it, greatly improve development efficiency.

Although the MagicGrid method has advantages over traditional requirement development methods in some aspects, there are still some drawbacks. One of the main problems is consistency, because the MagicGrid method is based on text and may suffer from semantic ambiguity and inconsistency. For example, different people may use different vocabulary to describe the same requirement, which may result in duplicated or unclear requirements in the model.

To address this issue, a synonym dictionary can be defined to map different expressions to the same concept, thus avoiding inconsistency caused by different expressions. For example, when processing requirements for controlling an plane's height, different expressions such as "controlling the plane attitude angle", "controlling the plane pitch angle", and "controlling the plane climb and descent angle" can be mapped to the same concept, such as "controlling the pitch angle". This can eliminate semantic ambiguity and inconsistency and better understand and process requirements. In building a synonym dictionary, techniques such as corpus and word vectors can be used to match and map words or phrases with similar semantics (Fig. 23).

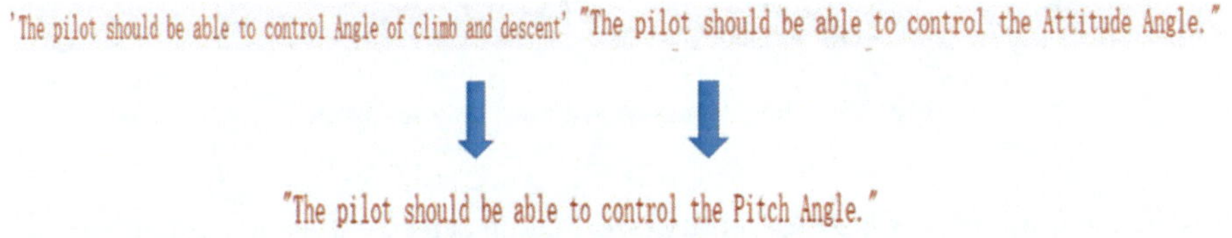

Fig. 23. Synonym replacement

Another approach is to use named entity recognition (NER) to identify professional terms, adopt a standardized terminology, and maintain it jointly by users and developers to ensure consistency of understanding. This can effectively eliminate the problem of duplication or ambiguity of requirements caused by different people using different words, and improve the accuracy and reliability of the model. For example, for the requirement "the aircraft transmits tower instructions to the pilot," NER technology can identify the subject (plane), object (pilot), and action (transmit tower instructions) in the requirement, and form a structured description to express the content and object of the requirement more clearly.

First, a certain amount of requirements data needs to be sorted out and labeled to establish a training set. Labeling requirements can use methods such as markup language to mark different parts of the requirements (such as subjects, objects, actions, etc.) separately (Fig. 24).

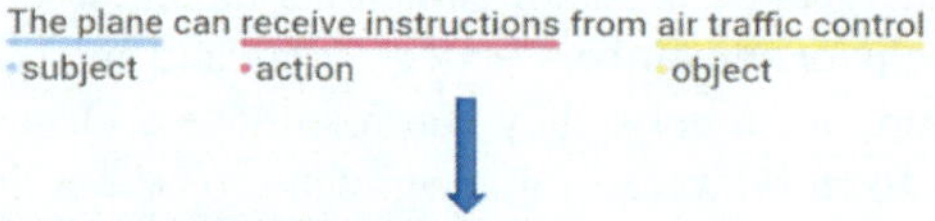

{"id":157,"text":"The plane can receive instructions from air traffic control","entities":[{"id":468,"label":"subject","start_offset":0,"end_offset":9},{"id":469,"label":"action","start_offset":14,"end_offset":34},{"id":470,"label":"object","start_offset":40,"end_offset":59}],"relations":[]}

Fig. 24. Label the data

Next, named entity recognition is used to train on the labeled data to learn how to recognize different entity types and their positions in sentences. During the training process, the accuracy of the model can be continuously optimized by adjusting algorithm parameters and increasing training data (Fig. 25).

```
global step 10, epoch: 5, loss: 0.00586, speed: 2.69 step/s
Evaluation precision: 0.65591, recall: 0.63542, F1: 0.64550
best F1 performence has been updated: 0.00000 --> 0.64550
global step 20, epoch: 10, loss: 0.00382, speed: 2.77 step/s
Evaluation precision: 0.92857, recall: 0.81250, F1: 0.86667
best F1 performence has been updated: 0.64550 --> 0.86667
global step 30, epoch: 15, loss: 0.00276, speed: 2.77 step/s
Evaluation precision: 0.91667, recall: 0.91667, F1: 0.91667
best F1 performence has been updated: 0.86667 --> 0.91667
global step 40, epoch: 20, loss: 0.00213, speed: 2.77 step/s
Evaluation precision: 0.91667, recall: 0.91667, F1: 0.91667
global step 50, epoch: 25, loss: 0.00173, speed: 2.76 step/s
Evaluation precision: 0.91667, recall: 0.91667, F1: 0.91667
global step 60, epoch: 30, loss: 0.00145, speed: 2.77 step/s
Evaluation precision: 0.91667, recall: 0.91667, F1: 0.91667
```

Fig. 25. The Process of Training Labeled Data

After the training is completed, entity information can be extracted from the requirement text by referencing the optimal model. For example, the requirement "The plane transmits tower instructions to the pilot" can extract entities "The palne", "the pilot" and "transmit tower instructions" and form a structured description, such as "the plane transmits tower instructions to the pilot". This structured description can express the content and object of the requirement more clearly, avoid unnecessary ambiguity and inconsistency, and improve the accuracy and reliability of the model (Fig. 26).

[{"subject": [{"text": "The plane", "probability": 0.9978847537644242}], "object": [{"text": "the pilot", "probability": 0.9992364275358696}], "action": [{"text": "transmit tower instructions", "probability": 0.9757586922001451}]}]

Fig. 26. The result of Named Entity Recognition

6 Conclusion

Requirements development, as the foundation of the aircraft development, runs through the entire life cycle of aircraft development. However, traditional requirements development method is based on text, which has problems such as incomplete information expression and ambiguity. Therefore, this paper proposes to introduce the requirements development method based on MagicGrid into the aircraft system design process. This paper demonstrates the specific process of using MagicGrid for requirements development in the flight control system example, successfully obtaining system-level, subsystem-level and component-level requirements, and establishing traceability to achieve requirement traceability. However, this method also has problems of some redundant and ambiguous requirements. Therefore, this paper proposes some improvement ideas, but it still needs further improvement and integration with the MagicGrid method in the future work, in order to better apply the MagicGrid method to solve the problrms in the aircraft system requirement development process and improve the quality of requirement development.

References

1. Chen, Y., Tian, B., Liu, Z., Xie, Y.: Philosophical thoughts on some basic concepts of civil aircraft development system engineering. Civil Aircraft Des. Res. **03**, 35–41 (2017)
2. Deng, Z.: Overview of the application of requirements engineering in civil aircraft life cycle project management. Technol. Vision **26**, 255 (2015)
3. Wu, Y., Liu, J., Zheng, D.: Analysis of model-based systems engineering technology. Aviat. Sci. Technol. **9**, 69–73 (2015)
4. Zhang, B., Qi, F., Xing, T., Liu, Y., Wang, W.: Research and practice of model-based manned spacecraft development method. Aeronaut. J. **41**(7), 72–80 (2020)
5. Estefan, J.A.: Survey of model-based systems engineering (MBSE) methodologies. Incose MBSE Focus Group **25**(8), 1–12 (2007)
6. Morkevicius, A., Aleksandraviciene, A., Armonas, A., Fanmuy, G.: Towards a common systems engineering methodology to cover a complete system development process. In: INCOSE International Symposium, vol. 30, no. 1, pp. 138–152 (2020)
7. Wang, L., Zhan, C.: Modeling and analysis of civil aircraft top level RNP system architecture. In: Journal of Physics: Conference Series, vol. 1827, no. 1, p. 012118. IOP Publishing (2021)
8. Lu Z, Liu X, Mao Y, Fan H, Zhao Y.: Application of model-based system engineering method in satellite overall design. Spacecraft Eng. (03), 7–16 (2018)
9. Weihao, L., Yuqiang, G., Qipeng, C., Hui, Z.: Model-based system engineering adoption for trade-off analysis of civil helicopter fuel supply system solutions. In: Krob, D., Li, L., Yao, J., Zhang, H., Zhang, X. (eds.) Complex Systems Design & Management, pp. 311–323. Springer, Cham (2021). https://doi.org/10.1007/978-3-030-73539-5_24
10. Morkevicius, A., Aleksandraviciene, A., Krisciuniene, G.: From UAF to SysML: transitioning from system of systems to systems architecture. In: INCOSE International Symposium, vol. 31, no. 1, pp. 585–598 (2021)
11. Cui, Z., et al.: MBSE for civil aircraft scaled demonstrator requirement analysis and architecting. IEEE Access **10**, 43112–43128 (2022)
12. Herfman, H.: Model-Based System Engineering Best Practices. Aviation Industry Press, Beijing (2014)
13. Yang, M., Qian, X., Wang, J.: Research on capture method of civil aircraft operational requirements based on DoDAF modeling. J. Armament Eng. (06), 81–85 (2018)

Airport Collaborative Decision-Making in Single Pilot Operations of Commercial Aircraft

Yue Luo[1], Miao Wang[1](✉), Yong Chen[1,2], Kelin Zhong[3], and Guoqing Wang[1]

[1] Shanghai Jiao Tong University, Shanghai, China
{luoyue,miaowang,wang_guoqing}@sjtu.edu.cn, chenyong@comac.cc
[2] Commercial Aircraft Corporation of China, Shanghai, China
[3] COMAC Shanghai Aircraft Design and Research Institute, Shanghai, China
zhongkelin@comac.cc

Abstract. As an important direction of modern aviation technology, single pilot operations (SPO) of commercial aircraft puts forward a strong demand for airport capability and efficiency, so it is necessary to establish the airport collaborative decision-making (A-CDM) capability under SPO. Based on the system architecture of A-CDM under current dual-pilot operation mode, this paper analyzes the core information sharing mechanism and milestone method. Then, based on the characteristics of SPO, SPO-A-CDM is proposed. Finally, a co-simulation platform of flight scenario and system model is constructed, and a typical scenario – fixed-point deicing is taken as an example to carry out the model-based simulation analysis of SPO-A-CDM.

Keywords: Airport Collaborative Decision-making · Single Pilot Operations · Fixed-point Deicing · Commercial Aircraft

1 Introduction

With the rapid development of civil aviation, the scale and business volume of airports are also expanding day by day. Many large airports are faced with limited number of resources such as runways, parking spaces and surface support facilities, and low rate of intelligent resource allocation, which leads to low operating efficiency and poor collaborative decision-making ability [1]. In order to solve the above problems, airport collaborative decision-making (A-CDM) was proposed, aiming at making the airport, ground handling, air traffic control (ATC), airline and aircraft participate in collaborative decision-making through information sharing, so as to optimize the utilization rate of airport resources, reduce flight delays and improve the operation efficiency of the airport [2, 3].

Single pilot operations (SPO) is one of the core technologies for the new generation of commercial aircraft, which is of great significance in the aspects of operation economy and flight safety [4, 5]. In SPO, there is only one pilot on board, who has limited cognition and perception, and requires collaborative interaction with the ground to obtain a more comprehensive situational awareness. According to the requirements of airport

Z. Jing and X. Zhan (Eds.): ICASSE 2023, LNEE 1153, pp. 52–60, 2024.
https://doi.org/10.1007/978-981-97-0550-4_4

development in the Aviation System Block Upgrade (ASBU) proposed by International Civil Aviation Organization (ICAO), it is necessary to establish the A-CDM capability under SPO.

Therefore, this paper first introduces the system architecture of A-CDM under the current dual-pilot operation mode, and analyzes the core information sharing mechanism and milestone method. Then, based on the characteristics of SPO, SPO-A-CDM is proposed. Finally, a co-simulation platform of flight scenario and system model is constructed, and a typical scenario – fixed-point deicing is taken as an example to carry out model-based simulation analysis of SPO-A-CDM.

2 A-CDM in Dual-Pilot Operation Mode

2.1 A-CDM System Architecture

Airport Collaborative Decision Making (A-CDM) is an operational mechanism led by the airport and jointly participated by aircraft, airline and Air Traffic Control (ATC), based on information sharing, guided by "improving the efficiency of surface operations", and centered on "collaborative decision making" [6, 7].

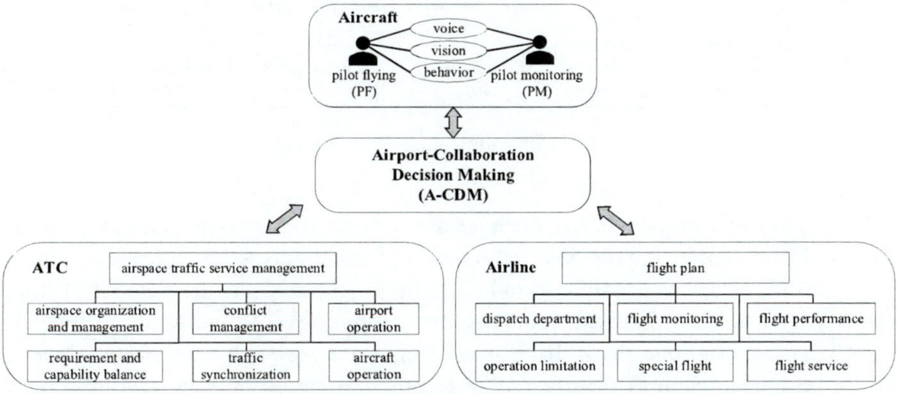

Fig. 1. System architecture of A-CDM.

As shown in Fig. 1, in the A-CDM system architecture, ATC is responsible for airspace traffic service management, including airspace organization and management, requirement and capability balance, conflict management, traffic synchronization, airport operation and aircraft operation; Airline is responsible for flight plan execution, including dispatch department, flight monitoring, flight performance, operation limitation, special flight and flight service. In the non-nominal flight, the PF and PM complete collaborative decision making through voice, vision and behavior and other face-to-face interactive ways. Ultimately, all parties can make a collaborative decision based on their own requirements, resulting in a favorable processing result that is beneficial to all parties involved.

2.2 Information Sharing

The prerequisite of collaboration is to solve the information interaction among all participants [8]. In A-CDM, real-time information sharing among air-ports, air traffic control, airlines and aircraft is the basis of collaborative decision-making. As shown in Fig. 2, all participants can perceive the common situation of airport and flight operation through the information sharing plat-form. Based on the shared information, the participating units can collaborate to make safer and more effective decision-making solutions and optimize their own decisions, which helps the airport effectively optimize the forecast of the overall operating situation, improve the operating efficiency, and then improves the accuracy of prediction and flight frequencies for aviation management.

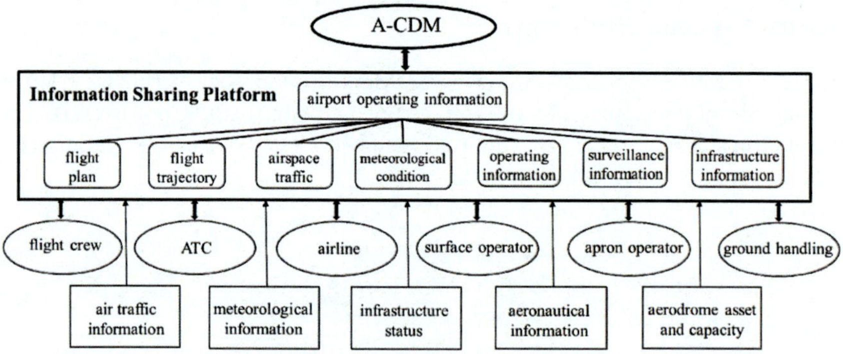

Fig. 2. Information sharing in A-CDM.

After the information is shared, the interests and demands of all parties can be further analyzed. For example, the resident airlines want their flights to be launched according to the schedule as far as possible to reduce delays; ATC expects the number of flights to be adjusted to the minimum, reducing the average takeoff and landing time; The airport hopes to improve the utilization rate of resources such as parking place and corridor bridge as much as possible, so as to obtain a larger throughput. Based on the requirements of all par-ties, a compromise that satisfies all parties can be reached through collaborative decision-making.

2.3 Milestone Method

Milestone events are significant events that occur during flight planning and flight operations. Completed milestone events will affect the decision-making process of the subsequent event, which will influence subsequent flight operations and accurate prediction of subsequent events. Defining mile-stones and tracking them helps improve each participating unit's awareness of flight situations, as well as anticipate potential flight delays and develop response plans in a timely manner. Figure 3 shows the key milestones of flight operations, which covers the entire airport operation from inbound, turnaround and outbound [9].

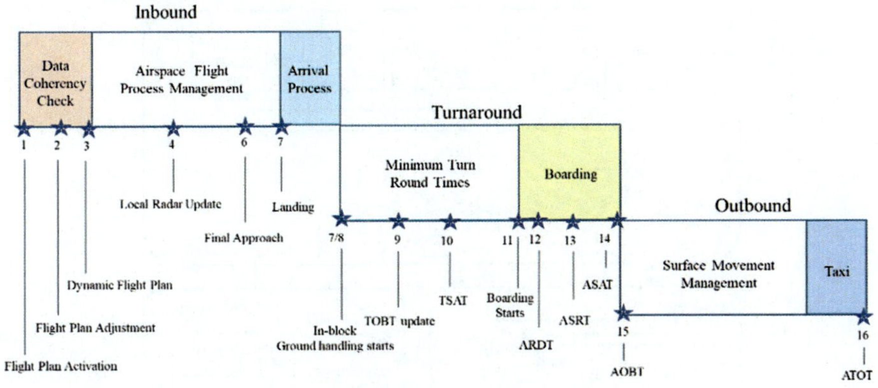

Fig. 3. Milestone method in A-CDM.

3 A-CDM in Single Pilot Operations Mode

3.1 Single Pilot Operations Mode

Single pilot operations (SPO) is a piloting mode in which there is only one pilot in the cockpit. To avoid excessive workload of the single pilot, more advanced airborne intelligent system and ground support are introduced, as shown in Fig. 4. The airborne intelligent system is divided into automatic, autonomous and intelligent modes. In the nominal flight, the airborne intelligent system works in the automatic mode and runs according to the preset automatic program. In the non-nominal flight, the airborne intelligent system works in the autonomous mode, which performs flight optimization processing based on input conditions and external environment. In the authorized airspace, the airborne intelligent system works in the intelligent mode. It can provide suggestions and support based on reasoning, learning, and mining, as a supplement to the single pilot's cognition.

The ground support system includes ground simulation system and a ground operator who is hired part-time by the airline dispatcher. The ground simulation system can visualize the received flight status for the convenience of the ground operator. The ground operator is responsible for monitoring the flight, interacting with ATC and, in non-nominal cases, coordinating with the single pilot on board. As a result, the decision-making mode of the cockpit has changed from the original face-to-face communication to human-machine collaborative decision-making and air-ground collaborative decision-making.

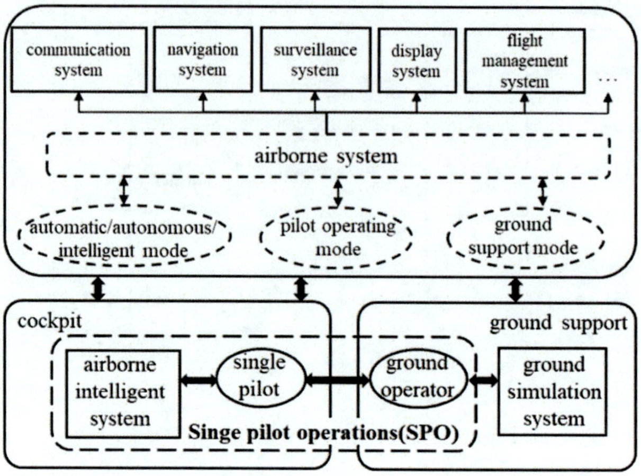

Fig. 4. Single pilot operations (SPO) mode.

3.2 Decision Process Organization of Single Pilot Operation

In SPO, the pilot on board makes flight decisions based on flight plan compliance, airspace environment adaptability, flight performance feasibility, air-craft/system capability availability, and flight safety effectiveness [10]. In different situations, the airborne intelligent system works in different modes to provide decision support for the single pilot on board and the operator on the ground.

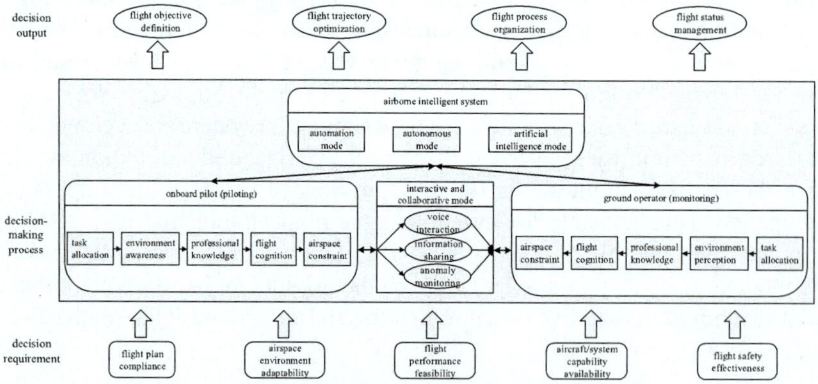

Fig. 5. Decision process organization of SPO.

Based on different task allocation, environmental awareness, professional knowledge, flight cognition and airspace constraint, the onboard pilot and the ground operator finally complete air-ground collaborative decision-making through voice interaction, information sharing, and anomaly monitoring, as shown in Fig. 5.

3.3 SPO-A-CDM

Due to the change of decision-making mode in SPO, SPO-A-CDM requires intelligent systems, including airborne intelligent system and ground intelligent system, to support human-machine collaborative decision-making and air-ground collaborative decision-making, so as to ensure flight safety. In addition, in view of the difficulty of information sharing and interaction existing in A-CDM, an information sharing platform need to be established to conduct intelligent processing and distribution of shared information, and to support intelligent decision making. The system architecture of SPO-A-CDM is shown in Fig. 6.

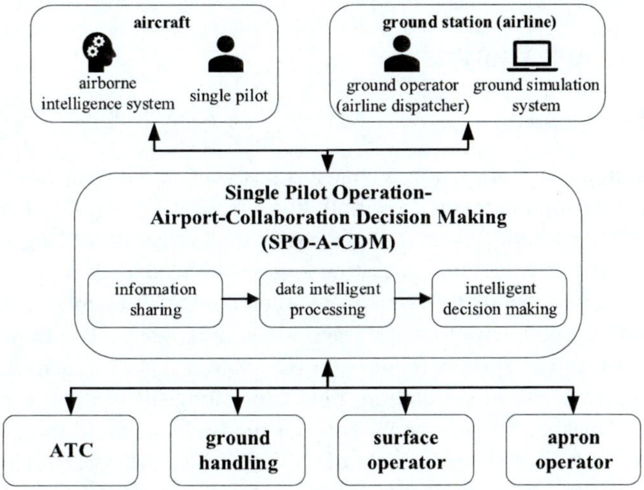

Fig. 6. The system architecture of SPO-A-CDM.

1) Intelligent system support
SPO-A-CDM cannot be separated from intelligent systems, including air-borne intelligent system and ground intelligent systems. Among them, the air-borne intelligent system can be divided into automatic, autonomous, and intelligent modes based on different situations. In intelligent mode, it can pro-vide solutions to pilots to support flight decisions based on reasoning, learning, and mining. The ground intelligent system relies on mobile Internet, geo-graphic information system (GIS), big data mining and other technologies to realize real-time perception of flight operation and vehicle activities, and intelligent allocation of ground resources to support airport collaborative decision-making [11].
2) Information sharing and intelligent processing
In SPO, in order to reduce the workload of the single pilot, many tasks will be handed over to the intelligent system and the ground. The consistency of flight and surface operating information obtained by all parties is the premise of collaborative decision-making, and information sharing is in fact the “glue” that ties these partners together. For airports, gate location, taxiing time from gate location to runway, expected start

time and length of deicing time should be provided. For ground handling, target guaranteed completion time, minimum layover time, flight plan and departure priority should be provided. For airlines, scheduling data should be provided. For ATC, estimated departure time, actual departure time, area control time, terminal control time, landing time of the flight, etc. should be provided. In addition, the aircraft needs to synchronize flight status to the ground station to support coordination and interaction between the ground operator and other parties.

Cloud computing is used to efficiently process and transmit the information of all parties through the information sharing platform to support the collaborative decision-making of all parties.

4 Modelling and Analysis

4.1 Simulation Platform

Firstly, the airport surface operation scenario is designed and the surface operating process is modelled through the scenario simulation software Prepar3D. Then, the internal logic of the surface operation is modelled through the system modelling software Magicdraw. The system architecture, operating organization, and information sharing of SPO-A-CDM will be modelled. A synchronization component between P3D and Magicdraw will be developed, with the flight scenario simulation module transmitting flight status information to the system model and the system model transmitting milestone trigger to the flight scenario simulation, thus controlling the operating process of the airport surface. Through the joint simulation of the flight scenario model and the system model, the analysis and verification of the SPO-A-CDM operational organization is completed. The simulation platform of SPO-A-CDM is shown as Fig. 7.

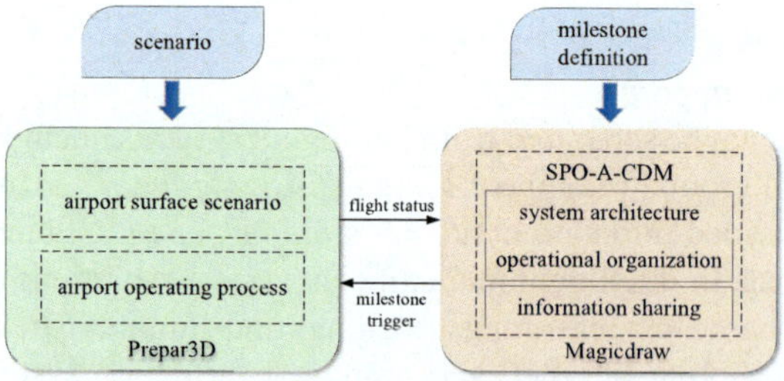

Fig. 7. The simulation platform of SPO-A-CDM.

4.2 Model-Based Simulation Analysis

In this section, a typical surface scenario – fixed-point deicing is taken as an example to carry out the model-based simulation analysis of SPO-A-CDM. At the beginning of the ice season, a deicing area is set up to provide deicing services for aircraft at a fixed location. In the event of ice and snow, after passengers have boarded the aircraft, the single pilot on board reports the ice buildup to the airline. Then, the ground operator applies to the airport for de-icing. The airport arranges the aircraft to proceed to the appropriate deicing area according to the progress of deicing. The ground operator receives the deicing information and synchronizes them with the airborne intelligent system. Then, the airborne intelligent system automatically loads the route and proceeds to the designated deicing area under the guidance of ATC. After de-icing, the ground operator applies to ATC for a taxi out, and then ATC issues a taxi clearance, the single pilot accepts the taxi clearance and starts taxiing [12].

Collaborative deicing under SPO-A-CDM involves interaction among the single piloted aircraft, the airline operation center, the ground operator, ATC and the airport. The information interaction model for collaborative deicing constructed through Magicdraw is shown in Fig. 8.

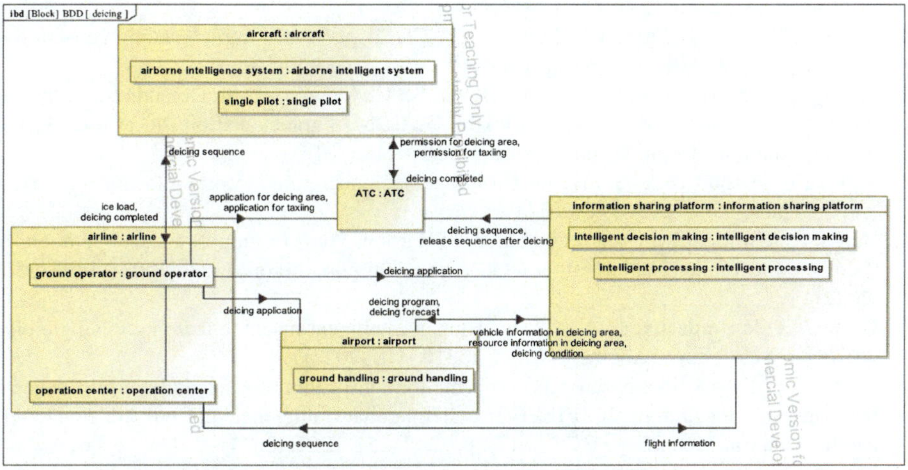

Fig. 8. The information interaction model for collaborative deicing.

The deicing process is combined with SPO-A-CDM to provide accurate flight information for deicing process through information sharing, ultimately facilitating deicing planning. By sharing deicing scheduling information, through intelligent processing and intelligent decision making, limited deicing resources can be planned and used in a rational manner to maximize the efficiency of deicing at the airport, enabling monitoring, scheduling and com-manding of deicing management, and optimizing the sequencing of flight departures to improve airport operating efficiency.

5 Conclusion

Based on the characteristics of SPO, SPO-A-CDM is proposed in this paper. Through the co-simulation platform of flight scenario and system model, model-based simulation analysis of SPO-A-CDM is carried out. By sharing in-formation, through intelligent processing and intelligent decision making, limited airport resources can be planned and used in a rational manner to maximize the efficiency of airport operation.

Acknowledgements. This work is partially supported by Natural Science Foundation of Shanghai (20ZR1427800). The authors also gratefully acknowledge the helpful comments and suggestions of the reviewers, which have improved the presentation.

References

1. Netto, O., Silva, J., Baltazar, M.: The airport A-CDM operational implementation description and challenges. J. Airline Airport Manag. **10**(1), 14–30 (2020)
2. Corrigan, S., Mårtensson, L., Kay, A., Okwir, S., Ulfvengren, P., McDonald, N.: Preparing for Airport Collaborative Decision Making (A-CDM) implementation: an evaluation and recommendations. Cogn. Technol. Work **17**, 207–218 (2015)
3. Joppart, P.: Airport collaborative decision making: Integrating airports into the network. J. Airport Manag. **5**(4), 295–297 (2011)
4. Harris, D.: Estimating the required number of Harbour Pilots to support airline operations of a single pilot commercial aircraft at a UK regional airport. Aeron. J. **126**(1303), 1497–1509 (2022)
5. Universit'e de Bordeaux, C. A. T. I. E.: Human intelligent machine teaming in single pilot operation: a case study. Augment. Cogn. **13310**, 348 (2022)
6. Zhang, X., Miao, X.: An optimization scheme for consortium blockchain in airport collaborative decision making. In: 2021 IEEE 3rd International Conference on Civil Aviation Safety and Information Technology (ICCASIT), pp. 247–251. IEEE (2021)
7. Verkerk, M.: Airport collaborative decision making: improving the information position of the turnaround coordinator in managing the target off block time (2018)
8. Eriksen, P.: Collaborative decision making information management in airports. In: Proceedings. The 21st Digital Avionics Systems Conference, vol. 2, pp. 7B4–7B4. IEEE (2002)
9. EUROCONTROL. Aiport CDM Implementation Manual, Edition 4 (2012)
10. Miao, W., Yue, L., Kai, H., Zhao, P., Guoqing, W.: Optimization and verification of single pilot operations model for commercial aircraft based on biclustering method. Chin. J. Aeron. **36**, 286–305 (2022)
11. Zheng, Y., Miao, J., Le, N., Jiang, Y., Li, Y.: Intelligent airport collaborative decision making (A-CDM) system. In: 2019 IEEE 1st International Conference on Civil Aviation Safety and Information Technology (ICCASIT), pp. 616–620. IEEE (2019)
12. Labeas, G.N., Diamantakos, I.D., Sunaric, M.M.: Simulation of the electroimpulse de-icing process of aircraft wings. J. Aircr. **43**(6), 1876–1885 (2006)

Research on Capability Catalog Generation of UAV Intelligent System Based on DoDAF

Yuexing Wang[1], Xiaoxu Dong[2], Jiazhi Yan[1], and Miao Wang[2](✉)

[1] AVIC (Chengdu) UAS Co., Ltd., Chengdu, China
[2] School of Aeronautics and Astronautics, Shanghai Jiao Tong University, Shanghai, China
miaowang@sjtu.edu.cn

Abstract. Unmanned aerial vehicles (UAVs) can carry out tasks such as border patrol and sea guard, which has a broad application prospect. How to apply intelligent technology in UAV system, so as to improve the efficiency of task execution in detection and route planning, has become a problem that needs to be studied. According to the application requirements, the intellectualization of UAV system can focus on three aspects, such as Intelligent detection and recognition, intelligent route planning and intelligent task planning. In this paper, the idea of systems engineering is applied and the modeling method based on DoDAF is adopted. First of all, the UAV task scenario model is established to specify the task activities that the UAV needs to complete in the scenario. Then, starting from the task scenario, the task architecture and system architecture DoDAF model of UAV intelligent system are established. Finally, the scenario model and DoDAF model are analyzed, and the capability catalog of UAV intelligent system is generated, which provides guidance for improving the intelligence of UAV system.

Keywords: DoDAF · UAV · Capacity demand

1 Introduction

In recent years, with the development of UAV technology, UAV has been widely used in border patrol and other industries [1]. UAV can carry out surveil-lance, reconnaissance, ground attack and other tasks, and can also be used for stability maintenance and anti-terrorism [2]. However, how to formulate the route of border patrol, and how to deal with emergencies such as detection of cross-border targets, are the difficulties facing UAV development. In order to solve the problems that UAV may encounter in patrol, it is necessary to improve the intelligence of UAV system.

The artificial intelligence of UAV system is now a hot research topic. Li[3] compared the effectiveness of UAV with intelligent system and UAV without intelligent system, but did not decompose the intelligent system into system-level analysis. Gucky [4] introduced AI into UAV control systems, but did not extend it to other systems. Hai [5] discusses how artificial intelligence (AI) can be used for unmanned aerial vehicle (UAV) monitoring, which is a good line of research, but not an intelligence upgrade within UAV systems. According to the existing research, to realize the intelligent UAV, we need to

Z. Jing and X. Zhan (Eds.): ICASSE 2023, LNEE 1153, pp. 61–74, 2024.
https://doi.org/10.1007/978-981-97-0550-4_5

pay attention to obstacle avoidance technology, interactive technology, communication technology and so on.

To provide a reference for improving the intelligence degree of unmanned aerial vehicle system, we propose a method of generating a capability catalog of unmanned caerial vehicle intelligent system based on DoDAF. The research idea of this paper is shown in Fig. 1. First, a scenario model is established to preliminarily confirm the requirements for capability analysis of the UAV intelligent system. Then, the multi-perspective requirement analysis method based on DoDAF is adopted to model and analyze the UAV intelligent system. Finally, the capability catalog of the UAV intelligent system is generated. In the multi-perspective requirement analysis method based on DoDAF, the basic requirements are obtained based on scenario model analysis, and the preliminary requirement confirmation of capability is realized. Then, based on scenario model analysis, task activities are obtained, and the mapping relationship between capability and operational activities is established. The relationship between units is obtained by analyzing task activities, and the mapping of capability and organization is completed. Activities are then decomposed into systems to establish a mapping relationship be-tween capabilities and systems. Finally, the system resource flow is determined according to the system activities, and the metric value of capability is determined.

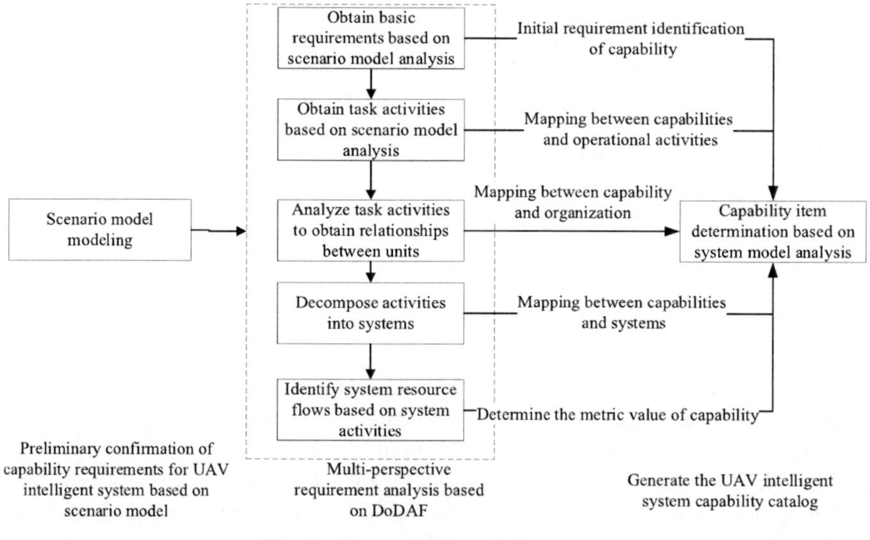

Fig. 1. Research idea of this paper.

2 Application Scenario Modeling and Demand Analysis of UAV Intelligent System

First of all, the content of the scenario is designed and the scenario model is established, which lays the application foundation for the subsequent model-based analysis and generation capability catalog. The blue objects represent our drones and the red objects represent possible targets. According to the application of UAV, the sea patrol scenario of UAV is designed. The target is to patrol the rectangular area composed of "rp-10", "rp-11", "rp-12" and "rp-13".The scenario model was built using a simulation scenario software named Origin, developed by Shareetech, as shown in Fig. 2. The specific scenario design is as follows: The ground control station determines the target of the sea patrol task, develops the task plan and uploads it to the UAV cluster. The drone receives the task plan and heads to the target area. Before reaching the patrol area, a drone detected a suspicious object. The UAV should be able to realize information transmission within the cluster and share information in the UAV cluster. In order to realize efficient detection, the UAV should have the capability of route change on board. Some UAVs rerouted their flight paths to detect and identify the target. The drone determines that the target is an unidentified flying object and transmits the detection information to the ground control station for further instructions. The UAV should be capable of identifying attack events and planning tasks on board. When the target is identified to attack us, complete the attack planning task on board.

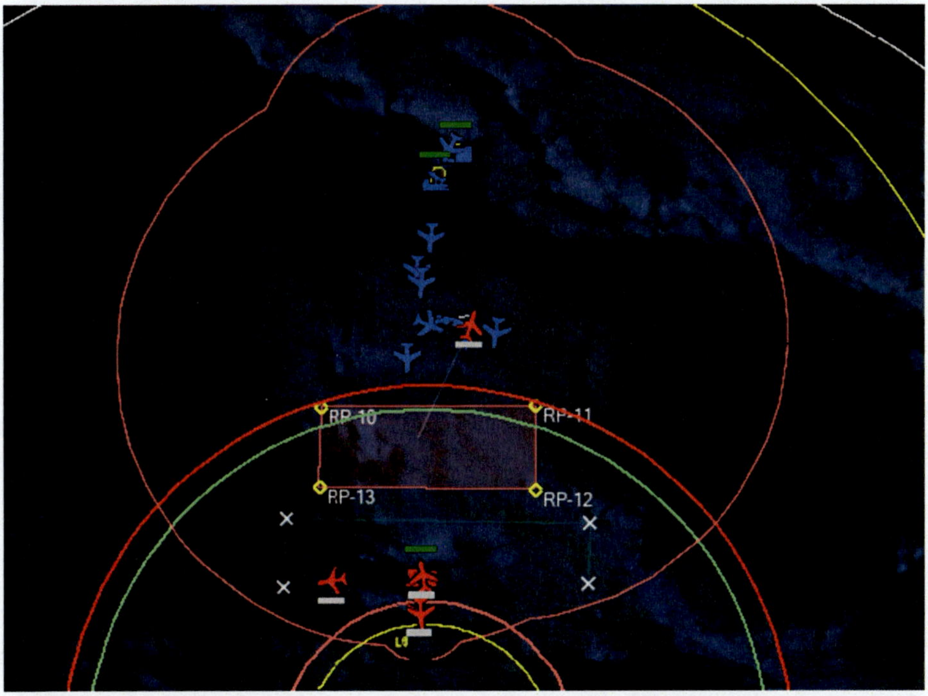

Fig. 2. Application scenario model of UAV intelligent system.

According to UAV application requirements, UAV system intelligence can be focused on three aspects: Intelligent detection and recognition, intelligent route planning and intelligent task planning.

(1) Intelligent detection and recognition

If the UAV has the disadvantage of poor shooting effect, the photos are not conducive to subsequent automatic analysis [6]. If there is a disadvantage of poor recognition capability, it is difficult to identify signals or targets, it will get half the result with less effort [7]. Therefore, to improve the intelligence of UAV system, we should pay attention to the intelligent detection and recognition of UAV. Specifically:

a) The UAV should be able to realize the discovery and identification of attack events;
b) The communication radiation source of the signal can be located and error estimation on the UAV;
c) The UAVs can collect, find direction and identify patterns of ultrashort wave communication radio and other signals.

The detection and identification function of UAV is the basis for the realization of various applications of UAV. Through the intelligent upgrade of the detection and identification function, the detection task of UAV can be better completed and the application range can be expanded.

(2) Intelligent route planning

UAV route planning is particularly important in the patrol process, because the UAV route represents whether the UAV patrol is comprehensive. Once there is a problem in the route planning, the UAV cannot efficiently complete the work [8]. Existing UAVs often need ground stations to control the route [9]. In order to achieve intelligence, UAVs should have the route reconnaissance capability on the UAV, and can carry out dynamic route planning according to reconnaissance situation information, such as automatically avoiding routes according to the range of threat targets and realizing autonomous obstacle avoidance of UAVs. If the route planning of UAV can realize the autonomous planning on board, the maneuverability of UAV patrol can be greatly improved.

(3) Intelligent task planning

Since UAVs can perform various tasks such as surveillance, reconnaissance and ground attack, it is more important to pay attention to the task planning of UAS to improve the intellectualization of UAS [10]. In terms of hardware, intelligent task manager and intelligent task data recorder should be installed in the system. For example, if a ground attack task is to be executed, the aircraft must have the attack planning capability and replan the task based on the situation information. UAV task planning is the basis of UAV task execution. By intelligentizing task planning, the efficiency of UAV task completion can be improved.

3 Multi-View Requirement Analysis Method Based on DoDAF

After the scenario model is established, typical views in DoDAF are selected for modeling in order to facilitate subsequent demand analysis. DoDAF is a common architecture modeling method [11]. By establishing capability view, Operation view and system view, and improving task architecture and system architecture at multiple levels, the requirements of UAV intelligent system are analyzed. The DoDAF architecture modeling idea is shown in Fig. 3. Firstly, the capability architecture analysis is realized through the capability view modeling, and then the operational view is established based on the scenario model to realize the task architecture analysis. Finally, the operational view is decomposed to the system level, and the system view is established to realize the analysis of system architecture and physical architecture.

From the perspective of architecture design, 12 perspectives such as DoDAF architecture operational view and system view are selected to describe the model and construct the conceptual model, which is divided into 12 construction steps:

1) CV-2 is constructed to describe capability and its classification relationship;
2) CV-4 is constructed to describe the traceability relationship between capability items;
3) OV-1 is constructed to describe operational concepts, including tasks, task implementers, task sequence and interaction with external environment;
4) OV-4 is constructed to describe organizational relationship;
5) OV-5 is constructed to define the flow of Operation resources;
6) SV-4a is constructed to identify and decompose system functions;
7) SV-7 is constructed to describe the system measurement value;
8) SV-6 is constructed to describe the data exchanged between system functions and its characteristics;
9) SV-10c is constructed and event tracking is drawn;
10) SV-10b is constructed and the analysis of system logical state transition is drawn;
11) SV-1 is constructed to draw the interface lines of the system;
12) SV-2a is constructed to describe the logical connection between system nodes and systems.

The process of DoDAF-based multi-perspective demand analysis method is explained by selecting typical models in typical detection scenarios. The process is as follows.

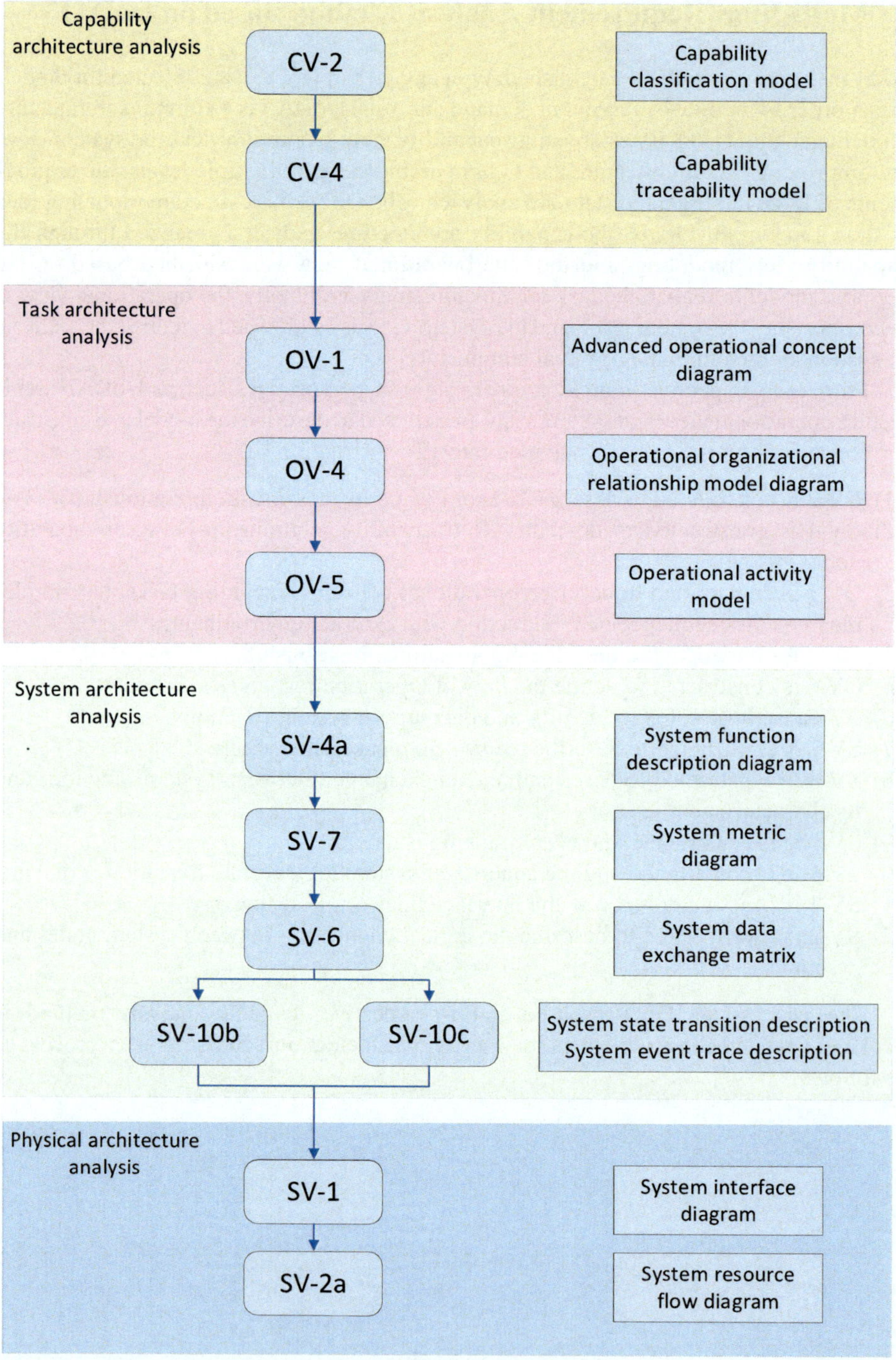

Fig. 3. DoDAF architecture modeling idea.

3.1 Basic Requirement Acquisition

Based on the scenario analysis of UAV intelligent system, capabilities are di-vided into six categories according to the classification idea of OODA ring [12], namely comprehensive perception capability, comprehensive identification capability, communication response capability, comprehensive defense capability, comprehensive strike capability and support capability. Establish the mapping relationship between intelligent system capabilities and OODA ring. Figure 4 shows the CV-2 capability classification model.

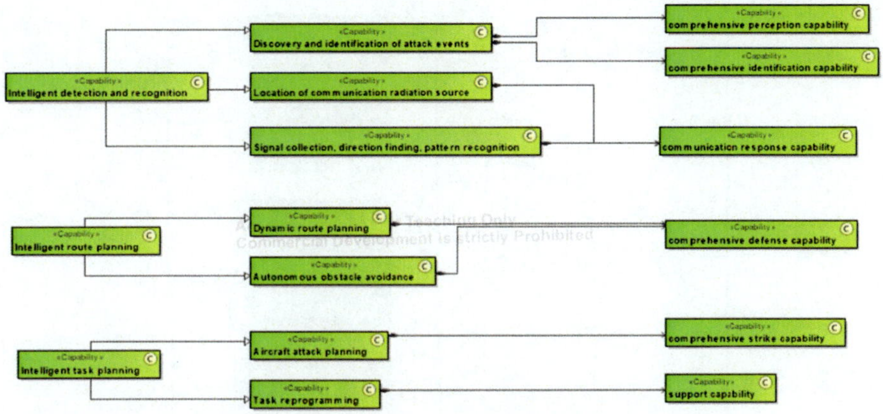

Fig. 4. CV-2 capability classification model.

According to CV-2 model, the basic requirements of capability can be defined, such as dynamic route planning, autonomous obstacle avoidance, etc. It is possible to specify the classification of capability items in the OODA ring.

3.2 Task Activity Acquisition

Then, according to typical application scenarios of UAV intelligent system (as described in Sect. 2, the activity transformation model is output, as shown in Fig. 5. The operational activity model describes the process of target detection through the cooperation of the ground station and UAV cluster.

After determining operational activities, the mapping model CV-6 between capability and operational activities is established, as shown in Fig. 6. As can be seen from the figure, dynamic route planning capability is related to flight route changing activities of UAV cluster. The capability to detect and identify attack events is related to the identification attack events of UAV cluster. Through the one-to-one correspondence between capabilities and operational activities, the specific application scenarios of each capability item can be defined, thus laying the foundation for determining the index value of capabilities.

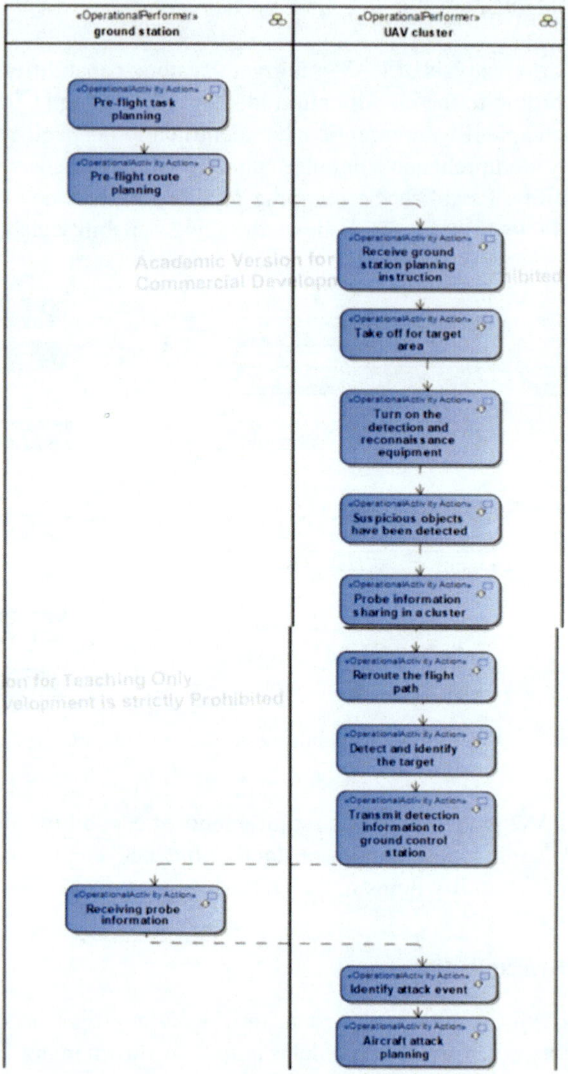

Fig. 5. OV-5 operational activity model.

3.3 Unit Organization Relationship Acquisition

In this step, the relationships between the units in the scenario are determined based on the Operation activity model. As can be seen from the analysis in Fig. 5, the ground station controls the UAV cluster and receives the detection information transmitted by the UAV cluster. And mapping the capabilities to the organization, all of which in this case is done by the UAV.

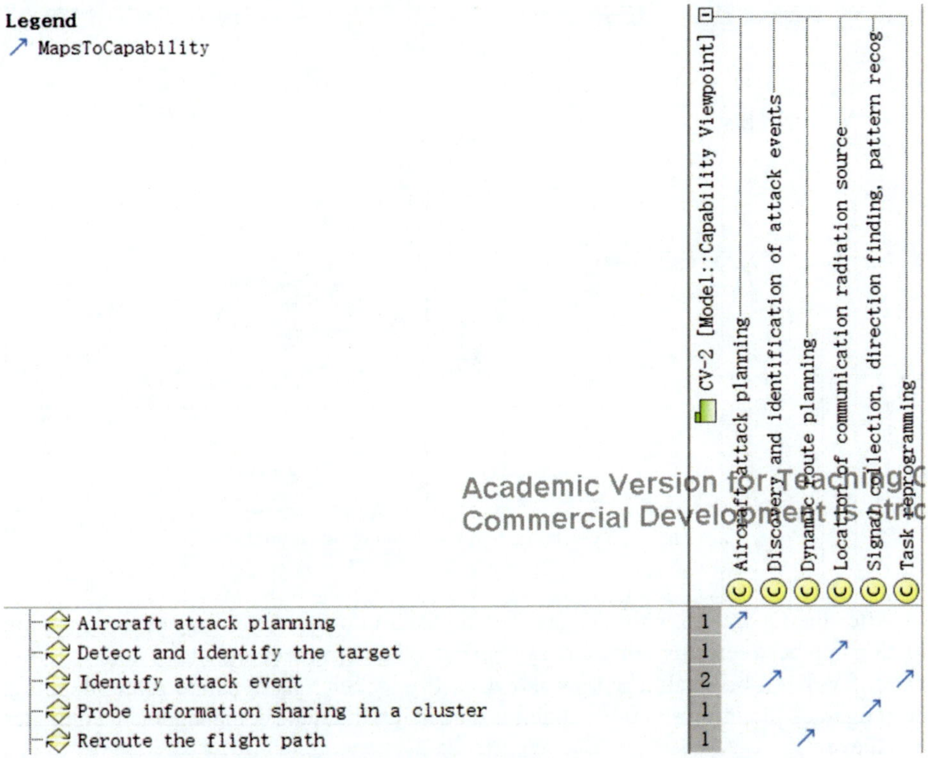

Fig. 6. CV-6 Mapping Model between Capability and Operational Activities (Part).

3.4 System Activity Decomposition

Operation activities are decomposed into various systems of UAVs so as to clarify the mapping relationship between capabilities and systems, as shown in Fig. 7. In the application scenario proposed in Sect. 3, the command is uploaded to the airborne link system of UAV after the ground station completes pre-takeoff task planning and pre-takeoff route planning. The drone takes off and turns on EO and SAR devices for detection. When the SAR device detects the suspicious target, the airborne link system transmits the detection information to realize the detection information sharing within the cluster. In the route planning module of the intelligent task Manager, the function of changing the flight route is realized. The Intelligent Task Data logger records relevant data. The detection related equipment carries out the detection task continuously, and transmits the detection information to the ground station by the airborne link system, and waits for the next instruction from the ground station. When the SAR recognizes that the target is attacking, the intelligent task management computer on the UAV completes the attack planning.

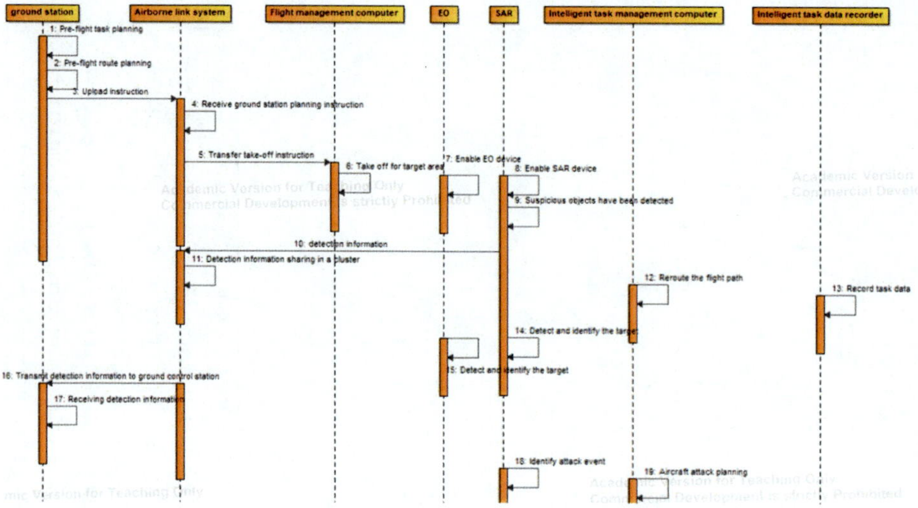

Fig. 7. SV-10c System timing diagram model.

After the Operation activities are decomposed into the system layer, the mapping relationship between capability and system can be determined, and the system function model SV-4 can be established, as shown in Fig. 8. Such as dynamic route planning, aircraft attack planning and other functions, need intelligent task management computer to achieve.

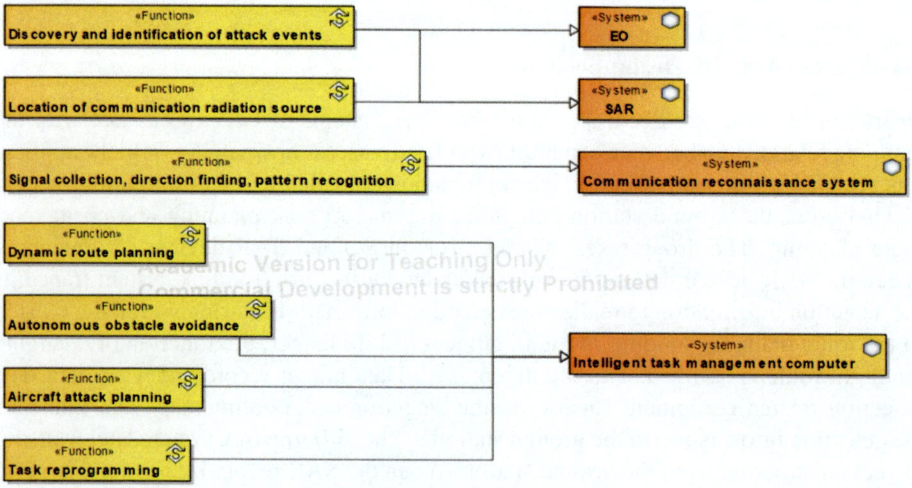

Fig. 8. SV-4 System function model.

3.5 System Resource Flow Identification

According to the activities between UAV systems, the typical resource flow between UAV intelligent systems is determined as shown in Fig. 9. The ground station uploads the control instructions to an intelligent task management computer in the drone. The intelligent task management computer sends the task data to the intelligent task data recorder for recording. EO transmits recognition information to the EO object recognition and detection module in the intelligent task data recorder. SAR transmits the detection information to the SAR target recognition detection module in the Intelligent task data logger. Communication reconnaissance system transmits identification information to the master software in the Intelligent Task data logger.

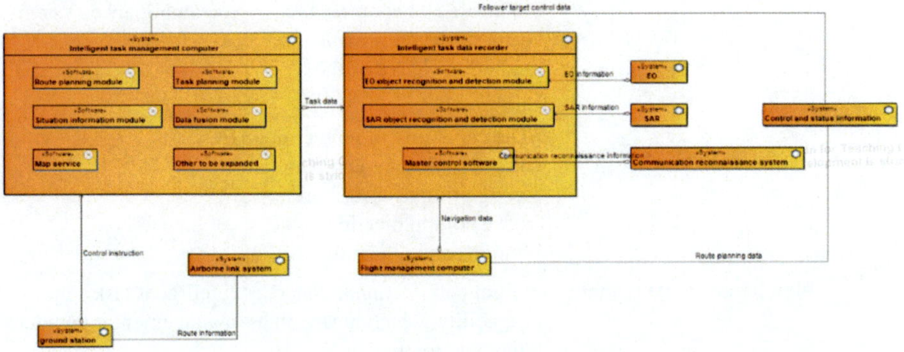

Fig. 9. SV-2a System resource flow model.

After determining the resource flow of the system, most of the modeling analysis process is completed. Finally determine a measure of capability, for example, to discover and identify attack events within 3 s.

4 UAV Intelligent System Capability Catalog Generation

According to the existing analysis results of scenario model and DoDAF model, the capability catalog of UAV intelligent system can be generated. Typical capability items are selected for description, as shown in Table 1. The capability catalog contains the capability classification, the specific content of the capability requirements and the matching relationship between the capability and the system. Take the capability to replan tasks based on situational information. It is a sub-item of intelligent task planning capability and related to intelligent task management computer system. Specifically, in order to achieve the capability of task replanning based on situation information, the task planning module in the UAV intelligent task management computer system needs to be upgraded to achieve the function of on-board task replanning.

By establishing the capability catalog of UAV intelligent system, the capability can be refined. The implementation of sub-level capability items to the system layer can provide reference for the research direction of system engineers and provide theoretical basis for the intelligent UAV system.

Table 1. UAV intelligent system capability catalog.

Number	Capability classification	Capability requirement	Related system
1	Intelligent detection and recognition	The capability to discover and identify attack events	SAR
2		The capability of the aircraft to locate and estimate the error of the communication radiation source of the signal	Communication reconnaissance system
3		The capability of collecting, direction finding and pattern recognition of ultrashort wave communication stations and other signals	EO
4	Intelligent route planning	Flight path reconnaissance capability, which means that dynamic route planning can be conducted based on reconnaissance posture information	Intelligent task management computer
5		Autonomous obstacle avoidance capability	Intelligent task management computer
6	Intelligent task planning	The capability to plan attacks	Intelligent task management computer
7		The capability to replan tasks based on situational information	Intelligent task management computer

5 UAV Intelligent System Capability Catalog Generation

In this paper, aiming at the requirements of system intelligence generated by UAV application scenarios, a typical application scenario of sea patrol is selected to complete the scenario description and establish the scenario model. Through the analysis of the scenario, the intelligent UAV system is determined in three directions: intelligent detection and recognition, intelligent route planning and intelligent task planning. Then, the multi-perspective requirement analysis method based on DoDAF is adopted to analyze and build the model step by step according to the capability view, operational view and system view, and improve the measurement requirement and mapping relationship of the capability items. Finally, by analyzing the scenario model and DoDAF model, the capability catalog of UAV intelligent system is generated, which provides a theoretical basis for the intelligent UAV system.

The innovations of this paper are as follows:

(1) Three directions of intelligent UAV system are proposed, and the specific capability items produced by each direction are analyzed.
(2) Combine scenario model with DoDAF architecture model. First, the scenario model is established to determine the scenario of intelligent application of UAV. Then, the DoDAF architecture model of intelligent system of UAV is
(3) By analyzing the scenario model and the DoDAF architecture model of the UAV intelligent system, the capability catalog of the UAV intelligent system is established to provide a theoretical basis for the intellectualization of the UAV system.

The future work is to load the proposed capability item onto the UAV, and conduct the same task scenario simulation with the UAV without intelligent capability to verify the effectiveness and feasibility of the intelligent capability item.

References

1. Jia, H., Wang, L., Fan, D.: The application of UAV lidar and tilt photography in the early identification of geo-hazards. Chin. J. Geol. Hazard Control **32**(2), 60–65 (2021)
2. Dwivedi, V.S., Patrikar, J., Addamane, A., Ghosh, A.K.: Maraal: a low altitude long endurance solar powered UAV for surveillance and mapping applications, pp. 449–454(2018)
3. Li, C.: Artificial intelligence technology in UAV equipment. In: 2021 IEEE/ACIS 20th International Fall Conference on Computer and Information Science (ICIS Fall), Xi'an, China, pp. 299–302 (2021)
4. Gúcky, J., Kováčiková, K., Novák, A., Bugaj, M.: Analysis of the introduction of artificial intelligence in the control of UAV. In: 2022 New Trends in Aviation Development (NTAD), Novy Smokovec, Slovakia, pp. 67–70 (2022)
5. Hai, D., Truong, H., Nguyen, T., Chien, F., Tran, T., Hua, T.: Energy-efficient unmanned aerial vehicle (UAV) surveillance utilizing artificial intelligence (AI). Wirel. Commun. Mob. Comput. **2021**, 1–11 (2021)
6. Ren, J., Jiang, X.: A three-step classification framework to handle complex data distribution for radar uav detection. Pattern Recogn. **111**, 107709 (2021)
7. Prihodova, K., Jech, J.: Gender recognition using thermal images from UAV. In: 2021 International Conference on Information and Digital Technologies (IDT) (2021)
8. Fu, Y., Ding, M., Zhou, C.: Phase angle-encoded and quantum-behaved particle swarm optimization applied to three-dimensional route planning for UAV. IEEE Trans. Syst. Man Cybern. Part A Syst. Humans **42**(2), 511–526 (2012)
9. Gu, Y., Cao, X., Sun, C.: A route planning algorithm for privacy protection of UAV states against eavesdropping. In: 2020 35th Youth Academic Annual Conference of Chinese Association of Automation (2020)
10. Atyabi, A., Mahmoudzadeh, S., Nefti-Meziani, S.: Current advancements on autonomous task planning and management systems: an AUV and UAV perspective (2020)

11. Griendling, K., Mavris, N.: Development of a DoDAF-based executable architecting approach to analyze system-of-systems alternatives. In: Aerospace Conference. IEEE (2011)
12. Sun, S., Meng, C., Zheng, G.: Assessment of combat capability of marine police equipment system based on ooda ring. J. Ordnan. Equip. Eng. **40**, 26–28 (2019)

Research on the Contribution Rate of Shipboard Manned/Unmanned Aerial Vehicle Cooperative Operation Based on Wargame Data Mining

Ling Peng[1,2], Xin Wang[2], Gechen Wang[3], Shitong Zhao[2], Miao Wang[3](✉), and Wenjun Xu[1]

[1] School of Information Engineering, Wuhan University of Technology, Wuhan, China
{pengling,xuwenjun}@whut.edu.cn
[2] China Ship Development and Design Center, Wuhan, China
[3] School of Aeronautics and Astronautics, Shanghai Jiao Tong University, Shanghai, China
{gechen.wang,miaowang}@sjtu.edu.cn

Abstract. With the rapid development of shipboard Unmanned aerial vehicles (UAVs), they will become the new normal for future maritime battlefield utilization. Current paper focuses on the following theme: how to allocate manned/unmanned aerial vehicle combat tasks in time sequence under specific combat scenarios according to the capability characteristics of shipboard unmanned UAVs, how to harness the role of shipboard UAVs in the system in a better way, and this is analyzed through the contribution rate of manned/unmanned aerial vehicle cooperative combat system. Current paper outlines the shortcomings of the current system contribution rate: (1) Lack of cooperative relationship between equipment; (2) Lack of objectivity and dynamism in the assessment process; (3) Lack of direct application of assessment results. Therefore, the following highlights are proposed in this paper to make up for the existing shortcomings Firstly, it is proposed to construct the manned/unmanned aerial vehicle cooperative system through DoDAF construction system framework and dynamic simulation of the wargame to meet the requirements of objectivity and dynamism. Secondly, an data mining algorithm is proposed to mine a large number of simulation output results based on the improved data structure of orthogonal list to analyze the temporal correlation relationship between the cooperative system. Finally, it provides a more objective basis for the allocation of capability requirement indicators of shipboard UAV in the cooperative combat system to optimize the existing cooperative combat model.

Keywords: DoDAF · Data mining · manned/unmanned aerial vehicle · contribution rate

1 Introduction

With the rapid development of the technical level of shipboard UAVs, they will become the new normal for future battlefield utilization. It can be carried on the aircraft carrier to perform various tasks in the deep sea and provide important support for maritime

Z. Jing and X. Zhan (Eds.): ICASSE 2023, LNEE 1153, pp. 75–88, 2024.
https://doi.org/10.1007/978-981-97-0550-4_6

operations [1]. Therefore, the introduction of shipboard UAVs into the existing maritime combat system will be a disruptive technology that will change the battlefield environment.

Compared to traditional shipboard manned aerial vehicle coordination or UAV clusters, manned/unmanned aerial vehicle coordination can create a new operational advantage with the following advantages [2]: (1) Flexible tactical warfare; (2) Making up for the shortcomings of the command and decision-making capabilities when dealing with unexpected situations during the execution of specific missions by UAVs; (3) Avoiding the operational costs of manned aerial vehicles performing high-risk missions, greatly shorten the weapons generation cycle and reduce the cost of combat attrition. Therefore, making full use of the respective advantages of manned and unmanned aerial vehicles to complement each other can achieve the combat effect of $1 + 1 > 2$ and improve the overall combat effectiveness.

However, how to allocate manned/unmanned aerial vehicle combat missions in a specific combat scenario by time sequence according to the characteristics of UAVs to make them work better in the system is critical. Therefore, it is important to establish a system contribution rate assessment method for shipboard manned/unmanned aerial vehicle.

In recent years, Chinese scholars have proposed the concept of equipment system contribution rate to assess the equipment system contribution rate from the dimension of system structure, system combat capability or system effectiveness [3]. Yu [4] constructed a system capability assessment model based on the correspondence between equipment system performance indicators and Observe-Orient-Decide-Act (OODA) loop capabilities, and then determined the equipment system contribution rate, the shortcoming is that it mainly constructs a tree indicator system for equipment system capability contribution rate assessment based on the hierarchical structure of the equipment system, without fully considering the relationship between capability indicators at the same level. Hua [5] introduces complex network theory into the study of combat systems, and constructs equipment contribution rate evaluation indexes based on the combat network model of the equipment system of the combat ring, the shortcoming is that the mapping relationship between the specific behaviors and capabilities of the observation nodes is difficult to reflect, while the determination of the weights may be subjective. Luo [6] established an assessment model of the contribution rate of the equipment system based on the quantity and capability indicators of the combat ring, but the shortcoming is that its assessment model is a static process and lacks dynamism.

The paper aims to evaluate manned/unmanned aerial vehicle cooperative operations through the system contribution rate assessment method. The existing assessment has the following shortcomings: (1) The existing assessment is mostly a static process, which is lack of dynamics, so it cannot display the dynamic manned/unmanned aerial vehicle cooperative process; (2) The existing evaluation methods almost focus on the equipment system itself, and the determination of weights may be subjective, lacking the corresponding application background to objectively evaluate it; (3) The existing evaluation methods mostly focus on the contribution of single equipment in the system, which leads to the lack of research on the relationship between equipment; (4) The existing research on the contribution rate of equipment system mainly focuses on its concept connotation understanding and evaluation method exploration. There is little direct application of the evaluation results.

The highlights of the paper:

(1) The cooperative architecture of shipboard manned/unmanned aerial vehicle is simulated by DoDAF2.0 model framework and Wargame [7], which meets the requirements of dynamic analysis. The simulation process is closer to reality, and the evaluation result is simple, intuitive and reliable [8];
(2) Mining dynamic manned/unmanned aerial vehicle cooperative model through data mining algorithm to meet the objective requirements;
(3) The relationship between manned and unmanned aerial vehicles is obtained through the analysis of the mined time series frequent patterns;
(4) Finally, the mining results provide a more objective basis for the allocation method of UAV's capability demand index in the cooperative system and optimize the existing cooperative combat mode.

2 Evaluation Process of Contribution Rate of Shipboard Manned/Unmanned Aerial Vehicle Cooperative System

As shown in Fig. 1, the contribution rate evaluation process of shipboard manned/unmanned aerial vehicle system is mainly composed of three parts: (1) A static model of manned/unmanned aerial vehicle architecture is constructed based on DoDAF2.0 framework; (2) The dynamic cooperative model of manned/unmanned aerial vehicle is constructed based on wargame. (3) Capturing and mining the cooperative time series data of manned/unmanned aircraft. Evaluating the contribution rate of the system according to the mining structure.

2.1 Constructing a Static Model Based on DoDAF

DoDAF2.0 framework [9] provides decision data through different viewpoints and models, which can be used as an effective tool to describe the cooperative combat mode of shipboard manned/unmanned aerial vehicles. Among them, the operational viewpoint (OV) is suitable for the macroscopic description of the cooperative combat mission of manned/unmanned aerial vehicles. The structure of the combat formation is constructed through the operational resource flow (OV-2), and then the interactive behavior between platforms is described through the operational activity model (OV-5b). The capability viewpoint (CV) is suitable for solving the problem of capability combination of shipboard manned/unmanned aerial vehicles in cooperative operations. The operational capabilities of shipboard manned aerial vehicles and shipboard unmanned aerial vehicles are obtained through the capability classification model (CV-2), and finally the mapping relationship between activities and capabilities is obtained through the mapping model of capabilities and operational activities (CV-6).

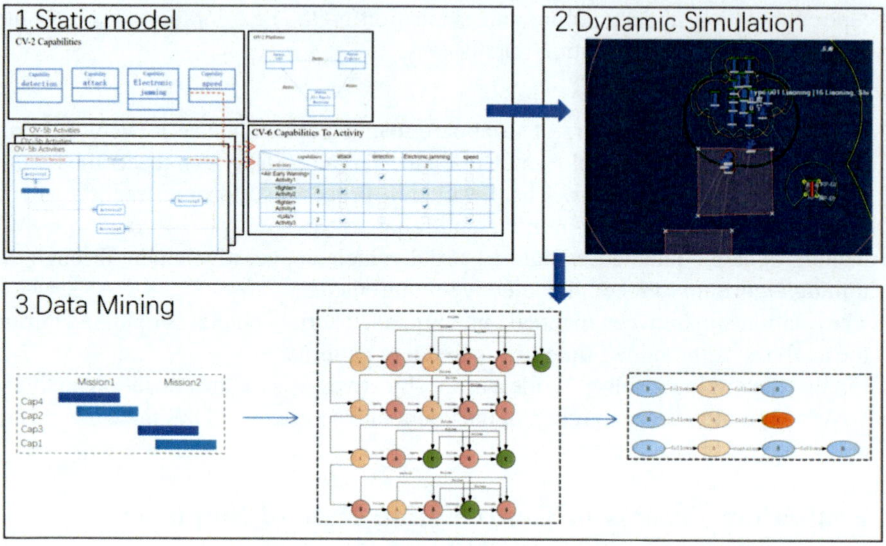

Fig. 1. Contribution rate evaluation process

First of all, CV-2 defines the capabilities that all units in the scenario need to complete the corresponding tasks and activities. Taking air-sea cooperative operations as an example, the capabilities of the shipboard UAV platform include detection capability, attack capability, electronic jamming capability and so on, as shown in Fig. 2.

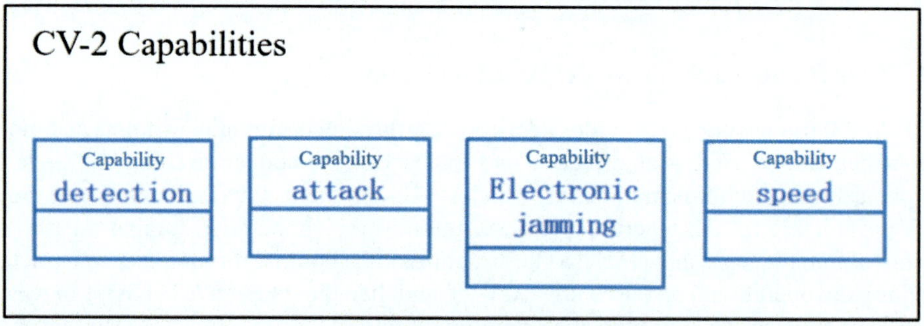

Fig. 2. CV-2 viewpoint

Secondly, OV-2 defines all the functional platforms in the scenario and the information transmitted between them. Taking air-sea cooperative operations as an example, the functional platform includes shipboard fighter, shipboard unmanned aerial vehicles and air early warning. The transmitted information includes detection information, situation information and attack instructions, which are described by the connection between platforms, as shown in Fig. 3.

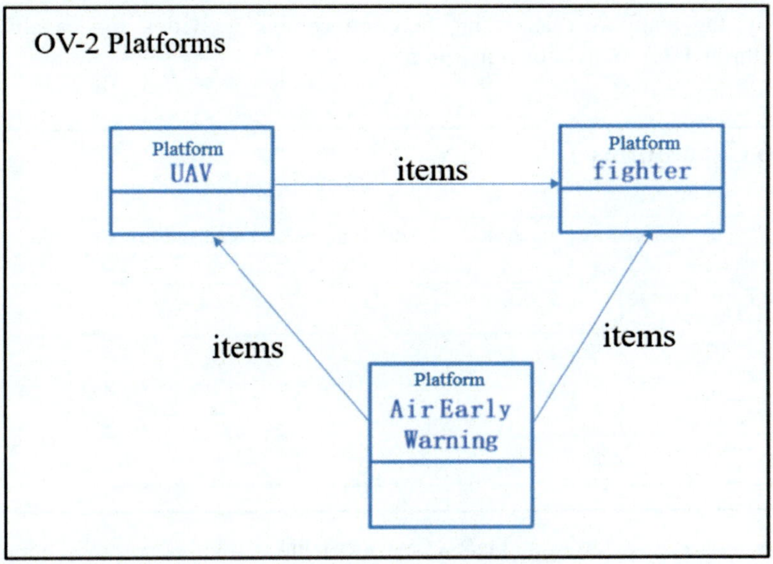

Fig. 3. OV-2 viewpoint

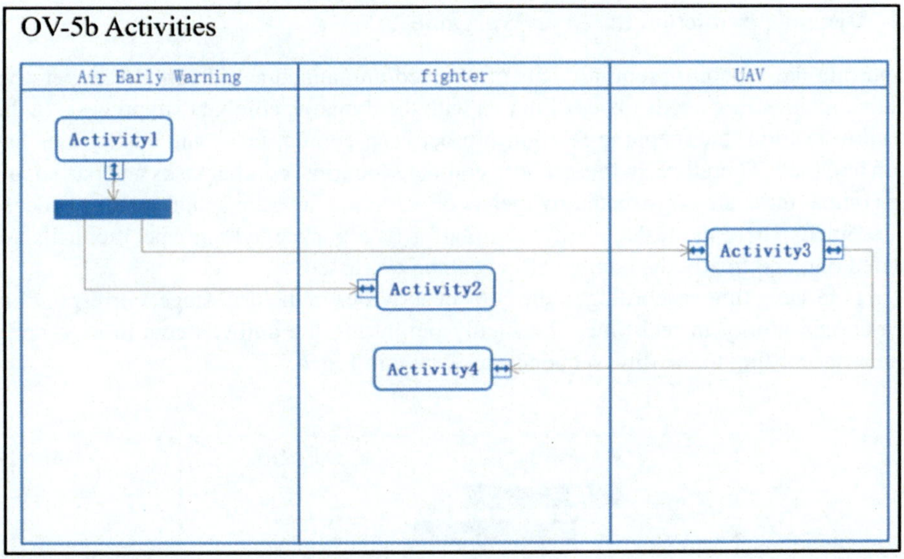

Fig. 4. OV-5b viewpoint

In addition, the cooperative combat mode of manned/unmanned aerial vehicle is combed and constructed through OV-5b. The use methods of shipboard UAV include reconnaissance, providing target indication, communication relay, bait, assault, feint and so on. These usage methods can flexibly form tactical tactics with manned aircraft according to operational requirements and application scenarios, as shown in Fig. 4.

Finally, the mapping relationship between combat activities and capabilities is obtained through CV-6, as shown in Fig. 5.

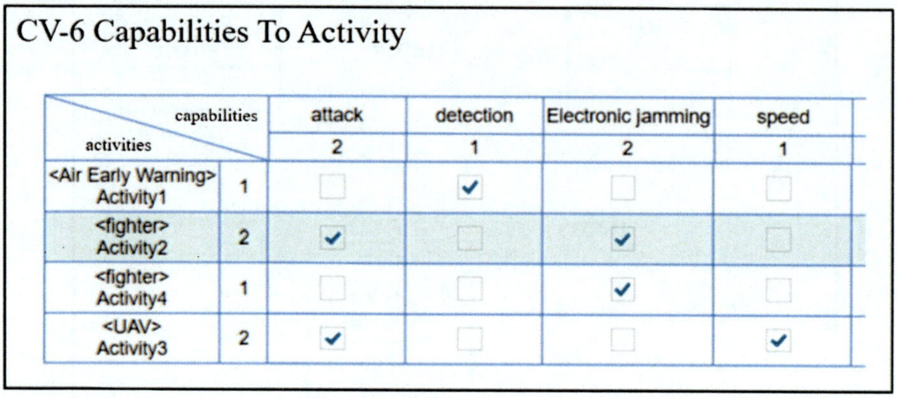

CV-6 Capabilities To Activity

capabilities / activities		attack	detection	Electronic jamming	speed
		2	1	2	1
<Air Early Warning> Activity1	1	☐	☑	☐	☐
<fighter> Activity2	2	☑	☐	☑	☐
<fighter> Activity4	1	☐	☐	☑	☐
<UAV> Activity3	2	☑	☐	☐	☑

Fig. 5. CV-6 viewpoint

2.2 Dynamic Deduction Based on Wargame

Capturing the capability requirements of manned/unmanned aerial vehicle cooperative combat architecture needs to be combined with the dynamic combat drill process. In the previous section, the mapping relationship between combat tasks and capabilities has been obtained. Therefore, in this section, combat scenarios, combat tasks and participating combat units are constructed by means of wargame, and the actual combat process is designed. Then, the tasks of different platforms are arranged and specific drills are carried out, and finally the combat effect can be evaluated.

At the same time, according to the current activities in the drill stage, corresponding to a certain ability in real time, and finally outputting the ability items in an orderly manner according to the time sequence, as shown in Fig. 6.

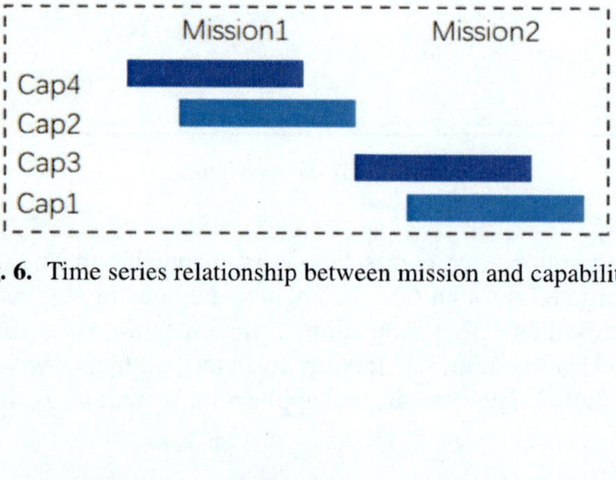

Fig. 6. Time series relationship between mission and capabilities

2.3 Data Mining of Shipboard Manned/Unmanned Aerial Vehicle Cooperative Operation Mode

Wargame in the last section contains a lot of data about manned/unmanned aerial vehicle cooperative combat mode. Therefore, the contribution rate of shipboard UAV capability under this system can be obtained by analyzing the task collocation of cooperative combat mode in different scenarios. On this basis, the dependence between the exercise results and the capability items of the shipboard UAV is evaluated, and the correlation between the capability items of the shipboard UAV and the operational results can be obtained.

In principle, there is a correlation between the execution of different combat missions, the capability items of drones and the strike effect. The problem of data acquisition related to capabilities, tasks and strike effects can be abstracted into a closed frequent pattern data mining problem, in which each deduction scenario is regarded as a transaction. The capabilities of manned/unmanned aerial vehicles and combat tasks appearing in the deduction process are regarded as discrete itemsets, and there is a time series correlation between them. Therefore, the correlation between these itemsets can be dealt with by time series closed frequent pattern data mining algorithm, and the contribution rate of UAV capability to manned/unmanned aerial vehicle cooperative system can be obtained.

3 Realization of Data Mining Algorithm

3.1 Closed Frequent Pattern Mining

Closed frequent pattern mining [10] is a machine learning method, which is used to mine the dependencies between transactions hidden in a large number of data, and has broad prospects for the contribution rate analysis of manned/unmanned aerial vehicle cooperative combat mode system. Its basic concepts are as follows.

Frequent Itemset (FI): Given a database $D = \{ d_1, d_2, \cdots, d_m \}$ with m transactions, where the full itemsets in a transaction is $I = \{ i_1, i_2, \cdots, i_n \}$ and the total number of attributes is n. Then each transaction d is a subset of the full itemset I. An important property of an itemset is the support value, mathematically defined as

$$\sup(S) = \left(\frac{Number\ of\ transactions\ containing{<}S{>}}{Total\ number\ of\ transactions\ in{<}D{>}} \right) \times 100\% \tag{1}$$

If the support of S is greater than or equal to a given minimum support, S is called FI.

Closed Frequent Pattern (CFP): The so-called closed itemset refers to an itemset X, and the support count of its direct superset (the smallest strict superset) is not equal to its own support count. If a closed itemset is frequent at the same time, that is, its support is greater than or equal to the minimum support threshold, it is called a CFP.

The extraction of CFP is the key of the algorithm. In the traditional algorithm, there is an algorithm to mine closed frequent patterns by designing orthogonal list structure [11]. In this paper, an improved data structure is proposed on the basis of orthogonal list, which further optimizes the analysis efficiency of the algorithm.

3.2 Algorithm Flow

The input of the algorithm is a transaction set composed of the capabilities of each unit in wargame. Each capability has three attributes: <capability name>, <start time> and <end time>. As shown in Table 1, take (A, 1, 3) as an example, where <A> stands for capability, <1> stands for start time and <3> stands for end time, and each row represents a combat mission. There are five time relationships between capabilities, namely<follows>, <matches>, <meets>, <overlaps>, <contains>.

Table 1. Transaction set input sample table

1	(A, 1, 3)	(B, 1, 3)	(A, 6, 12)	(B, 8, 11)	(B, 14, 16)	(C, 17, 20)
2	(A, 1, 2)	(B, 2, 6)	(A, 10, 12)	(B, 11, 15)	(C, 14, 17)	
3	(A, 4, 7)	(B, 11, 12)	(C, 12, 14)	(B, 17, 17)	(C, 18, 19)	
4	(B, 1, 5)	(A, 6, 14)	(B, 7, 10)	(C, 8, 9)	(B, 16, 18)	

Because of the need to mine the relationship between capabilities, it will greatly increase the time consumption to traverse and judge the relationship in the process of mining with exponential complexity. Therefore, the data structure is designed in advance, and each transaction is stored by the combination of linked list and array. Between each capability, the time sequence relationship information between two activities is stored in advance when the data is read in, as shown in Fig. 7.

The input transaction set is called transaction table. In the process of mining a single relationship R, there is a sequence $Y_k(X_1, X_2, \ldots, X_i)$ being mined for each row, where Y_k represents a row of the transaction table, X_i represents an activity in the transaction. Therefore, a corresponding table is needed to maintain and store the corresponding information for each relationship mining, and update it synchronously during the mining process.

The algorithm maintains three data sets at the same time, which are:

<Complete>, the linked list stores the excavated set of relationships, which is a table that stores the corresponding activity subscripts in the transaction table;

<Mining>, which maintains the information of the relationship being mined and corresponds to the input transaction table;

<Candidate>, the linked list stores the candidate set of a certain stage in mining;

This algorithm is based on depth-first. In the process of Mining, a mining linked list is initialized for multiple relationships being mined once. The linked list stores the subscripts of the corresponding activities in the transaction table, and the head node stores and maintains the attribute information of the row in mining, which determines whether the row can be searched for the next candidate set in mining. Find a relationship in depth first at a time, then rollback to the node information.

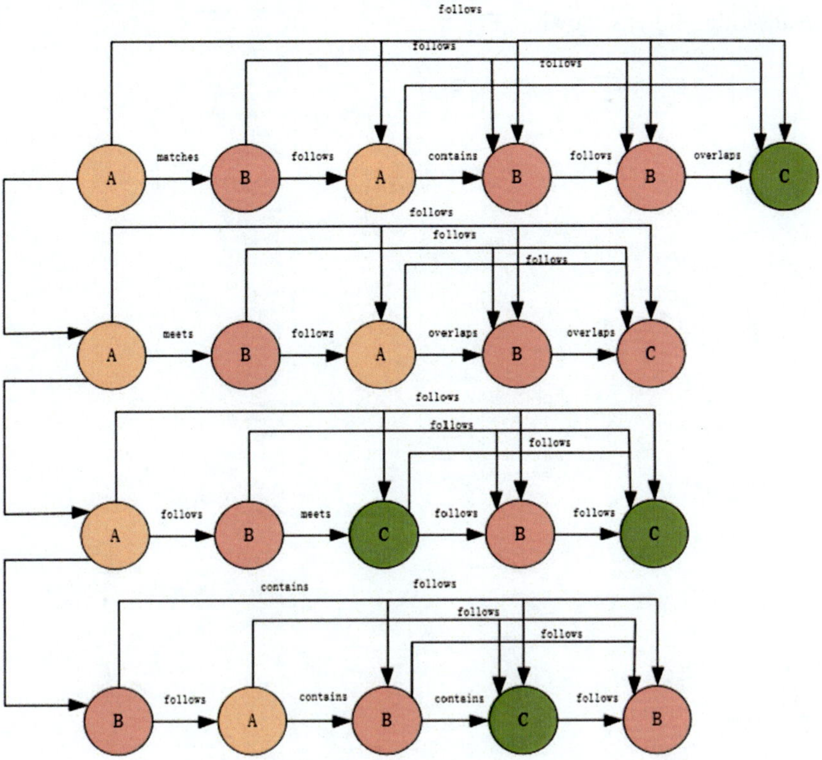

Fig. 7. Data structure

Rollback node refers to: after finding a relation <r>, in a row, the corresponding activity subscript is (X_1, X_2, X_3, X_4) and a node is rolled back to delete X4, that is, from the sequence of (X_1, X_2, X_3), the corresponding candidate set of $X_1 \rightarrow X_2 \rightarrow X_3$ sequence is searched, and the mining of the next relation is continued.

Find the first activity, and record its subscripts in <Mining>. For the first activity, manually call the linked list to generate its <Candidate> set, and then traverse the <Candidate> in a circular way. For each activity in <Candidate>, record the subscript in <Mining>, that is, a node with depth priority. For each of these nodes, <Mining> is carried out recursively. Every time a <Candidate> set is generated, it will traverse the mined relationships in <Complete>. Pruning if the excavation paths are the same by comparing the subscripts. After this relationship mining, it will be stored in <Complete>, and the nodes in Mining will be rolled back to mine the next relationship. That is to say, there is a <findNextItem> method to receive the candidate set generated by manual call.

The algorithm flow is shown in Fig. 8.

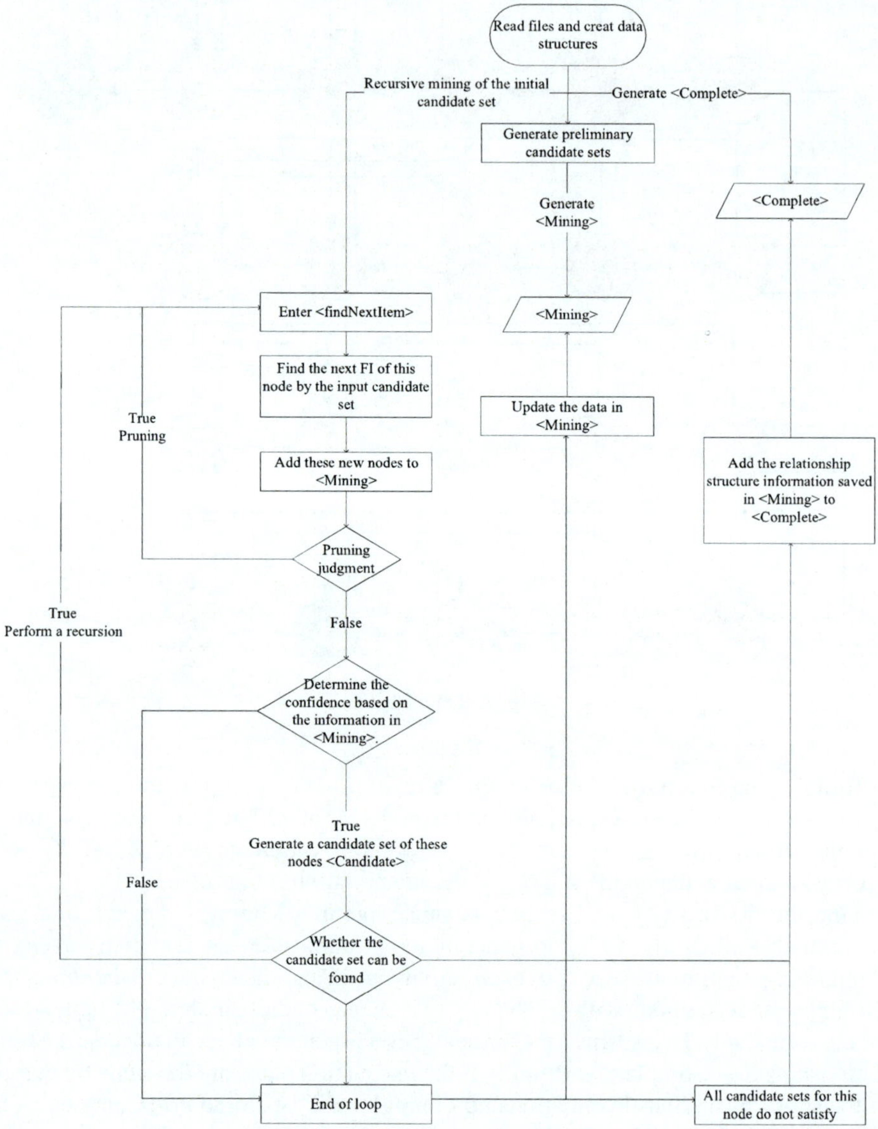

Fig. 8. The algorithm flow

Taking Table 1 as an example, when the support threshold is set to 0. 67, the output is shown in Fig. 9.

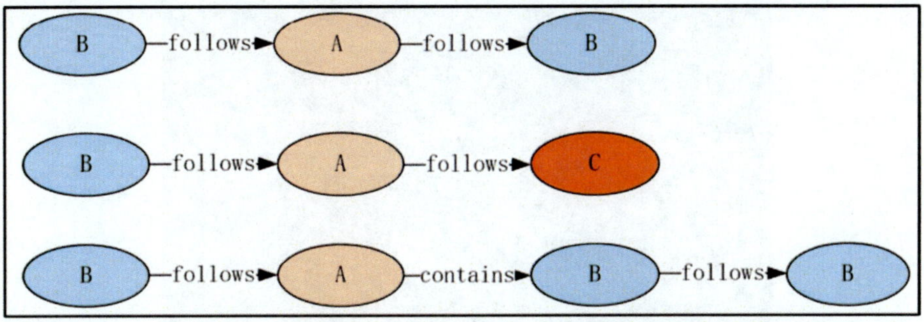

Fig. 9. Table 1 Mining Results

4 Experimental Verification

4.1 Scenario Description

In this paper, an air-to-sea attack scenario is designed as a verification to analyze the contribution rate of UAV capability to manned/unmanned aerial vehicle cooperative operations and verify the effectiveness of the proposed algorithm. The red side is a flight formation composed of shipboard UAVs and manned aircraft, including attack aircraft and early warning aircraft, with the intention of breaking through the four patrol fighters of the blue side and destroying a large destroyer of the blue side. The weapon of Red UAV can be mounted flexibly, and it has the function of inspection and combat. It can change the mounting type and weight flexibly according to the combat mission, and it is equipped with electronic warfare function. They take off from a nearby aircraft carrier. Fighters take off from the nearby red aircraft carrier and teamed up with UAVs.

The capabilities of shipboard UAVs include electronic jamming capability, cruising speed capability, decoy capability, detection capability, attack capability, communication relay capability and target indication capability. The capabilities of shipboard manned attack aircraft include strike capability, electronic jamming capability, detection capability, situation display capability and command capability of UAVs.

4.2 Simulation Result Analysis

The simulation was automatically run by Origin, a simulation platform of Shareetech. 700 deduction simulation data were collected, as shown in Fig. 10. The deduction results include various task matching modes of manned/unmanned aerial vehicle cooperative operations, the victory and defeat of corresponding scenes and the loss of both units. Summarize the simulation scenes and distinguish the scenes with high winning rate and small red loss. The proposed algorithm is used to mine frequent closed patterns, and the contribution of UAV capability to equipment system is obtained.

The mining results contain a series of ability items, which are further visualized by linked list diagram. Different nodes with different colors in the diagram represent different capability items, and the time series relationship between capabilities is connected by arrows.

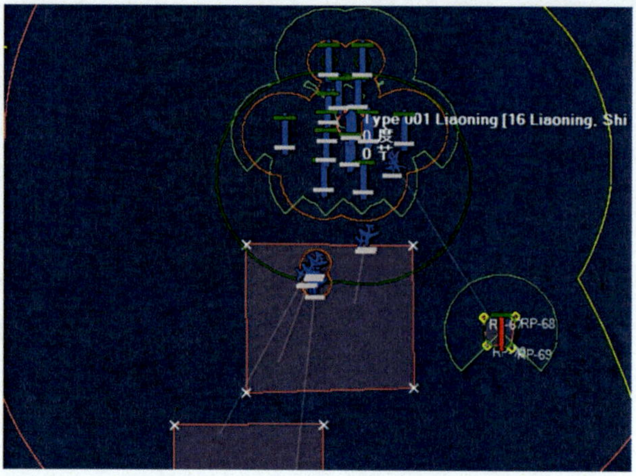

Fig. 10. Application Scenario Model of Air-to-Sea attack

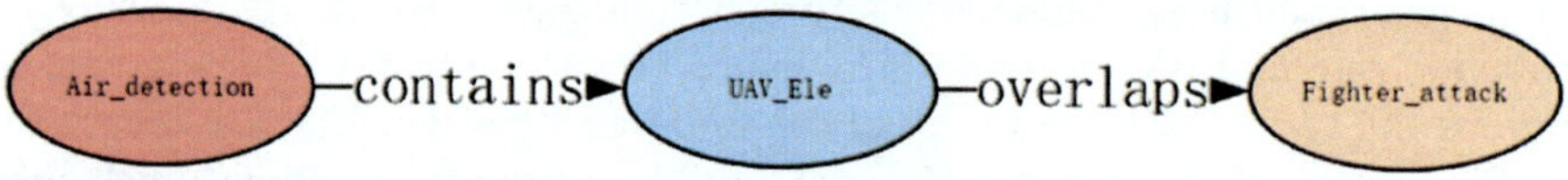

(a) UAVs electronic jamming, fighters attack

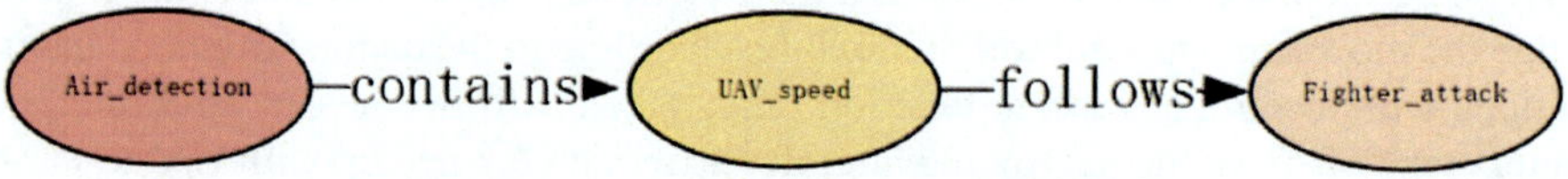

(b) UAVs feint, fighters attack

(c) Long-range attack by fighters , assault by UAVs

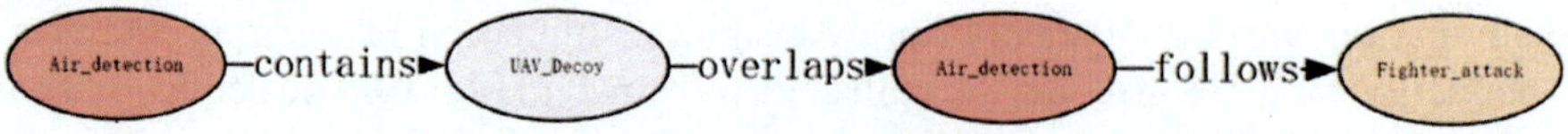

(d) UAVs bait, manned aerials vehicles detect and attack

Fig. 11. Closed Frequent Patterns in Winning Scenarios

As shown in Fig. 11(a), UAVs first carries out the electronic jamming task, which can effectively shield the enemy detection and cause interference to the enemy defense system, which can effectively reduce the Loss of red units. But the damage to the blue ship is average, indicating that the target ship cannot be completely destroyed and the effect is relatively mild.

As shown in Fig. 11(b), UAVs attacks in other directions, luring the blue side to pursue, attracting the blue side's firepower and consuming its weapons. Then the fighters of red side went to the target point to bomb, so that the attack effect was better and the loss was less.

As shown in Fig. 11(c), fighters first strikes with long-range missiles. UAVs approach and attack with low-altitude blind area, which can achieve good strike effect and the loss of red personnel is not high.

As shown in Fig. 11(d), UAVs radar is actively turned on, which used as bait to lure the defense aircrafts of blue side to track and attack. The air early warning of red side waits for an opportunity to detect the blue defense trend and command the fighters to strike. But the bait effect is average, and the strike effect is average.

Therefore, it is best to use fighters to carry out long-range fire attack first, then UAVs to carry out close attack; the effect is good, but the loss of UAVs is large by using UAVs as feint to lure the enemy to attack, then using manned aircraft to attack,; Through the electronic jamming of UAVs, manned aerial vehicles can effectively reduce the detection ability of defenders, but the attack effect is average; Deploying UAVs as bait can protect manned aerial vehicles formation, but the deception effect and attack effect are average.

5 Conclusion

Aiming at the lack of objective and dynamic analysis method for the contribution rate of UAV capability used in manned/unmanned aerial vehicles cooperative combat system, this paper firstly constructs the static framework of manned/unmanned cooperative combat system. Then the relationship between activity and ability is obtained through dynamic deduction. This paper puts forward an algorithm for mining manned/unmanned cooperative patterns based on time series closed frequent pattern mining, which mines effective time series relationship of shipboard manned/unmanned capability collocation from a large number of wargame data, which can give better play to the role of shipboard UAVs in joint combat system, providing guidance for the capability development of shipboard UAVs.

The highlights of the paper:

(1) The cooperative architecture of shipboard manned/unmanned aerial vehicle is simulated by DoDAF2.0 model framework and Wargame, which meets the requirements of dynamic analysis. The simulation process is closer to reality, and the evaluation result is simple, intuitive and reliable;
(2) Mining dynamic manned/unmanned aerial vehicle cooperative model through data mining algorithm to meet the objective requirements;
(3) The relationship between manned and unmanned aerial vehicles is obtained through the analysis of the mined time series frequent patterns;
(4) Finally, the mining results provide a more objective basis for the allocation method of UAV's capability demand index in the cooperative system and optimize the existing cooperative combat mode.

The future work is to quantify the contribution of UAV's capability items through analytic hierarchy process (AHP) and other distribution methods, and confirm the capability demand indicators and their importance.

Acknowledgements. This work is partially supported by Natural Science Foundation of Shanghai (20ZR1427800), New 945 Young Teachers Launch Program of Shanghai Jiao Tong University (20X100040036).

References

1. Fan, B., Li, Y., Zhang, R., Fu, Q.: Review on the technological development and application of UAV systems. Chin. J. Electron. **29**(2), 199–207 (2020)
2. Fan, J., Li, D., Li, R., Yang, T., Wang, Q.: Analysis for cooperative combat system of manned-unmanned aerial vehicles and combat simulation. In: 2017 IEEE International Conference on Unmanned Systems (ICUS), pp. 204–209. IEEE (2017)
3. Yin, X., Hu, X., Rong, M., Liu, H., Yang, Y.: Review of evaluation methods of contribution rate to system of systems. J. Syst. Simul. **31**(6), 1027 (2019)
4. Zhang, Y., Huang, J.: Study on contribution degree of equipment to system of systems using OODA loop model. Mod. Def. Technol. **42**(2), 177–182 (2017)
5. Hua, B., Fan, Z., Lin, C., Wang, Z., Zang, Y.: Evaluation of contribution rate of armaments to system-of-systems based on complex network and improved information entropy. In: 2021 IEEE International Conference on Advances in Electrical Engineering and Computer Applications (AEECA), pp. 632–635. IEEE (2021)
6. Luo, X., Zhu, Y., He, R.: Research on evaluation method of contribution to system war fighting for weapons and equipment system based on complex network. Fire Control Command Control **42**(2), 83–87 (2017)
7. Hernandez, A.: Advancements in scenario-based learning through quantification of wargaming techniques. In: ECRM 2017 16th European Conference on Research Methods in Business and Management, p. 154. Academic Conferences and Publishing Limited (2017)
8. Yajie, D.O.U., Zhexuan, Z.H.O.U., Danling, Z.H.A.O., Yong, W.E.I.: Weapons system portfolio selection based on the contribution rate evaluation of system of systems. J. Syst. Eng. Electron. **30**(5), 905–919 (2019)
9. Piaszczyk, C.: Model based systems engineering with department of defense architectural framework. Syst. Eng. **14**(3), 305–326 (2011)
10. Prabha, S., Shanmugapriya, S., Duraiswamy, K.: A survey on closed frequent pattern mining. Int. J. Comput. Appl. **63**(14) (2013)
11. Chekanin, V.A., Chekanin, A.V.: Multilevel linked data structure for the multidimensional orthogonal packing problem. Appl. Mech. Mater. **598**, 387–391 (2014)

Cooperative Organization and Application Mechanism Based on Intention Environment Target for Maritime Ship-Aircraft Cooperation

Jingshi Wang[1,2], Zehua Zou[1], Miao Wang[1](✉), and Guoqing Wang[1]

[1] Shanghai Jiao Tong University, Shanghai, China
miaowang@sjtu.edu.cn
[2] Jiangsu Automation Research Institute, Lianyungang, China

Abstract. As the complexity of the maritime application environment increases, the traditional fleet-based application system relying solely on manned ship is inadequate to meet growing demands. To address this issue, the collaborative application between ship-borne unmanned aerial vehicles (UAVs) and manned ships is necessary. However, as the degree of intelligence of ship-borne UAVs cannot make decisions fully by itself, an effective ship-aircraft cooperative decision-making architecture mechanism is required. This paper proposes a collaborative decision-making mechanism and organization based on the strategy-tactics-mission architecture to enable effective ship-aircraft cooperative decision-making in modern maritime ship-aircraft cooperative application. This paper clarifies the concept of modern maritime ship-aircraft cooperative decision-making and analyzes the objectives and demands of this approach based on intention, environment, and objectives. Based on these requirements, a collaborative decision-making mechanism and organization are established using a strategy-tactics-mission architecture. A collaborative decision-making goal system is proposed, and a collaborative decision-making phase is divided into unilateral and multilateral decision-making processes. A collaborative operational decision-making model mechanism is constructed, which includes both the unilateral and multilateral decision-making processes. Finally, a typical maritime multi-ship aircraft collaborative scenario is established based on the proposed collaborative decision-making mechanism. The paper models the six stages of collaborative decision-making based on the strategy-tactics-mission framework, including both the unilateral and multilateral collaborative decision-making processes between ship-borne UAVs and manned ships.

Keywords: Ship-aircraft cooperation · Collaborative decision-making · Cooperative application

1 Introduction

The current architecture of the world fleet system primarily relies on large manned ships formations, utilizing a centralized command and control mode to achieve high command efficiency. However, this structure concentrates a large number of forces in one location,

Z. Jing and X. Zhan (Eds.): ICASSE 2023, LNEE 1153, pp. 89–103, 2024.
https://doi.org/10.1007/978-981-97-0550-4_7

increasing the likelihood of being targeted by the opponent and potentially causing the entire system to be paralyzed by a single fault. To address this issue, cross-domain distributed collaborative application at sea can integrate platforms across various domains, including land, sea, air, space, electricity, and network, to construct an all-around three-dimensional situational awareness network [1]. This network can be achieved through communication networks and integrated command and control systems distributed across different domains, thereby implementing system cluster/distributed application forms to enhance overall operational efficiency.

Additionally, a collaborative system of manned ships and unmanned platforms can achieve an efficient situational awareness network [2], a powerful and resilient network, fundamentally realizing cross-domain distributed collaborative application ideas [3]. Shipborne UAVs have the advantages of low cost, simple structure, and strong reconnaissance capabilities and also possess significant advantages in electronic application, air defense application, air application, anti-ship application, anti-submarine application, and land application when conducting sea application [4]. However, current UAVs are primarily used for simple tasks such as reconnaissance and search and rescue in areas with weak air defense capabilities. Facing the complex changes in future fields, a single UAV may not meet applicational requirements. Therefore, research on collaboration between multiple UAVs as well as between multiple UAVs and ground vehicles, surface ships, and satellites is essential for future applicational mode development [5–7].

This paper discusses the changing needs of maritime application by examining the transformation process of operation. It also explores the requirements of ships and aircraft in the process of maritime cooperative application a and investigates the needs of a modern ship-aircraft cooperative application mode, the needs of ship-aircraft cooperative application process decision-making, and the needs of cooperative decision-making mode based on intention environment and objectives. Additionally, this paper proposes an operational mechanism of ship-aircraft cooperative decision-making, constructs a cooperative decision-making goal system, divides the cooperative decision-making phase, and proposes both the unilateral cooperative decision-making mechanism and the multilateral cooperative decision-making mechanism. Finally, a typical maritime ship-aircraft cooperative scenario is constructed, and the unilateral ship-aircraft cooperative decision-making process in this scenario is modeled based on the source diagram of domestic system engineering modeling software, and the multilateral ship-aircraft cooperative decision-making process is verified.

2 Intent, Environment, and Target Requirements and Concepts

2.1 Requirements for Modern Ship-Aircraft Cooperative Mode

In response to the modern ship-aircraft cooperation needs, clear research development direction, analysis of maritime cross-domain manned and unmanned ship-aircraft needs, mainly in the following areas [8, 9].

(1) Cross-domain dynamic field scenario posture real-time and consistent organization needs. Domain is a space to describe the role of missions. The traditional ship domain is based on the physical space role of shipboard physical domain. Modern maritime

application to build a multi-domain collaborative mission, the establishment of cross-domain dynamic field scenario posture in real time and consistency, to provide manned ships at sea and shipboard UAV collaborative posture display to meet the needs of the modern maritime domain.

(2) Information quality, scope and timeliness requirements for different mission domains. Applicational range is a description of the area covered by the mission. The traditional ship range to ship navigation, the role of the ship's physical distance and range to determine the area. Modern maritime application to build a full space collaborative mission to establish the quality, scope and timeliness of information needs of different areas to meet the modern manned ships at sea and shipboard unmanned aircraft collaborative decision-making capabilities, performance and effectiveness needs.

(3) The unilateral and multilateral collaborative decision-making mechanism needs between ship and aircraft. The model is a way to describe the implementation of the mission. Traditional ship models are built with how the shipboard capabilities are utilized. Modern maritime operations build integrated capabilities to collaborate on missions, establish unilateral and multilateral collaborative decision-making mechanisms for multiple platforms such as main ships, ships and drones, and meet the organization and application requirements for different environments mode.

(4) Based on cross-domain information defects of missions and target monitoring and evaluation needs. Mission operation state is to describe the gains obtained from missions. The traditional ship domain defines effectiveness in the form of the effect of action on targets. Modern maritime operations to the capability preservation state, network operation organization and mission composition as the mission target results, the construction of target results of the collaborative mission, the establishment of cross-domain information-based mission operational state monitoring and evaluation, effective perception and determination of mission operational state and enhance the effectiveness of the collaborative mission to meet the modern manned ships at sea and shipboard UAV collaborative mission control and management needs of modern manned ships and UAVs.

2.2 Decision-Making Mode of Ship and Aircraft Cooperative Process

For the ship-aircraft collaborative complex process, a collaborative mission decision-making approach must be used to eliminate the offset. Therefore, for the multi-organizational, multi-dimensional, and multi-domain collaborative model of manned ships and shipboard UAVs, multi-organizational knowledge, cognitive and intention-based collaborative decision- making must be established as shown in Fig. 1.

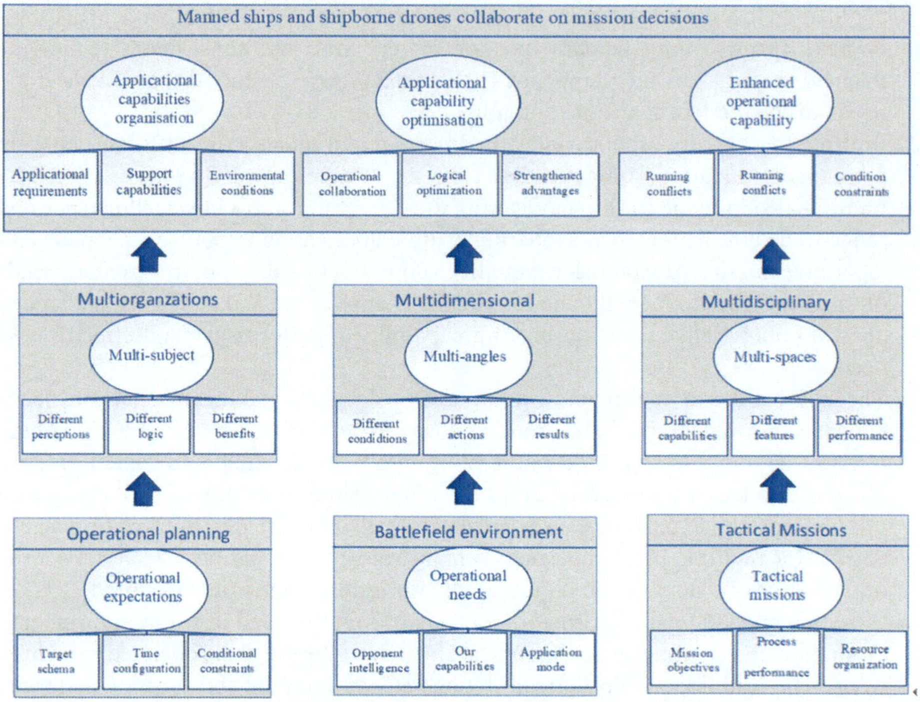

Fig. 1. Cooperative decision-making mode in operational process

2.3 Decision-Making Mode of Ship and Aircraft Cooperative Process

Process situational awareness is the basis and guarantee of ship-aircraft cooperative process decision-making. Situational awareness is a description of the role subject based on its own initial intention, oriented to the role environment, determine the time period, perceive the associated objects, identify the constraints and associated and unassociated factors, construct the elimination of factors intended to amend, provide and guide the role subject's decision- making, as shown in Fig. 2. It mainly includes:

(1) Strategic situational awareness: by establishing strategic mission requirements, implementing constraint, correlation and uncorrelated factor analysis based on acquired opponents' intent and capability intelligence, establishing strategic development pressure impacts, identifying strategic development constraints, and forming the final strategic organizational intent - strategic goal expectations.
(2) Field situational awareness: In response to strategic expectations and performance indicator definitions, implement constraint, correlation, and uncorrelated factor analysis based on field object and condition perceptions based on applicational planning, field direct threat, potential threat, and no threat classification, and form the final field organizational intent - field operational objectives in the face of complex, opaque field environment, cope with complex coupling such as electromagnetic interference, complete consistent situational awareness of manned ships and shipboard UAVs, and

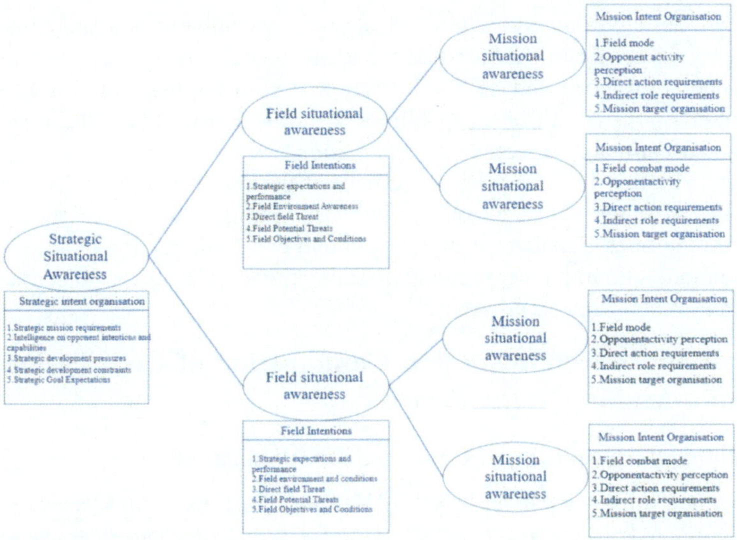

Fig. 2. Organizational structure of situation awareness in operational process

achieve collaborative decision-making is the next noteworthy direction in the field of situational awareness.

(3) Mission situational awareness: for field operational objectives and environment, according to the field operational model organization, based on units activity perception, implementation of constraint, correlation and uncorrelated factors analysis, field direct role and indirect role, forming the final mission organization intention unit mission objectives.

2.4 Collaborative Decision-Making Model Based on Intention, Environment and Objectives

The modern cooperative application process is based on the cooperative application of multiple individuals, each of which has its own independent target space, operating environment, behavior mode and result state; The effectiveness of modern coordinated results is based on the target space, operating environment, behavior pattern and result state, and achieves balanced and optimized results through integrated technology. Therefore, the decision-making model based on intention, environment and target is based on the decision-making situation of operational results that supports balance and optimization, mainly including:

(1) Intent: Applicational intent is based on the real-time embodiment of operational strategy, that is, on the basis of the initial definition of Applicational strategy, with the change of time (passage), according to the changes in the environment and the state of the mission, real-time revision and improvement of the Applicational strategic objectives.

(2) Applicational environment: The applicational environment is a specified applicational process environment based on the applicational strategy, that is, the applicational operation mode and field objectives that are established and revised in real time based on the conditions, capabilities and time constraints of the Applicational strategic mission and expected implementation.
(3) Mission: The mission is a mission activity of the unit formed by determining the mode based on the field and forming the task configuration of the mode. That is, based on the operational strategic mission and expectations, the field environment and the mode, the implementation and establishment of mission objectives and the operation of coordinated missions.

3 Operational Mechanism and Composition of Coordinated Decision-Making in the Process

3.1 Operational Mechanism of Coordinated Organization

Modern manned ships and carrier-based UAVs coordinate operations based on strategic mission planning, field environment perception, tactical mission organization, multi-organization, multi-dimensional, multi-field composition, according to the needs of capabilities, performance and efficiency targets, implement a decision-making mode based on intention, environment and objectives, and realize the decision-making, operation, control and management process of missions. The main ones include:

(1) Event-based collaborative decision-making mechanism: Event-based collaborative decision-making is a multi-master decision-making coordination mechanism for traditional operations. Event-based collaborative decision-making focuses on the process, key events, and chronological sequence. Such as intelligence, interception, etc., through information perception, operation mode organization and result goal definition, the event-oriented first-come-first-served mechanism is adopted.
(2) Collaborative decision-making mechanism based on strategy-tactical-mission framework: Collaborative decision-making based on strategy-tactical-mission architecture is a multi-master decision-making coordination mechanism of modern operations, which mainly focuses on the guidance, constraint and role of mission activity configuration decision-making guidance, constraints and roles of mission activity configuration of intention, environment and conditions. That is, to establish multi-master collaborative decision-making oriented to strategic guidance, field environmental constraints, and mission control, form an integrated operation mode of all individuals in the process, and maximize the realization of strategic goals, adapt to field conditions, and give play to the capabilities of units.

The ship-aircraft coordination process organization is based on the comprehensive process organization of maritime needs, field environment and distributed platform, that is, according to the environment, according to the status of each platform of distributed collaborative operation, based on the current capabilities and performance of each platform, and according to the mode of each platform, the route planning, operation efficiency, operational constraints and guidance of each unit in distributed coordination are realized, and the distributed collaborative process organization is completed, as shown in Fig. 3.

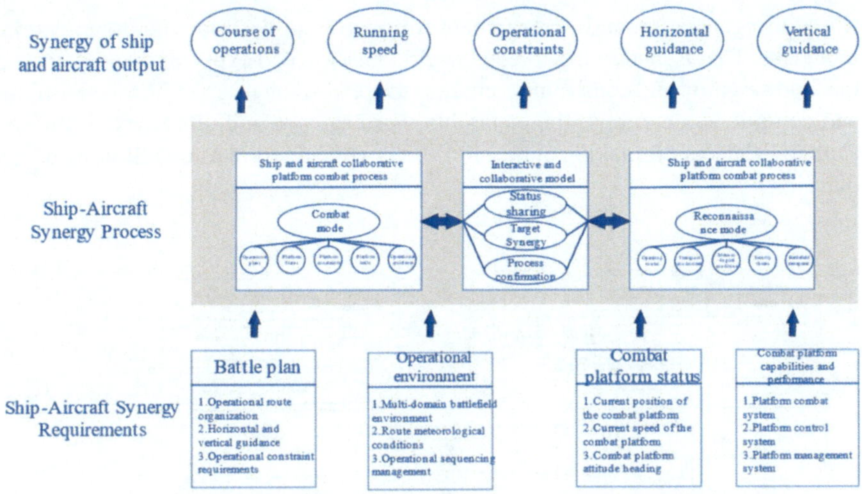

Fig. 3. Organizational mechanism of collaborative process

3.2 Operational Organization of Operational Coordinated Decision-Making

The operational organization of operational coordinated operational decision-making is to determine the expectations and benefits of operational results by constructing the operational cycle time, regional location, and scale range, and guiding the objectives of operational coordinated decision-making; Determine mission capability, performance, and effectiveness through its own capabilities, environment, and conditions, and support operational mission target decision-making, as shown in Fig. 4. The main ones include:

(1) Strategic planning organization and management: Strategic planning is the planning organization that describes the process. That is, based on strategic mission, based on strategic intelligence, according to its own capabilities, build the operational cycle time, regional location, scale and range, determine the expectations and benefits of operational results, guide operational coordination decision-making goals, and form operational strategic plans.
(2) Field environment organization and management: The field environment is a constraint that describes the process. That is, according to the strategic planning and result expectations, in view of the opponent's activities and advantages and disadvantages, and based on its own capabilities and advantages and disadvantages, it builds operational needs, organizational models, role fields, and conditions, determines the objectives of the process and the field situation, constrains the conditions for operational coordination decision-making, and forms an operational organization plan.

(3) Mission organization and management: mission is to describe the organization of unit tasks. That is, according to the strategic mission and expectations, according to the field situation and constraints, and in the face of its own capabilities, environment and conditions, determine the capability, performance and efficiency of missions, support the decision-making of mission objectives, and form a unit mission operation plan.

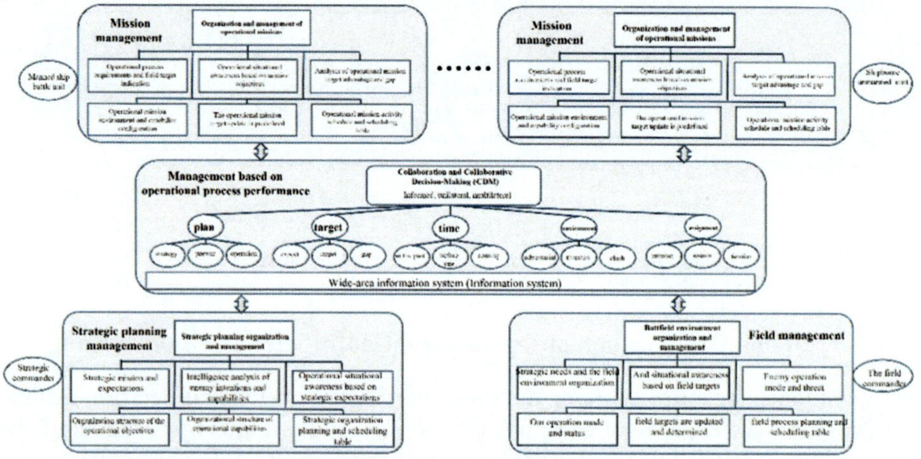

Fig. 4. Operation organization of operational coordination decision-making

The cooperative decision-making organization mechanism of ships and aircraft is based on the plan, according to the environment, based on the current capabilities and performance of each platform, to meet the needs of security, adopt an automated, autonomous and intelligent interactive coordination mode, and realize the definition of coordinated objectives, optimization of tracks, organization of processes and management of operational status, and realize collaborative decision-making organization according to the task division of labor, environmental perception, professional knowledge, operational cognition and environmental constraints of each platform, as shown in Fig. 5.

3.3 Operational Collaborative Decision-Making Target System

Coordinated decision-making between manned ships and carrier-based UAVs must establish a clear, relevant and measurable target system. The applicational collaborative decision-making goal system is divided into: applicational strategy and application strategy expectations, performance domain requirements for reflecting the effectiveness of strategic expectations on the field, field-oriented mode specified field result objectives, mission-oriented logic and measurable indicators, mission implementation path status and mission operation objectives, operational collaborative decision-making target organization system shown in Fig. 6, and collaborative decision-making processing process as shown in Fig. 7.

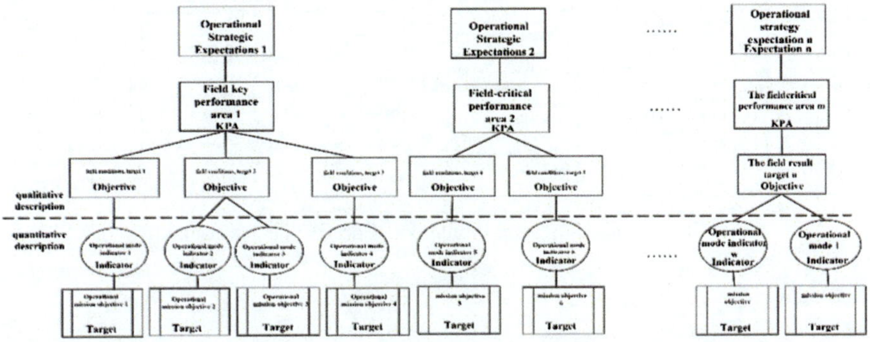

Fig. 5. Organizational structure of cooperative decision-making objective system in operational process

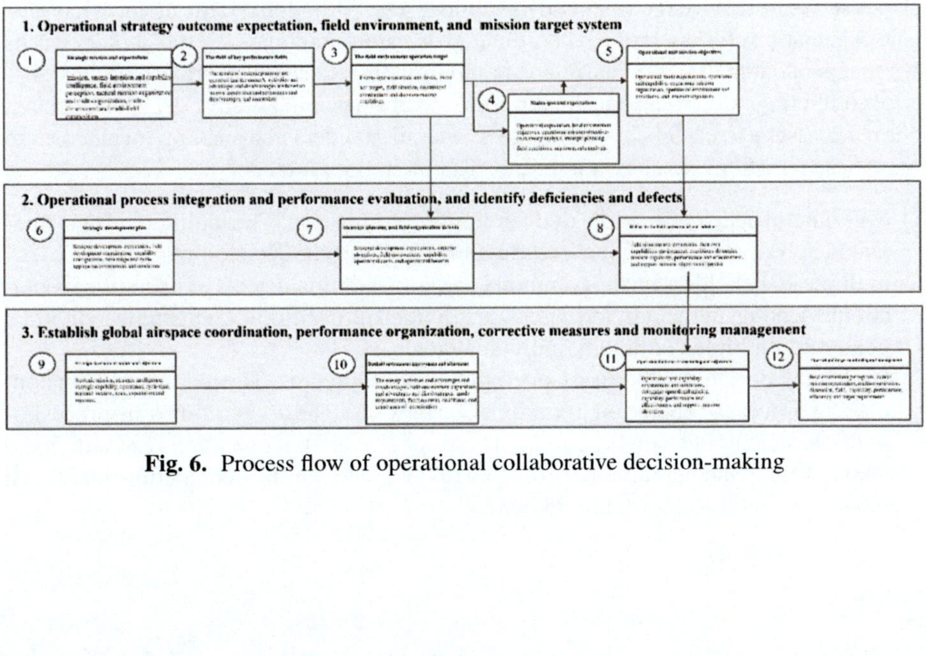

Fig. 6. Process flow of operational collaborative decision-making

3.4 Phases of Operational Coordinated Decision-Making

The applicational coordination decision-making process is completed through six stages, starting from the identification of applicational coordination requirements in the applicational process, as shown in Fig. 7. Through the analysis of applicational coordinated operation, standardize and verify the organization of operational coordinated operation, determine the requirements of operational collaborative performance, confirm the mode of operational coordinated decision-making, implement the decision-making process of operational coordination, and finally implement coordinated operation, maintenance and improvement.

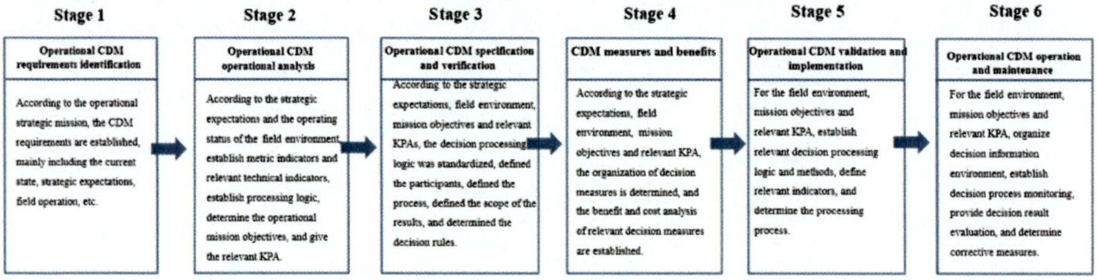

Fig. 7. Six phases of operational collaborative decision-making

3.5 Operational Collaborative Decision-Making Mode Mechanism

The collaborative decision-making process for manned ships and carrier-based UAVs is known to involve multiple operational stakeholders. Such as commanders, aircraft pilots, various services, service facilities and institutions, etc. Although all stakeholders have the same expectations for operational coordinated decision-making, each stakeholder has its own independent mission, goals, interests and limitations, and there are unanimous, separate, conflicting, and irrelevant situations in operational collaborative decision-making, so it is necessary to establish a cooperative operational decision-making mechanism to cover these situations, as shown in Fig. 8. The main ones include:

(1) Multilateral decision-making of operational coordination: The multilateral decision-making process of operational coordination describes the decision-making process of multiple stakeholders on the common goals and individual goals of operational operations. Among them, individual goals are divided into multiple compatible individual goals and multiple conflicting individual goals.
(2) Unilateral decision-making of operational coordination: The unilateral decision-making process of operational coordination describes the decision-making process of multiple stakeholders on the common goals of operational operations and individual goals. Among them, individual goals are divided into multiple compatible individual goals and multiple conflicting individual goals.

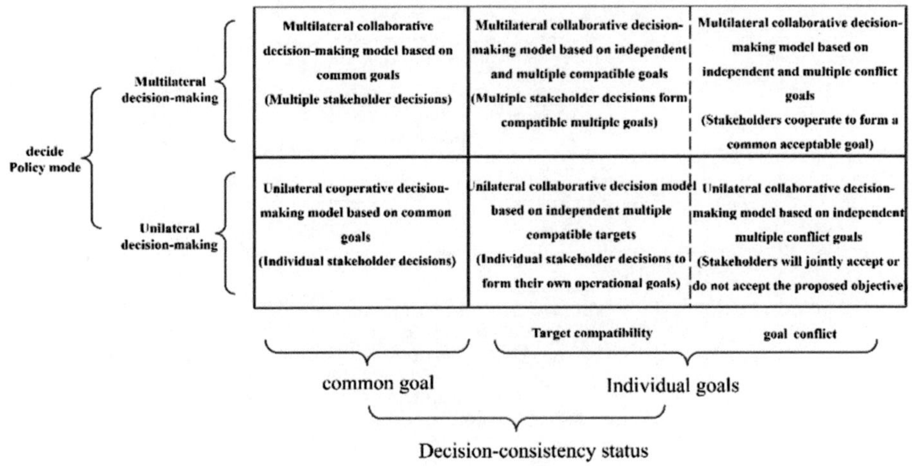

Fig. 8. Operational Collaborative Decision Mode Mechanism

(3) Commonality, compatibility and conflict of operational collaborative decision-making goals: The commonality of operational collaborative decision-making with stakeholder participation is to describe the common goal that everyone shares, compatibility is to describe the acceptability of the goals of other participants, and conflict is to describe the conflict or confrontation between everyone's independent goals.

4 Scenario Modeling and Analysis

In order to realize the strategic intention of controlling sea power in a certain sea area, according to this strategic mission, the red side has planned a operation plan against the port infrastructure of a certain island and reef controlled by the blue side, and unmanned ship groups, electronic manned aircraft and reconnaissance by dispatching manned ships According to the specific objectives and field environment of destroying port infrastructure, air defense positions and radars on the blue islands, the red side formulated a multi-platform coordinated operation strategy, as shown in Fig. 9, and formulated a action plan according to the six stages from the perspective of operational coordinated decision-making. The specific process of the operation is: first, through manned electronic aircraft to electronically suppress the island where the blue side is located, and then unmanned reconnaissance aircraft are dispatched to reconnoiter the blue The specific situation of the radar station of the port facilities on the island, and the unmanned reconnaissance aircraft guide the guns of manned ships to carry out ships' action on the blue islands, destroying the blue radars and air defense and coastal defense positions. Finally, the unmanned boat group enters the island port area, destroys the blue port infrastructure, and completes the operation objective.

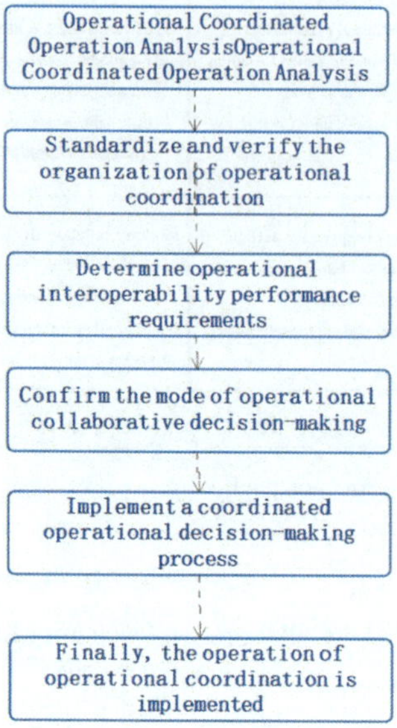

Fig. 9. Six stages of operational coordinated decision-making

In the course of operations, units need to make unilateral decisions in coordination. For example, reconnaissance UAVs need to make unilateral decisions on coordination based on the information they obtain and reconnaissance targets in the process of reconnaissance of specific conditions on blue side islands. If the reconnaissance UAV finds that the blue's radar or communication system needs to notify the manned electronic jammer for jamming or suppression, the reconnaissance UAV finds the ground, air defense or coastal defense position needs to notify the manned ship to action, and at the same time, according to the manned ship action demand, the unmanned reconnaissance aircraft cooperate to guide the action. Due to the limitation of manned ship action capability, unmanned reconnaissance aircraft need to temporarily abandon further reconnaissance and cooperate with it to provide guidance action, unmanned reconnaissance aircraft need to make decisions according to the intelligence obtained by themselves and target conditions, adjust their own reconnaissance plans, give priority to meeting the guidance needs of manned ships, and the unilateral decision-making process is shown in Fig. 10.

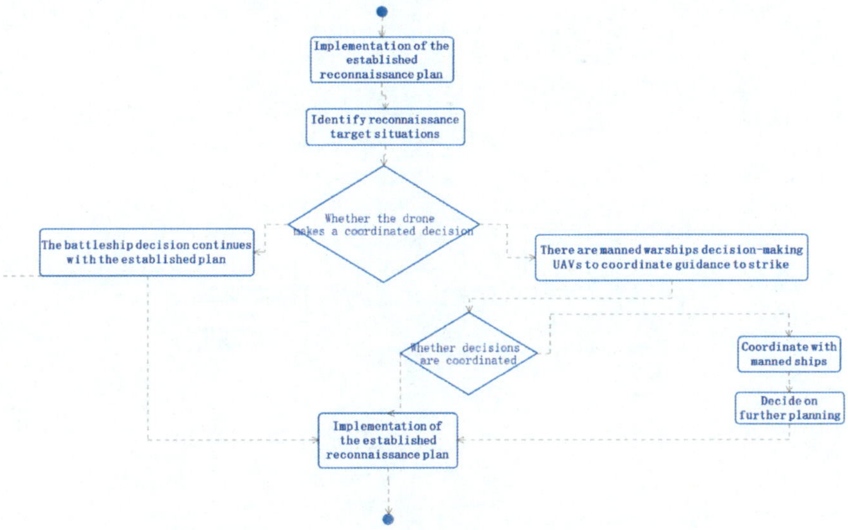

Fig. 10. Unilateral collaborative decision-making process

In the process of operation, units need to coordinate multilateral decision-making, for example, when unmanned boat groups enter the port area and destroy port facilities, they need to cooperate with electronic jamming drones and surface unmanned boats, in order to maximize the action effect, multilateral decision-making is required, which is led by the command and control center located on the manned ship, which is limited by the reconnaissance information of manned aircraft and according to its own action capability. It was determined that the target needed to be hit by an unmanned boat group. The unmanned boat group determines the action plan according to its own capability limits and location, and at the same time, electronic warfare aircraft need to timely carry out corresponding electronic suppression according to the needs of unmanned reconnaissance aircraft and unmanned boats. The unmanned reconnaissance aircraft determines the reconnaissance target according to its own reconnaissance target, directly notifies the unmanned boat of the action target to carry out unmanned coordination with the unmanned boat, or reports the action situation to the command center of the manned ship, and the command center decides whether to continue to launch the action by the unmanned boat, and models it in the source diagram of the system engineering modeling software The multilateral decision-making process is shown in Fig. 11.

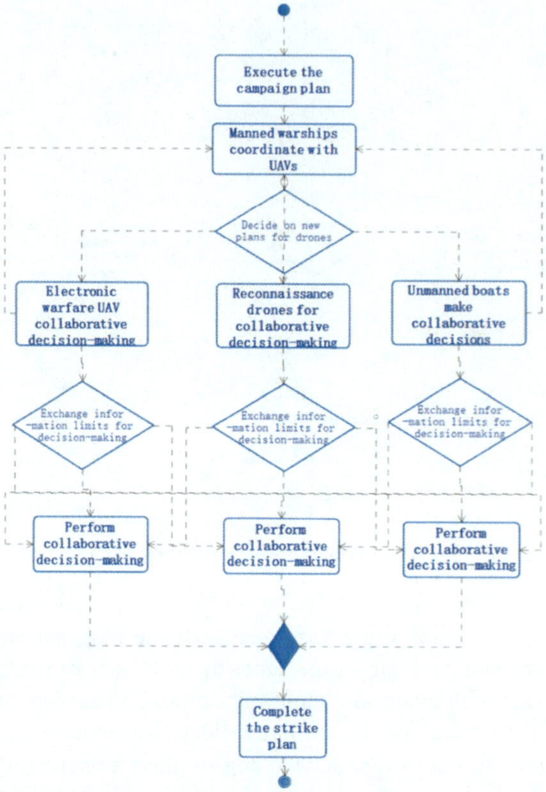

Fig. 11. Multilateral collaborative decision-making processes

5 Conclusion

This paper analyzes the model, organization and applicational requirements of modern ship-aircraft cooperative application based on strategic mission planning, field environment perception and tactical mission organization of ship-aircraft cooperative applications. The paper also validates the proposed unilateral and multilateral ship-aircraft cooperative decision-making mechanism by constructing a typical maritime ship-aircraft system scenario to realize the manned ship and ship-aircraft UAV cooperation, organization and operation process design.

Acknowledgements. This work is partially supported by Natural Science Foundation of Shanghai (20ZR1427800), New Young Teachers Launch Program of Shanghai Jiao Tong University (20X100040036).

References

1. Zhu, Z., Du, Q., Wang, Z., Li, G.: A survey of multi-agent cross domain cooperative perception. Electronics (2022). https://doi.org/10.3390/electronics11071091

2. Huckleberry, S.: Cross domain synergy: using artillery in the fight for sea control. Fires 51–59 (2017). https://www.proquest.com/trade-journals/cross-domain-synergy-using-artillery-fight-sea/docview/2101842214/se-2. Accessed 2 Apr 2023
3. Sun, H., Xu, J., Yang, L.: An UCAV intelligent combat system for distributed combat network. In: 2021 40th Chinese Control Conference (CCC), Shanghai, China, pp. 7840–7844 (2021). https://doi.org/10.23919/CCC52363.2021.9549535
4. Smith, G.: Organization and innovation: integrating carrier-launched UAVs. Naval War Coll. Rev. **70**(3), 79–100 (2017). JSTOR. http://www.jstor.org/stable/26398042. Accessed 2 Apr 2023
5. Li, R., Ma, H.: Research on UAV swarm cooperative reconnaissance and combat technology. In: 2020 3rd International Conference on Unmanned Systems (ICUS), Harbin, China, 2020, pp. 996–999 (2020). https://doi.org/10.1109/ICUS50048.2020.9274902
6. Luo, J., Su, B., Yuan, G.: Development trend of unmanned combat system and its application in intelligent war. In: 2020 International Conference on Robots & Intelligent System (ICRIS), Sanya, China, 2020, pp. 275–278 (2020). https://doi.org/10.1109/ICRIS52159.2020.00075
7. Zhou, Y., Rao, B., Wang, W.: UAV swarm intelligence: recent advances and future trends. IEEE Access **8**, 183856–183878 (2020). https://doi.org/10.1109/ACCESS.2020.3028865
8. COMMAND AFS: Resiliency and Disaggregated Space Architectures: White Paper (2013)
9. Sujit, P., Saripalli, S.: An empirical evaluation of co-ordination strategies for an AUV and UAV. J. Intell. Robot. Syst. **70**(1), 373–384 (2013)

LPV Robust Filter Based Fault Diagnosis Method for Aeroengine Control System

Huihui Li(✉), Linfeng Gou, and Zongting Jiang

Northwestern Polytechnical University, Xi'an, China
lihuihui@mail.nwpu.edu.cn

Abstract. As the most typical component of an aeroengine control system, the sensor operates under high temperature, high pressure and high speed for a long period of time. It is exposed to a high rate of faults occurring. An analysis and design study of LPV robust filter-based fault diagnosis method for aeroengine control system is conducted based on the idea of multi-sensor fusion estimation under unknown disturbance and model uncertainty. Simulation experiments verify the effectiveness of the filter is applied to sensor fault diagnosis of the aeroengine control system.

Keywords: Aeroengine control system; Sensor fault diagnosis · LPV · Robust filter

1 Introduction

As the most typical component of an aeroengine control system, the sensor operates under high temperature, high pressure and high speed for a long period of time [1, 2]. It is exposed to a high rate of faults occurring. Fault can result in unsatisfactory engine control system performance or even system instability [3]. To avoid serious losses due to engine control system faults, it is necessary to monitor and diagnose the engine control system in order to meet the requirements for efficiency, reliability and safety of the engine and the aircraft [4].

Fault diagnosis of aeroengine control systems is a typical multi-sensor fusion problem. Kalman filter-based multi-sensor information fusion is one of the most important methods [5, 6]. It is well known that the engine is very complex, modelling the physics involves making many assumptions and simplifying [7]. The linear model obtained from it contains a lot of uncertainty, so the application of Kalman filtering is somewhat limited. One of the solutions is to design a robust filter. This means to consider the effect of model uncertainties on the filtering performance when designing the filter, so that for all allowed uncertainties there is a minimum upper bound on the robust filter [8].

The above analysis is based on a Linear time-invariant (LTI) models, which can only describe the engine control system close to a given stationary operating point of the engine and lacks dynamic information about the model [9]. The Linear Parameter Variation (LPV) framework is used to diagnose faults in changing operating states of a highly nonlinear system. In summary, this paper proposes a robust LPV filter approach for aeroengine fault diagnosis to improve the accuracy of fault diagnosis method.

Z. Jing and X. Zhan (Eds.): ICASSE 2023, LNEE 1153, pp. 104–110, 2024.
https://doi.org/10.1007/978-981-97-0550-4_8

2 Multi-sensor Information Fusion

2.1 Theoretical Background

Multi-sensor information fusion state estimation is a combination of traditional estimation theory and data fusion theory for estimation problems. It is the study of how best to use the useful information contained in multiple data sets to estimate unknown quantities, usually from multiple sensors. Obviously, estimation fusion has many applications. The reason for this is that many design estimation problems have data from multiple information sources.

Two basic fusion methods exist: central and distributed fusion. The centralised fusion approach will be the focus of this paper. In central fusion, all sensor data are transmitted to a central processor for processing and fusion, as shown in Fig. 1. The fusion centre is globally optimal in terms of unbiased linear minimum variance estimation because it can use the raw measurement data from all sensors without any loss of information [10].

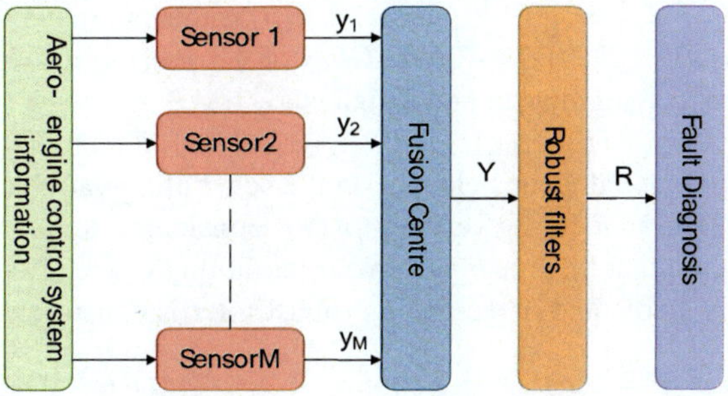

Fig. 1. Fault diagnosis based on a centralised fusion approach

2.2 Description of the Problem

Let the state equation of the multi-sensor fusion system be as follows.

$$\dot{x}(t) = Ax(t) + Bw(t) \tag{1}$$

where $x(t) \in R^n$ is the n-dimensional system state, A is the system matrix, B is the noise distribution matrix, and $w(t) \in R^h$ is the noise vector.

The measurement equation is described as follows.

$$y_i(t) = C_i x(t) + v_i(t) \tag{2}$$

where $y(t) \in R^m$ is the measurement output of the i_{th} sensor, C is the corresponding measurement matrix, $i = 1, 2, ..., M$. $w(t)$, $v_i(t)$ are all zero-mean Gaussian white noise, and are uncorrelated.

$$E\left\{\begin{bmatrix} w(t) \\ v_i(t) \end{bmatrix}\begin{bmatrix} w'(t) & v_i'(t) \end{bmatrix}\right\} = \begin{bmatrix} Q(t) & 0 \\ 0 & R_i(t) \end{bmatrix}\delta_i$$

where the symmetric semi-positive definite array Q,R is the variance array of $w(t)$, $v_i(t)$, respectively. δ_i is the Kronrecker function.

The state estimation problem of the multi-sensor fusion system is defined as the problem of finding the best estimate $\hat{x}(t)$ of the system state variable $x(t)$ on the basis of the observations of the observation system $\{y_1(t), y_2(t), \ldots, y_N(t)\}$.

3 Fault Diagnosis Based on LPV Robust Filters

An aeroengine model with sensor faults, unknown disturbances and model uncertainty can be expressed as follows.

$$\begin{aligned}\dot{x}(t) &= (A(\rho) + \Delta A(\rho))x(t) + (B(\rho) + \Delta B(\rho))u(t) + E_d(\rho)d(t) + E_f(\rho)f(t)\\ y(t) &= C(\rho)x(t) + F_d(\rho)d(t) + F_f(\rho)f(t)\end{aligned} \tag{3}$$

where $x(t) \in R^n$ is the state variable, $u(t)$ is the input variable, $d(t)$ is an external disturbance bounded signal, satisfying $||d(t)|| \le ||d(t)||_\infty$, $f(t)$ represents the fault variable,$E_f(\rho) = 0$, $F_f(\rho) = \begin{bmatrix} 0 \; I \end{bmatrix}$. $\Delta A(\rho)$, $\Delta B(\rho)$ is the parameter uncertainty matrix of the system, $\Delta A(\rho) = E_A(\rho)\Sigma_A F_A(\rho)$, $\Delta B = E_B(\rho)\Sigma_B F_B(\rho)$, where $\Sigma_A(t)$, $\Sigma_B(t)$ is an unknown time-varying matrix and satisfies $\Sigma_A^T(t)\Sigma_A(t) \le I$, $\Sigma_B^T(t)\Sigma_B(t) \le I$.

Assumption 1: $A(\rho)$, $B(\rho)$, $C(\rho)$, $D(\rho)$, $E_d(\rho)$, $E_f(\rho)$, $F_d(\rho)$, $F_f(\rho)$ is a continuous and bounded function and depends closely on the scheduling variable $\rho(t)$.

Assumption 2: The scheduling variable $\rho(t)$ is considered to be measurable.

On the assumption that the linear parameter variation system (3) at the control input $u(t)$ is quadratically stable and detectable, a robust filter is constructed.

$$\begin{cases}\dot{\hat{x}}(t) = A(\rho)\hat{x}(t) + B(\rho)u(t) + K(\rho)[y(t) - C\hat{x}(t)]\\ \hat{y}(t) = C\hat{x}(t)\\ e(t) = x(t) - \hat{x}(t)\\ r(t) = y(t) - \hat{y}(t)\end{cases} \tag{4}$$

where K is the fault detection filter gain to be found, and varies with scheduling parameters,$K(\rho) = \sum_{i=1}^{r} \rho_i(\theta(t))K_i.\hat{x}(t)$ is the estimated state vector, $e(t)$ is the state estimation error, $z(t)$ is the introduced intermediate vector and $r(t)$ is the residual vector.

Then, the following error dynamic system is obtained:

$$\begin{cases}\dot{e}(t) = (A(\rho) - K(\rho)C)e(t) + \Delta A(\rho)x(t) + \Delta B(\rho)u(t)\\ +(Ef(\rho) - K(\rho)Ef)f(t) + (E_d(\rho) - K(\rho)F_d)d(t)\\ r(t) = Ce(t) + F_d d(t) + F_f f(t)\end{cases} \tag{5}$$

From a computational point of view, the filter gain design problem can be characterised by Linear matrix inequalities (LMI) [11, 12]. From the Schur Complementary Lemma, the Cross Term Amplification Lemma and the Bounded Real Lemma, the following conclusions can be drawn.

Given the scalar $\gamma > 0$, the LPV robust filter (4) is asymptotically stable and satisfies $\|G(s)\|_\infty < \gamma$, if there exists a symmetric positive definite matrix $P > 0$, and a scalar $\varepsilon_i > 0$, $i = 1, 2$, such that for all scheduling variables $\rho(t)$, Eq. (6) holds.

$$\begin{bmatrix} \begin{matrix}P(A(\rho)-K(\rho)C)+\\(A(\rho)-K(\rho)C(\rho))^TP+C(\rho)^TC(\rho)\end{matrix} & 0 & 0 & \begin{matrix}P(E_d(\rho)-KF_d(\rho))\\+C(\rho)^TF_d(\rho)\end{matrix} & PE_A(\rho) & PE_B(\rho) \\ * & -\gamma^2I+\varepsilon_1F_A^T(\rho)F_A(\rho) & 0 & 0 & 0 & 0 \\ * & * & -\gamma^2I+\varepsilon_2F_B^T(\rho)F_B(\rho) & 0 & 0 & 0 \\ * & * & * & -\gamma^2I+F_d(\rho)^TF_d(\rho) & 0 & 0 \\ * & * & * & * & -\varepsilon_1I & 0 \\ * & * & * & * & * & -\varepsilon_2I \end{bmatrix} < 0 \tag{6}$$

The problem of solving the filter can be equated to finding a gain matrix $K(\rho) = \sum_{i=1}^{r} \rho_i(\theta(t))K_i$ such that the estimation error is dynamically stable and the following objective functions are minimized:

$$\begin{gathered}\min \gamma \\ \text{s.t. equation}(6).\end{gathered} \tag{7}$$

The convex multi-cell technique can be applied to transform the parameter-dependent LMI equation into a finite-dimensional LMI under multi-cell vertices for solution, resulting in a filter that can meet global performance requirements.

4 Experiment and Discussion

In this section, simulation tests are performed using the LPV engine control model to verify the effectiveness of the proposed method. The architecture of the aeroengine that is used is shown in Fig. 2 [13].

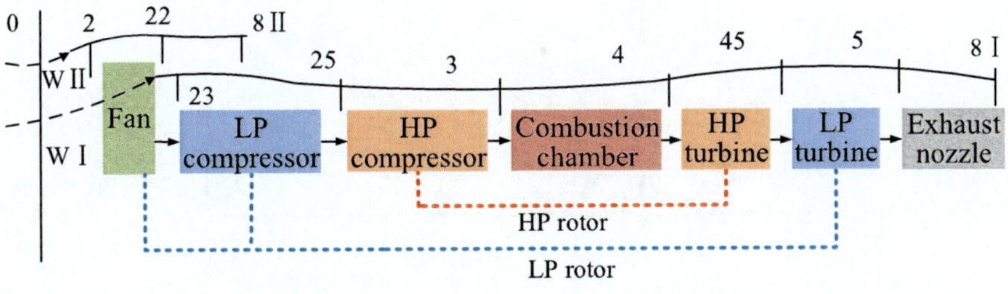

Fig. 2. The architecture of the aeroengine

Based on the analysis of the thermodynamic relationships between the engine components as shown above, a non-linear component-level model of the engine is developed.

Several steady states are then selected for Jacobi linearisation under different scheduling parameters. These linearised models form a system of equations which are solved to obtain the affine LPV model. Finally, the multi-cell LPV is obtained by transforming the affine LPV model.

The selected scheduling parameter is flight altitude (H), flight Mach number (Ma), High pressure rotor speed (NH). The simulation is verified for steady-state conditions with a scheduling parameter of $H = 0km, Ma = 0, N_H = 97.26\%$.

The following fault scenarios are considered: in the flight condition of $H = 0km, Ma = 0, N_H = 97.26\%$, at $t = 5$–10 s, the low pressure rotor speed sensor has suddenly developed a hard fault, with a fault amplitude of 5%.

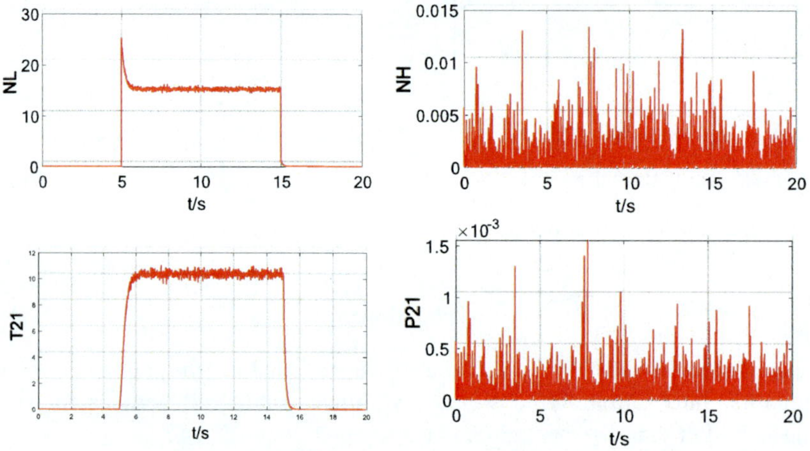

Fig. 3. The residuals generated by the designed robust LPV filter

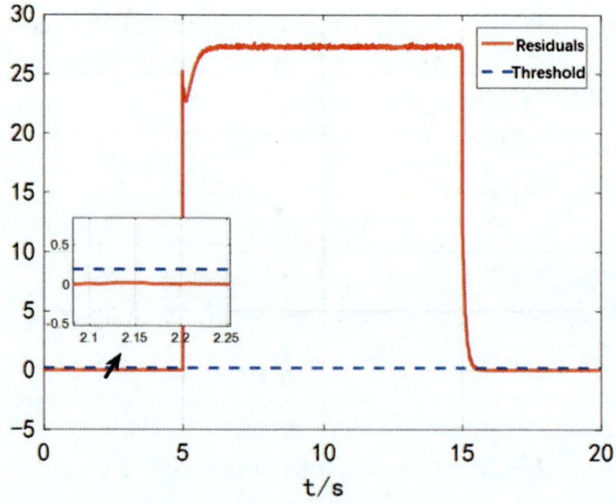

Fig. 4. The Fault diagnosis results from the designed robust LPV filter

Figure 3 and Fig. 4 show the simulation results for fault diagnosis. The residual signal obtained by comparing the predicted output of the fault detection filter with the actual measured output of the aeroengine control system is shown in the curve in Fig. 3. The theoretical adequacy criterion to ensure the performance of H_∞ is given above based on linear matrix inequalities. Since the effect of the uncertainties is effectively separated from the estimation error, the robust filter converges quickly and tracks the engine output well when no faults occur, so the residuals are stable in the vicinity. The residuals are evaluated as shown in Fig. 4, where the red solid line indicates the residuals and the blue dashed line indicates the threshold. When no faults occur, the residuals are within the threshold. When a fault occurs, the residuals rapidly exceed the threshold range, at which point it can be determined that a fault has occurred in the control system. In addition, false alarms can be avoided by selecting predefined thresholds. The simulation results verify the effectiveness of the proposed method in fault detection.

5 Conclusion Experimental Verification

This chapter focuses on the analysis and design study of the LPV robust filter-based fault diagnosis method for aeroengine control systems based on the idea of multi-sensor fusion estimation. First, the main structural form of the multi-sensor fusion estimation system and the problem description of multi-sensor fusion estimation are systematically introduced. Then, the H_∞ theory is introduced into multi-sensor system state fusion estimation and applied to fault diagnosis of aeroengine control systems. Simulation experiments verify the effectiveness of the filter applied to sensor fault diagnosis of aeroengine control systems.

Acknowledgements. This work is partially supported by National Major Science and Technology Project (J2019-V-0010-0105), Science Center for Gas Turbine Project (P2022- DB-V-001-001), the National Natural Science Foundation of China (Grant Number 62203364), the Fundamental Research Funds for the Central Universities (The authors also gratefully acknowledge the helpful comments and suggestions of the reviewers, which have improved the presentation.

References

1. Li, H., Gou, L., Chen, Y., et al.: Fault diagnosis of aeroengine control system sensor based on optimized and fused multidomain feature. IEEE Access **10**, 96967–96983 (2022)
2. Sun, R.-Q., et al.: Hyperelliptic Kalman filter-based aeroengine sensor fault FDIA system under multi-source uncertainty. Aerosp. Sci. Technol. **132**, 108058 (2023)
3. Shen, Q., Yue, C., Yu, X., et al.: Fault modeling, estimation, and fault-tolerant steering logic design for single-gimbal control moment gyro. IEEE Trans. Control Syst. Technol. **29**(1), 428–435 (2020)
4. Huihui, L., Linfeng, G., Yingxue, C., et al.: Intelligent fault diagnosis of aero-engine sensors based on multi-domain feature optimization. Propul. Technol. **44**(02), 272–284 (2023)
5. Xiaojun, S., Han, Z., Binhai, S., et al.: Weighted fusion robust incremental Kalman filter. J. Electron. Inf. **43**(12), 3680–3686 (2021)
6. Wang, H.: Research on robust estimation theory and its application in aerospace. Northwestern Polytechnic University (2019)

7. Cai, C., Zheng, Q., Zhang, H.: A new method to improve the real-time performance of aero-engine component level model. Int. J. Turbo Jet-Engines **40**(1), 101–109 (2023)
8. Qi, W.J.: Information Fusion Robust Kalman Filter. Heilongjiang University (2015)
9. Gómez, S., López, R., Valencia, G., et al.: Actuator and sensor fault estimation based on a proportional multiple-integral sliding mode observer for linear parameter varying systems with inexact scheduling parameters. Int. J. Robust Nonlinear Control **31**(17), 8420–8441 (2021)
10. Li, Q.: Research on the application of $H\infty$ filtering theory in multi-sensor information fusion state estimation. Shandong University (2009)
11. Tan, J., Zheng, H., Meng, D., et al.: Active input design for simultaneous fault estimation and fault tolerant control of LPV systems. Automatica **151**, 110903 (2023)
12. Qiusheng, J., Xinxing, S., Huacong, L., Xiaobao, H., Yan, L.: Robust variable parameter controller design for aero-engine full envelope based on hybrid region pole configuration. Propul. Technol. **41**(02), 431–438 (2020)
13. Li, H., Gou, L., Zheng, H., et al.: Intelligent fault diagnosis of aeroengine sensors using improved pattern gradient spectrum entropy. Int. J. Aerosp. Eng. **2021**(1), 1–20 (2021)

A Research on the Intelligent Flight Deck Development Trend for the Civil Aircraft

Hua Meng(✉), Qinpeng Dong, Chunling Zhao, Xianchao Ma, and Dayong Dong

Shanghai Aircraft Design and Research Institute, Shanghai 201315, China
{menghua,dongqinpeng,zhaochunling,maxianchao,
dongdayong}@comac.cc

Abstract. The Flight Deck is the brain of the aircraft and is the key to controlling the operation of the aircraft. With the development of industrial technology and information technology, the Flight Deck of civil aircraft has roughly experienced two main stages of electronicization and informatization, and in recent years, with the rapid development of intelligent technology, the concept of intelligent Flight Deck has received more and more attention and favor, and has also become the focus of the development of the next generation of aircraft. This paper mainly studies the application scenarios and development trends of future advanced technologies in intelligent Flight Decks, takes the three aspects of intelligent "information perception, collaborative decision-making and processing execution" as the main line, and explores the application prospects of cutting-edge technologies such as artificial intelligence, big data, multi-channel interaction, machine vision/infrared detection, 5G/ATG/high-speed satellite communication/navigation technology in the Flight Deck of civil aircraft, so as to analyze the main directions of future intelligent Flight Deck design and formulate a roadmap for the development of intelligent Flight Deck technology of civil aircraft.

Keywords: Intelligent · Flight Deck · Development trend · Advanced · Automated Cockpit · Road map · civil aircraft

1 Introduction

Reviewing the development history of civil aviation in the past 100 years, "The evolution of aircraft cockpit can be roughly divided into two stages. Before the 1970s, along the path from simplicity to complexity, various instruments and switches in the cockpit continued to increase and gradually became saturated; Starting from the late 1970s, with the continuous development of avionics technology, it began to shift from complexity to simplicity." (Zhou Qihuan. 1998).

"The civil aircraft cockpit design have grown through generations, and the characteristic of cockpit have evolved from the original simple to complex integration, then to the direction of simple and intelligent." (Feng Zhixiang et al. 2021).

Since the beginning of the 20th century, human industrial civilization has experienced a development process of mechanization/electrification, electronic/informatization and

Z. Jing and X. Zhan (Eds.): ICASSE 2023, LNEE 1153, pp. 111–124, 2024.
https://doi.org/10.1007/978-981-97-0550-4_9

intelligence/interconnection, and the civil aircraft cockpit accompanying its development has also undergone obvious upgrades in three aspects: "information perception, collaborative decision-making and execution processing".

So, this paper holds that the civil aircraft cockpit has roughly experienced two stages: the Electrified Cockpit (the first stage) and Informative Cockpit (the second stage), and is now in the early stage of Intelligent Cockpit (the third stage).

The information perception referred to in this article mainly refers to the perception of aircraft system status (normal, abnormal) and the acquisition of external environmental information (terrain, meteorology, air traffic control, routes, terminals, and other aircraft, etc.).

The collaborative decision-making referred to in this article mainly refers to the processing of information by human or aircraft systems, as well as the relevant tools needed in this process.

The processing execution referred to in this article mainly refers to the interaction between humans and machines, as well as the execution subjects and methods for decision instructions.

2 Characteristics of Electrified Cockpit (The First Stage)

The period from 1900 to 1980s was a period of rapid development of the electrified cockpit. "For a considerable period of time, with the rapid, long-range, and large-scale development of aircraft, airborne equipment has become increasingly complex. The cockpit equipment has evolved from fully mechanical instruments and handle switches to more complex electromechanical instruments, lighting and audio signals, and knobs. In addition to the pilot, the crew members will also be equipped with random mechanics, and remote aircraft will also be equipped with random telegraph operators, navigators, and ultimately a 5-person crew. Due to the fact that each instrument panel can only provide one piece of information, dozens of instruments and over a hundred switches are distributed throughout the cockpit, requiring the driver to operate in all directions, forming a crowded and chaotic maze like cockpit." (Zhou Qihuan. 1998).

During this period, the main characteristics of the civil aircraft cockpit in the three aspects of "information perception, collaborative decision-making and processing execution" were as follows:

a) **Information Perception:**

System status awareness: In the cockpit at this stage, the flight crew obtains the status information of the aircraft system through various instrument panels and complex and diverse mechanical dials installed in the cockpit. They need to concentrate highly to deal with complex and diverse information; When the system fails, it mainly depends on the human response to the aircraft itself and the independent distributed single system warning (warning lights and warning sounds) to perceive the aircraft status.

External environment perception: in the cockpit at this stage, the ground-based radio navigation represented by visual landmark navigation, PAPI light and NDB/VOR/DME/ILS is the main navigation means for the flight crew at this stage;

the flight crew obtains terminal and route information through paper charts and meteorological reports (TAF/METAR, etc.); The voice communication using HF/VHF is the main way for the cockpit to communicate with the outside world; The conflict between aircraft mainly depends on the visual guidance of the crew and the ground radar.

b) **Collaborative decision making:**

At this stage, the cockpit, because of the complex and diverse information perceived, is not intuitive enough, and the relevant decision-making basically depends on the flight crew itself.

Because the aircraft system is difficult to provide effective decision-making assistance, the crew of piloting, navigation, communication and system management has a high workload. The cockpit is composed of a five-person/three-person crew composed of pilots, co-pilots, mechanics, communicators and navigators. The crew cooperates closely to complete relevant collaborative decision-making.

c) **Process execution:**

At this stage, the cockpit has a low degree of automation, and the pilot needs to operate "two levers and one rudder" for a long time to drive manually; the mechanical control handle, switch and knob are the main human-computer interaction modes, and the flying quality and safety are highly dependent on the pilot's "driving skill level" and the mechanic's "troubleshooting level".

During the execution of the mission, the flight crew mainly relies on the paper manuals such as AFM/FCOM/QRH to execute the corresponding operating procedures. In order to avoid human errors, the paper Checklist is gradually introduced into the cockpit and becomes an important means to ensure flight safety.

On the whole, the first stage of the electrified cockpit has a high workload, and the safety of aircraft take-off and landing and route planning are greatly affected by the external environment, which greatly limits the safety, efficiency and economy of civil aviation operation (Fig. 1).

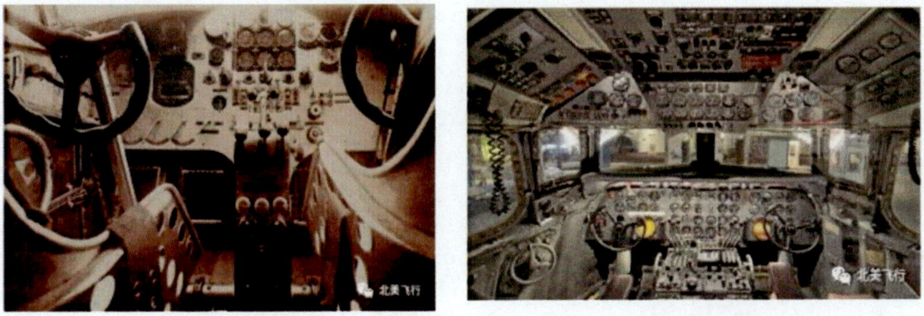

1900-1940 1940-1980

Fig. 1. Typical representation of the first stage cockpit

3 Characteristics of Informative Cockpit (The Second Stage)

Since the 1980s, with the introduction of electronic technology into the aviation industry, the transformation of the cockpit has entered a new stage, evolving into the mainstream glassy cockpit today. Electronic instruments have replaced traditional mechanical instruments, and various parameter information of the aircraft has been digitized. A single screen can integrate and display multiple information, and the functions of various equipment are integrated with each other. The number of instrument equipment in the cabin has been greatly simplified, and pilots have transformed from a multi person crew to the standard two person crew today. (Feng Zhixiang et al. 2021).

Thus, From the 1980s to the 2010s, with the vigorous development of electronic information technology, especially computers, The cockpit has ushered in a significant historical revolution.

a) **Information Perception:**

System status awareness: In the cockpit of the second stage, the cathode ray tube (CRT)/flat panel liquid crystal display (LED) quickly replaced the mechanical instrument, and the cockpit entered the era of "glass cockpit". The flight crew can obtain digital and graphical aircraft status information through the integrated display; In addition, the concept of "quiet and dark cockpit" is introduced, which greatly liberates the workload of the crew under normal conditions, especially for the abnormal state under system failure, the central electronic integrated crew warning system helps the crew to quickly achieve fault location through CAS message/warning light/warning tone/PBA/electronic checklist and other related elements. Greatly enhance the situational awareness of the crew.

External environment perception: In the cockpit at this stage, due to the popularity of electronic computer, high-precision inertial navigation and GNSS satellite-based navigation technology, VOR/DME/ILS/GLS/FLS integrated navigation has become the mainstream; the data link communication technology based on ACARS and broadband satellite communication has been promoted on a large scale; Pilots can also obtain electronic charts, weather reports and other terminal and route information in time through EFB (Electronic Flight Bag) and CPDLC. In addition, with the continuous upgrading of airborne equipment, Doppler color weather radar, EGPWS (Enhanced Ground Proximity Warning System), TCASII, ADS-B and other technologies have greatly enhanced the integrated monitoring capability of the crew to deal with weather, terrain and other aircraft.

b) **Collaborative decision making:**

At this stage of the cockpit, the large screen display replaces the traditional mechanical instruments, the cockpit is greatly simplified, the cockpit information is concise and intuitive, and the flight crew can make relevant decisions more easily.

At the same time, with the improvement of automation of key systems represented by flight control computer and flight control computer, some functions (such as flight guidance FD, electronic checklist ECL, flight plan, performance calculation) have become effective tools for pilots to make decisions, which greatly reduces the workload of the flight crew, and also allows the pilot and co-pilot to have enough. The "mechanic, communicator and navigator" became history, and the two-person crew became standard. Although the pilot's driving ability is decreasing day by day, the requirements for the pilot's situational awareness and the ability to deal with unexpected situations are greatly increased. Therefore, the current civil aircraft cockpit still needs a standard two-man pilot, relying on the good crew resource management level of PF and PNF (PM). Correspondingly, in addition to the conventional aircraft knowledge training and skill training, in order to meet the procedural rules of various new navigation technologies, Airlines need to spend a lot of money to maintain the qualifications of the crew (such as CATIIIb, HUD, etc.).

c) **Process execution:**

In the second stage of the cockpit, the civil aircraft cockpit began to have a real sense of "man-machine function allocation". Due to the improvement of automation level, Fly by wire operation, FADEC engine full authority digital control, Autopilot autopilot, AutoThrust automatic thrust management and other technologies were introduced into the cockpit by leaps and bounds. The role of the flight crew has gradually changed from "controller" to "monitor" of the aircraft. In addition to critical stages such as aircraft takeoff and landing or some abnormal situations, most of the energy of the crew is released. Of course, the current automation is still a conventional automation application that needs the crew to lead. At this stage, new human-computer interaction methods such as PBA, CCD, MKB and touch screen have been promoted, and flight safety in the "quiet and dark cockpit" has evolved from relying on the pilot's "driving skill level" to "situational awareness level".

In the process of mission execution, electronic checklists and FCOMs are gradually introduced into the cockpit, which improves the efficiency of the crew in mission execution and becomes an important means to ensure flight safety. Airlines gradually accept and promote the "paperless" cockpit. At the same time, with the introduction of IMA (Integrated Modular Avionics System)/CCR (Common Computing Resource), the cockpit has gradually begun to modularize integrated avionics.

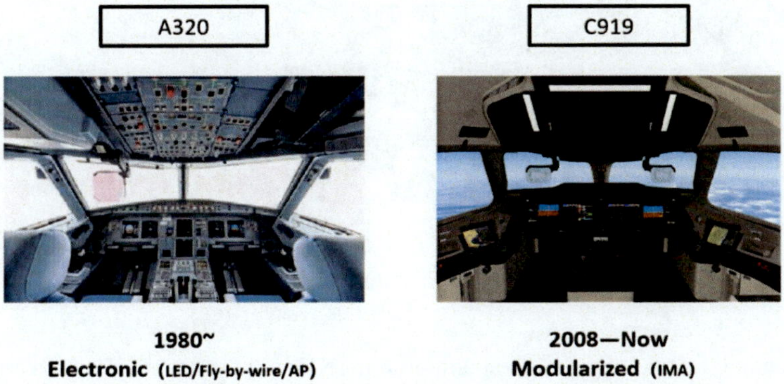

Fig. 2. Typical representation of the second stage cockpit

To sum up, thanks to the development and application of computer technology, the second stage of the cockpit has greatly liberated the attention of pilots, greatly guaranteed the safety of civil aviation operation, and greatly improved the efficiency of flight operation (Fig. 2).

4 Latest Technologies and Research Status

The design of civil aircraft cockpit has evolved from the concept of "function centered" to the design concept of "people centered". Today, when the new scientific and technological revolution breaks out, it will bring a wide imagination space to the design of civil aircraft cockpit. After fully considering the characteristics and use needs of pilots, the cockpit with new technology will achieve cross generational development. (Feng Zhixiang et al. 2021).

In the era of intelligence, intelligent systems based on artificial intelligence (AI) technology are gradually entering people's daily work and life, and entering the cockpit of large commercial aircraft is no exception (Bailey et al., 2017).

Looking forward to the future aircraft cockpit, a new generation of technological revolution led by 5G technology, artificial intelligence, new display and control technology is breaking out. When 5G technology is integrated with artificial intelligence, new display and control technology, the future cockpit will be more intelligent, equipment and functions will be more integrated, and the space will be more spacious and comfortable.

In the 21st century, especially from 2010 to now, companies and institutions represented by Boeing, Airbus, Gulfstream, Thales and NASA have carried out a number of studies on the future intelligent cockpit, such as the Gulf 650 cockpit and the concept cockpit proposed by Thales. Compared with the current mainstream second stage cockpit, it has made a bold partial breakthrough. This paper analyzes some of the key achievements (Fig. 3).

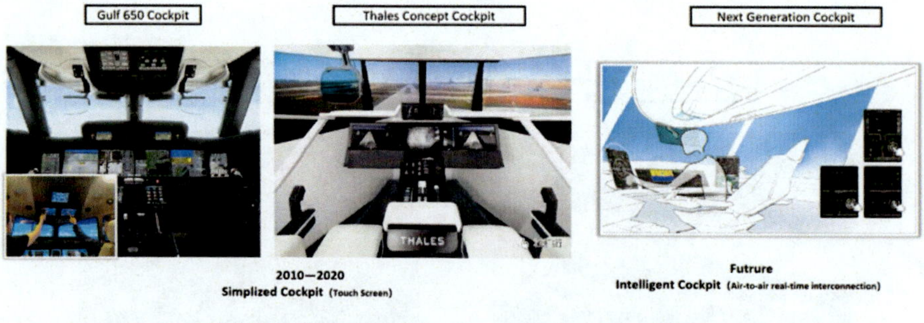

Fig. 3. Concept of the future third stage cockpit

Unmanned technology has great potential in reducing the cost of research and development/use of civil aviation airliners, improving flight safety and improving flexibility of use, so it has always been highly concerned by the aviation industry, airlines and even the general public in various countries. As one of the two giants in the world's civil aircraft manufacturing industry, Boeing has invested a lot of manpower, material and financial resources in the field of unmanned airliners for many years, and has made significant progress in the long-term research and testing of unmanned airliner related technologies. At present, Boeing is preparing to adopt single-pilot technology in the next generation of airliners before the technology of unmanned airliners is fully practical, which can be used as a necessary link and transition before the final realization of unmanned airliners, and can also br e economic benefits to airlines. At present, Boeing's related research demonstrates a single-person driving mode, which is intended to be applied to its latest generation of Boeing 797 airliners, hoping to serve as a link and transition before civil airliners are fully unmanned. In this mode, there will be only one pilot in the cockpit of the airliner flying in the air, who can provide backup and monitoring for the airborne unmanned system, especially responsible for dealing with various unexpected situations during the flight. At the same time, there will be another "pilot" on the ground, who is somewhat similar to the ground operator in the current UAV system, who can share part of the work for the pilot on board and the UAV system, and can monitor multiple aircraft at the same time. In addition, in view of the increasingly mature touch technology, Boeing has fully used touch screen technology in its latest B777X model, and touch screen technology is in the stage of transformation from research to application.

Since 2014, under the "Clean Sky" program of the European Union SESAR program (Tatjana Bolic et al. 2021), Airbus has carried out a study on the subversive cockpit technology "DISCO", aiming at studying the driving technology of a single pilot, the head-mounted display with specific SVS function can improve the situational awareness of a single pilot, and the sensor based on lidar can be used in the icing environment of traditional air hostels. It can also sense clouds, wind shear or icing environment, etc. AI speech recognition technology will convert voice to steady, reduce the workload of manual input by pilots, and enhance the efficiency of communication with air traffic controllers. It is expected to reach TRL3 in 2021 (to be updated) and TRL5 in 2023. In June 2020, based on A350-1000, Airbus completed the test flight of the ATTOL project, successfully verified the fully automatic flight based on machine vision, and expanded the research direction of unmanned aircraft. The technology is derived from the Wayfinder autonomous flight control system at Airbus A3 Innovation Center. The team developed software based on computer vision and machine learning to examine the environment around the aircraft and calculate the best navigation scheme by using sensors (including cameras, lidar) and more powerful computers. Autonomous technology frees pilots from routine tasks such as take-off and landing, cruise and so on, and puts more energy on decision-making and mission management, which helps to make the best decisions in emergencies, thus improving flight safety. At the same time, only airborne systems are used to achieve autonomous flight, which gets rid of the dependence on external equipment, thus reducing the cost of airport infrastructure construction.

From 2012 to 2014, through FDDLR laboratory, NASA carried out special research on single-person crew cockpit, including crew cooperation in operating procedures, use of automation, responsibilities and division of labor between cockpit and ATC, etc. In 2015, the relevant report was released, but the report only gave the results of the principle test. (Cynthia A. Wolter et al. 2015).

In 2018, the Display and Decision Support of Intelligent Cockpit System, published by China Institute of Aeronautical Radio Electronics, focuses on the integrated intelligent cockpit technology proposed by NASA, and discusses the display and operation capabilities, information processing and collaborative decision support of the next generation cockpit from the conceptual level, in order to better cope with future changes in air traffic management.

In 2019, "Research on Efficient Human-Computer Interaction Methods in Future Aircraft Cockpit Design" proposed a new type of human-computer interaction method for aircraft cockpit, this "data glove" interaction system constructs a highly reliable and efficient human-computer interaction platform between pilots and aircraft, which provides flight data acquisition and analysis. It is an important part of the efficient man-machine integration of the future cockpit.

In February 2020, the team of Wang Zhao and Xiao Gang of Shanghai Jiaotong University applied for the "Single Pilot Driving System and Control Method", and proposed the SPO mode organization and operation mode for four different scenarios. (Wang Zhao 2020).

PBN is one of the core technologies for building a new generation air transportation system, which can accurately guide aircraft and improve flight operation safety; Provide vertical guidance, implement continuous and stable descent procedures, and reduce the risk of controllable ground collision; Improve round-the-clock operations, improve flight normality, and ensure the safety of airport operations with complex terrain; Realize flexible and optimized flight paths, increase aircraft load, reduce flight time, and save fuel; Avoid noise sensitive areas, reduce emissions, and improve environmental protection level; By implementing parallel routes and increasing the positioning points of inbound and outbound routes within the terminal area, traffic flow can be increased; Reduce the horizontal and vertical spacing between aircraft and increase airspace capacity; Reduce the demand for ground to air communication and radar guidance, facilitate command, and reduce the workload of pilots and controllers; Reduce investment in navigation infrastructure and operational costs, and improve the overall economic efficiency of operations. (Huang Jing. 2014).

Speech recognition technology, as part of human-computer interaction, is essential for machine intelligence. The utilization of robots as the co-pilot in civil aircraft is a major breakthrough and innovation direction in the civil aviation industry. The application of speech recognition technology to the co-pilot of the robot can make the command of the captain directly to the co-pilot program, making it possible to cooperate between the captain and the robot pilot. In 2020, a standard yelling speech database was established by Civil Aviation Flight University of China. (He Liqing. 2020).

Based on the above research on the three stages of cockpit development, with the third stage of intelligent cockpit, it will achieve leapfrog development in information perception, collaborative decision-making and execution processing, change the current human-computer collaborative work mode, reduce the workload/qualification requirements/quantity of the unit, and improve system management and fault handling. Improve the adaptability of terminals and routes, and realize the development of civil aviation airliners from "low visibility operation" to "all-weather operation".

5 Development Trend of Intelligent Cockpit (The Third Stage)

There is currently no specific and complete solution for the future intelligent cockpit, but researchers have looked forward to the development of intelligent cockpit based on the advantages of intelligent technology, the limitations of automation technology, flight safety, and the needs of airlines. (XU Wei, 2022).

From 2010 to now, with the vigorous development of advanced technologies such as artificial intelligence, big data, multi-channel interaction, machine vision/infrared detection/airborne image recognition technology, 5G/ATG/high-speed satellite communication/navigation technology, and the continuous innovation of civil aviation authorities and main manufacturers in various countries, new technologies of civil aircraft cockpit have sprung up like mushrooms. The cockpit is ushering in a new historical revolution. This paper is briefly summarized as follows:

a) **Information Perception:**

System state perception: At the current stage, "glass cockpit" will be further developed with the application of touch screen technology, and "display and control integrated super-large screen cockpit" may become the next trend. At the same time, on the basis of the concept of "quiet and dark cockpit", more intelligent algorithm rules will be used to further reduce unnecessary redundant information. Only the key information required in the current stage is provided, and the situational awareness of the crew will be further enhanced; on the basis of ECAM/EICAS + ECL, the aircraft will monitor and even predict the state change of the aircraft in combination with the real-time health state management system, and realize automatic isolation and disposal in case of system failure (Fig. 4).

Fig. 4. Conceptual Diagram of Future Cockpit (Picture from Internet)

Perception of external environment: In the past decade, with the maturity of infrared detection technology, airborne image recognition technology and GNSS technology and the increasingly powerful function of airborne navigation database, HUD/EVS/SVS/CVS/PBN (especially high-precision RNP) technology has been partially promoted and will gradually become the "standard configuration" of future intelligent cockpits; At the same time, the further maturity of data link communication technology based on 5G/ATG/high-speed satellite communication (high bandwidth/low latency/high availability and integrity) will support the real-time detection, sharing and avoidance of meteorological risks between aircraft and aircraft, aircraft and the ground. At the same time, the real-time interconnection of aircraft status information with ATC/AOC/other aircraft information (including the real-time uplink and downlink of airborne database) is realized; in addition, with the maturity of machine vision technology, the importance of "human eye recognition" will be further reduced.

b) **Collaborative decision making:**

At present, with the rapid development of artificial intelligence technology, big data and multi-channel interactive technology represented by automobile automatic driving, combined with the development of the new generation of communication/navigation technology and flight control computer mentioned above, the automation level of civil aircraft cockpit is expected to evolve to the direction of "intelligent AI", on the basis of obtaining sufficient internal and external information. Automation is upgraded from an "effective assistant tool" to a "reliable assistant" that can replace most of the pilot's duties by means of "intelligent algorithm and autonomous learning of pilot's behavior patterns" to help pilots automatically perform tasks at critical stages. For example, TCAS automatic avoidance function under AP connection and automatic emergency descent function under pilot disability have been realized on A350 and other aircraft. In addition, when the communication technology is mature enough, real-time remote control on the ground will also become possible.

Based on the above prediction, the man-machine cooperation mode of the third stage cockpit may also undergo a subversive change, from the mainstream pilot-co-pilot-automation (aircraft system) to the multi-mode driving mode of "pilot-automation (AI), pilot-ground controller, automation (AI)-ground controller".

c) **Process execution:**

Based on the above, the third stage of intelligent cockpit, human and automation (AI) will be complementary/alternative roles, therefore, based on artificial intelligence, big data, multi-channel interaction, machine vision, 5G/ATG/high-speed satellite communication technology based on 4D track "intelligent flight" will become an intelligent cockpit. Including but not limited to "automatic taxi/automatic takeoff/automatic interval management/automatic descent/autonomous route planning/wake prediction and active avoidance/automatic landing".

Among them, the multi-channel interaction technology represented by touch control, voice control and gesture control will further simplify and improve the efficiency of interaction between the crew and the aircraft.

At the same time, the intelligent management of the system under normal conditions (such as intelligent environmental control) and knowledge-based autonomous fault isolation and disposal under abnormal conditions will also greatly reduce the workload of the crew and reduce the requirements for the qualification/quantity of the flight crew.

6 Development Route

Based on the above analysis and summary of the development stage of civil aircraft cockpit, combined with the current development direction of future cockpit at home and abroad, we can see that the main direction of future civil aircraft intelligent cockpit is the following three points:

a) Intelligent information perception: study how to improve the accuracy, timeliness and reliability of aircraft perception of internal and external information such as external environment (weather, terrain, airport, ATC, other aircraft, etc.), aircraft system status, personnel status and operation intention. Including but not limited to: touch screen technology, machine vision technology, information detection and sharing technology based on data link communication, CVS technology and high-precision PBN technology.
b) Intelligent collaborative decision-making: After obtaining sufficient perception information and authorization, based on technologies such as big data analysis model, high-speed computing system, specific algorithm and deep learning, intelligent AI assistant decision-making instructions are provided. Some instructions can try to control the aircraft directly instead of people, so as to better cope with interference, special cases or unexpected situations. Including but not limited to: research on multi-mode man-machine interaction decision-making mode between "pilot-automation (AI), pilot-ground controller, automation (AI)-ground controller".
c) Intelligent disposal execution: research to improve the automation level of the existing aircraft system, and execute relative procedures or actions according to the set rules to cope with the conventional and normal operation process. In addition, through big data and machine autonomous learning technology, specific procedures or actions can be completed to meet abnormal operational needs under special circumstances. Including but not limited to: "intelligent flight" based on 4D track, such as "automatic taxi/automatic takeoff/automatic interval management/automatic descent/autonomous route planning/wake prediction and active avoidance/automatic landing"; multi-channel interactive technology represented by touch control, voice control and gesture control; Knowledge-based system intelligent management, autonomous fault isolation and disposal technology under abnormal system conditions (Fig. 5).

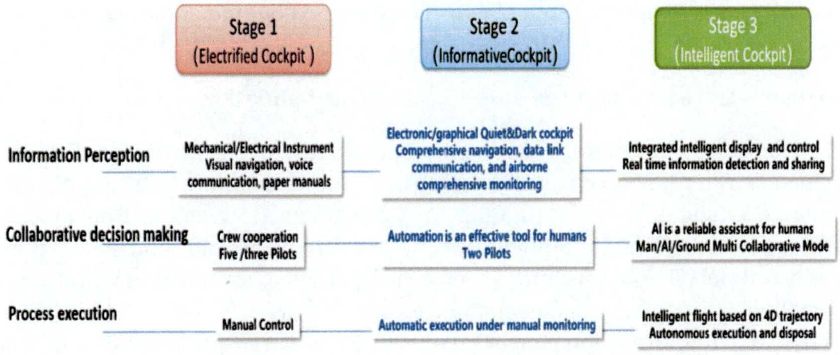

Fig. 5. Summary of Cockpit Development History

Through the above three aspects of research, ultimately, the intelligent cockpit can be upgraded synchronously with the future air traffic control, dispatch, maintenance, airport and other interactive subjects, and deeply integrated with the next generation of the entire civil aviation transport system, so that the elements of the system play their respective maximum efficiency, so that the cockpit's operational safety, human-computer interaction comfort, operational economy, etc. All of them have been greatly improved to achieve high-density, all-weather, single or even unmanned driving, and enhance the operational efficiency and benefits of the company, air traffic control, airport and other main bodies.

7 Apply Strategies and Recommendations

The Intelligent Flight Deck technology is a future project for domestic OEMs, It will take the subsystem or prototype as the carrier, and complete the laboratory test and typical real environment verification by setting the typical operating environment and working conditions. Some technologies can be combined with future test aircraft and real flights to test and improve the real operating environment.

In terms of strategy, it is suggested to start from two aspects of "difficulty": on the one hand, fully tap the potential of existing mature technologies and carry out creative applications (such as comprehensive utilization of GLS/SVS/EVS/laser radar/camera and other technologies to achieve automatic take-off and landing), in order to further improve flight safety and efficiency; On the other hand, we should make full use of new technologies such as artificial intelligence, large data analysis, high-bandwidth air-ground integrated communication, touch screen technology, machine vision technology, information detection and sharing technology based on data link communication to upgrade key systems intelligently in an all-round way and to drive, navigate, communicate and manage systems. To realize "intelligent flight" based on 4D track, such as "automatic taxiing/automatic takeoff/automatic interval management/automatic descent/autonomous route planning/wake prediction and active avoidance/automatic landing", as well as knowledge-based intelligent system management, so as to solve the difficulties and pain points in the current civil aviation operation mode with "intelligence".

The key technical points listed above can be carried out in parallel, and the overall research process can be divided into six steps:

a) **Intelligent Cockpit Requirements Capture and Validation**

Based on the current operation system, the advanced driving, navigation and communication functions that have been used in the most advanced aircraft or will be applied in the near future are studied. Including: intelligent braking mode BTV, anti-run-out protection, take-off safety monitoring, intelligent taxiing and airport ground guidance, automatic landing based on PBN or machine vision, intelligent display and control information push, intelligent monitoring status monitoring, etc.

To study the future development trend of civil aviation transportation system at home and abroad, such as the Single Sky Program in Europe, the NEXGen Program in the United States, Airbus DISCO (Single Pilot) and other market operation requirements expected by airlines.

Through investigation and research, combined with the adaptability and maturity of existing advanced technology and potential new technology, especially combined with artificial intelligence technology, big data technology and air-ground integration technology, the relevant technical characteristics of future intelligent cockpit are analyzed and demonstrated. Including but not limited to intelligent interaction technology, intelligent driving technology, intelligent navigation technology, intelligent communication and monitoring technology, and intelligent state management technology. Through the trade-off, the conclusive analysis and prediction are given.

b) **Research on Intelligent Driving Mode Technology**

According to the driving characteristics of intelligent cockpit, advanced control technology, navigation technology, automation function and auxiliary control technology of ground equipment are considered to determine the man-machine cooperative decision-making mode of supporting flight piloting under normal and abnormal flight conditions, to construct the normal and abnormal abilities and behaviors of pilots, and to support the health of pilots. Set up a driving mode oriented to the flight process and the pilot's ability.

c) **Research on Human-Machine Interaction Decision-Making Mode of Intelligent Cockpit**

According to the flight process status and intelligent driving characteristics (such as single-person SPO), the organization regulates the flight and non-standard flight conditions and flight process requirements, determines the man-machine function allocation, determines the normal and abnormal behavior and ability modes of pilots, and constructs the responsibilities and objectives of pilots, automatic systems and other roles (such as ground operators).The interactive decision models of knowledge and cognition, rules and logic, and events and conditions are established.

d) **Research on Intelligent Technology and Credibility Evaluation Technology of Pilot Interaction**

Establish the validity area analysis of driving knowledge, cognition and rules, logic coordination of intelligent cockpit (such as single-person system), construct the conformity analysis of flight process events and conditions with process and state coordination, determine the process of knowledge, rules, events and conditions coordination in decision-making process, and establish the driving mode ability of intelligent cockpit. And an interactive credibility evaluation index system is realize.

e) **Research on Safety Analysis Technology of Flight Process under Human-Aircraft Cooperative Decision-Making**

Based on the possible functions of intelligent cockpit, such as intelligent interaction, intelligent driving, intelligent navigation, intelligent communication and surveillance, and intelligent state management, and aiming at the safety problems caused by system environment perception deviation and pilot's situational awareness understanding error, the operation scenario of cockpit system is analyzed, and the failure modeling and analysis method of decision support function are studied. Construct the verification environment of the expected functional safety requirements to support the safety verification of the simulation system.

f) **Development and Test of Intelligent Cockpit Simulation System**

Build the intelligent cockpit simulation system and platform, and carry out the mission simulation of the whole flight process based on the engineering simulator. In the process, it is necessary to complete the corresponding layout of the intelligent cockpit, the preliminary design of the human-computer interface and operation procedures.

8 Conclusion

This article studies the application scenarios and development trends of future advanced technologies in intelligent cockpit, with the main focus on intelligent "information perception, collaborative decision-making, and processing execution". It analyzes the three development stages of civil aircraft cockpit, and combines current research on the development direction of future cockpit at home and abroad to explore artificial intelligence, big data, multi-channel interaction, machine vision/infrared detection The application prospects of cutting-edge technologies such as 5G/ATG/high-speed satellite communication/navigation technology in civil aircraft cockpit were analyzed to analyze the main directions of future intelligent cockpit design, and a preliminary roadmap for the development of domestic civil aircraft intelligent cockpit technology was formulated.

References

Bailey, R.E., Kramer, L.J., Kennedy, K.D., Stephens, C.L., Etherington, T.J.: An assessment of reduced crew and single pilot operations in commercial transport aircraft operations. In: 2017 IEEE/AIAA 36th Digital Avionics Systems Conference (DASC), pp. 1–15. IEEE (2017)

Wolter, C.A., Gore, B.F.: A Validated Task Analysis of the Single Pilot Operations Concept National. Aeronautics and Space Administration (2015). https://ntrs.nasa.gov/search.jsp?R=20150001421. Accessed 24 Jan 2019

Feng, Z., Bai, Y.: The cockpit design evolution of civil aircraft. Civil Aircraft Design Research, no. 1 (2021)

Huang, J., Jiang, W.: Introduction to the Current Situation and Future Development of NEXTGEN Construction in the United States, pp. 202–203 (2014)

He, L.: Study of Artificial Intelligence Flight Co-Pilot Speech Recognition Technology. Civil Aviation Flight University of Chinas (2020)

Bolic, T., Ravenhill, P.: SESAR: the past, present, and future of european air traffic management research. Engineering **7**(2021), 448–451 (2021)

Xu, W.: User centered design (VII): from automated to intelligent flight deck. Chin. J. Appl. Psychol. **28**(4), 291–313 (2022)

Zhou, Q.: Development course of commercial Aircraft's Flight deck. Aerosp. Avionics 66–69 (1998)

Check for updates

Angle of Attack of the Aircraft Calibration

Hui Yang[1(✉)], Chao Xi[2], Zhengdi Zhu[1], and Yang Fang[1]

[1] Shanghai Aircraft Design and Research Institute, Shanghai, China
ya_ben_cool@126.com
[2] Chinese Flight Test Establishment, Xi'an, China

Abstract. The installation position for the AOA sensor is discussed. The results indicated that best position for AOA sensor is near the largest width part of the fuselage. Two flight test methods for AOA calibration are described, one is fuselage-nose-boom method, and the other is attitude angle or attitude-flight-path angle method. An example of AOA calibration using attitude angle method is presented, and the vertical wind has affected the flight test results. Each flight test condition should be repeated at least twice to ensure consistent results for attitude angle method.

Keywords: Installation position · AOA · Flight test · Calibration

1 Introduction

Some airframe manufacturers will define angle of attack (AOA or α) as the angle between the flight path axis and the wing axis. However, most of them will define α as the angle between the flight path axis and the x-body axis, which is the definition used in this paper as illustrated in Fig. 1.

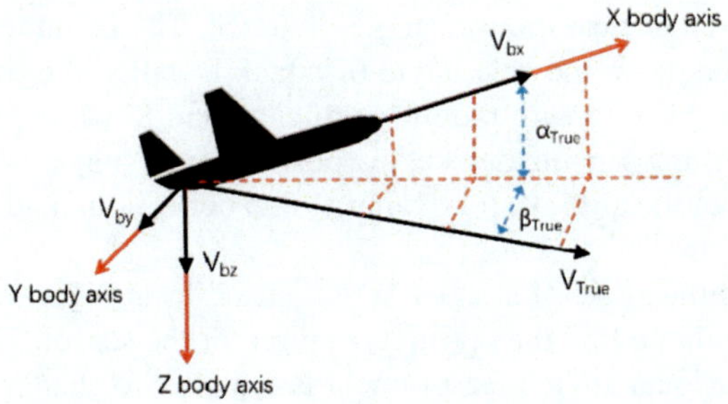

Fig. 1. AOA definition

AOA data is belong to airdata system of the aircraft, and it's a very critical parameter for the aircraft's safety, erroneous AOA data will possibly lead to an aircraft crash. For example, the AOA sensor sends data to a 737 Max software system that pushes the nose

Z. Jing and X. Zhan (Eds.): ICASSE 2023, LNEE 1153, pp. 125–133, 2024.
https://doi.org/10.1007/978-981-97-0550-4_10

of the aircraft down if it senses an imminent stall. That software, triggered by erroneous data from AOA sensors, is believed to have played a role in the crashes of Lion Air and Ethiopian Airlines jets on October 29, 2018, in which 189 people died.

The pivoted vane is generally used as sensing device for measuring the angle of attack for an aircraft, and typically be mounted on the forebody of the subsonic transport aircraft. The pivoted vane shown in Fig. 2, which is a mass-balanced wind vane free to align itself with the direction of the airflow, has been used primarily in aircraft.

Fig. 2. A pivoted vane

2 Installation Position for AOA Sensor

However, the sensing device installed on any position of the aircraft will only measure the local angle (αi) rather than the true angle of attack. The amount by which the local angle of attack differs from the true angle of attack is called the position error of the installation. This error varies with the lift coefficient and Mach number of the aircraft. For some locations of the sensing device, the position error may also vary with changes in the configuration of the aircraft, for example, flap deflection, landing-gear extension, and so forth.

If the position of the angle of attack (AOA) sensor has not yet been determined, the first objective should be to find the optimal position for the sensor. This position should be chosen so that the local angle measurement is sensitive to changes in AOA, and less being affected by changes in angle of sideslip (AOS) or velocity. The citation 1 has used CFD method to find the position for AOA sensor and citation 2 has used fundamentals of airflow around a body of revolution to choose the position. Both [1] and [2] show that the best position for AOA sensor is near the largest width part of the fuselage. The contour show in Fig. 3 presents the sensitivity of AOA to AOS for an aircraft. In this figure, the lighter color indicate better performance for AOA sensing device. Alternatively, the AOS effect could be eliminated by using the averaged signal of two detectors, one on each side of the fuselage. Generally, for subsonic transport aircraft, the local angle of attack changes by approximately 1.5 to 1.8° for every one-degree change in true angle of attack.

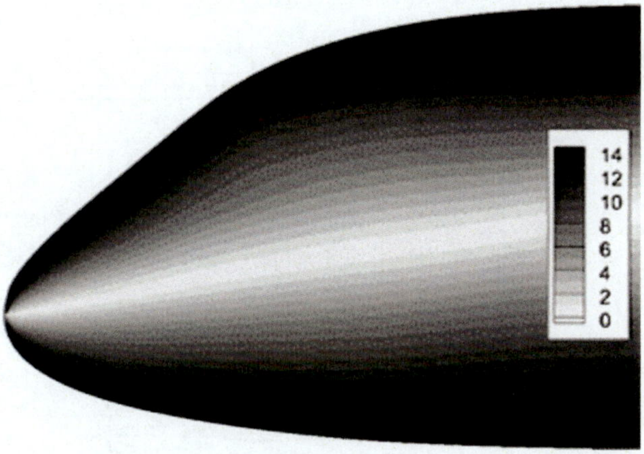

Fig. 3. The sensitivity of AOA to AOS for a subsonic transport aircraft

3 Flight Test Methods for Calibrating AOA

Once the AOA sensor position is established, the calibrations of angles of attack should be determined to account for the disturbing presence of the aircraft in the flow field. The calibrations must be performed under various flight conditions of airspeeds and altitude as well as aircraft gravity and configurations (combinations of flaps, gear, and external stores).

An angle-of-attack installation may be calibrated in flight by any one of a number of methods. There are two kinds of method, which have been used primarily in the flight testing of aircraft, will be given here.

3.1 Fuselage-Nose-Boom Method

In the fuselage-nose-boom (as shown in Fig. 4, from reference [3]) method the local angle of attack decreases as the distance from the nose increases (as shown in Fig. 5, from reference [4]) and the local angle of attack at a point with considerable distance ahead of the fuselage is considered to represent a close approximation to the true angle of attack of the airplane. Structural considerations and boom bending prevent a noseboom from being very long. Tests to determine the variation of local angle of attack with distance ahead of the fuselages of an airplane (reference [5]) showed that, with increasing distance ahead of the fuselage, the local angle of attack approached the true angle of attack asymptotically and that at a distance of at least 1.5 fuselage diameters the local angle of attack was essentially equal to the true angle of attack. In accelerated maneuvers, additional errors may be introduced by such items as the pitching velocity of the aircraft, boom bending, fuselage flexibility, and lag in the transmission system of the angle of attack instrumentation. Reference [5] to [8] have presented more details about this method.

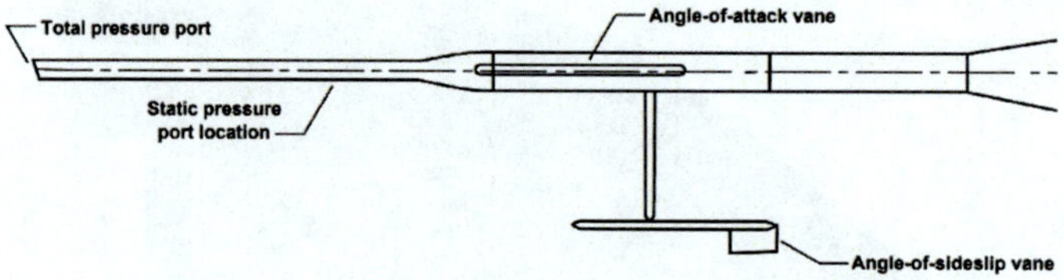

Fig. 4. A typical flight test noseboom

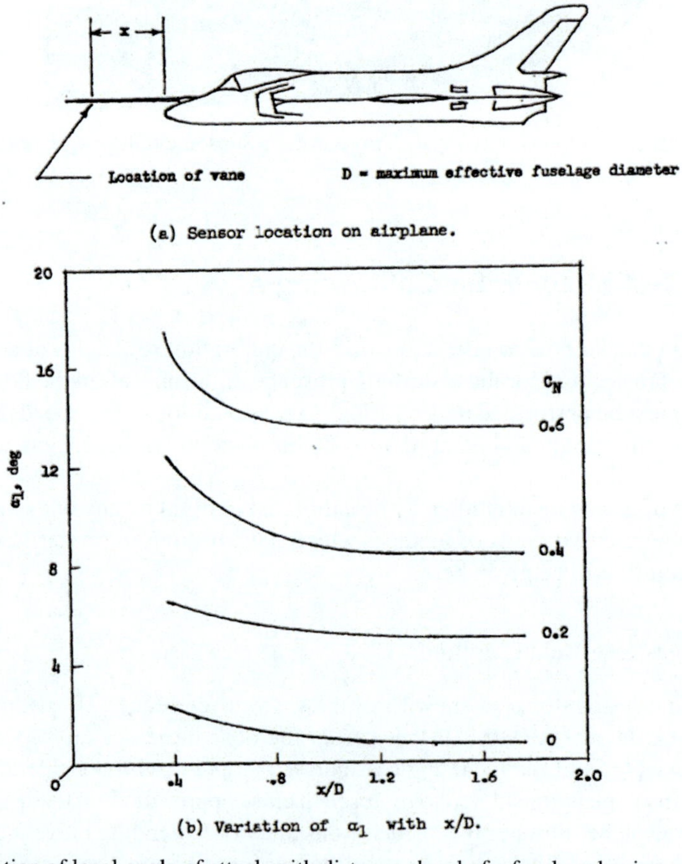

Fig. 5. Variation of local angle of attack with distance ahead of a fuselage having an ogival nose with rounded tip (data from Ames Aero. Lab.). M = 0.80

3.2 Attitude Angle or Attitude-Flight-Path Angle Method

True angle of attack can be determined during steady flight as the difference between the pitch attitude angle and flightpath climb angle of the airplane. This analysis requires minimum effort, but the result may not be valid during unsteady flight.

From reference [9], the matrix summary form of the transformation from the flight path axis true airspeed to the true airspeed in the earth axis (N, E, D) is as follows:

$$\begin{Bmatrix} V_{gN} \\ V_{gE} \\ V_{gD} \end{Bmatrix} = [\psi] \cdot [\theta] \cdot [\phi] \cdot [\alpha] \cdot [\beta] \cdot \begin{Bmatrix} V_t \\ 0 \\ 0 \end{Bmatrix} - \begin{Bmatrix} (V_{wN}) \\ (V_{wE}) \\ (V_{wD}) \end{Bmatrix} \tag{1}$$

where, V_{gN}, V_{gE} and V_{gD} are north, east and down ground speed, V_{wN}, V_{wE} and V_{wD} are north, east and down wind speed. ψ is heading angle (degrees from true north), θ is pitch attitude, ϕ is roll attitude, α is angle of attack, β is angle of sideslip and V_t is true airspeed. The matrices are used to transform the true airspeed from the flight path axis to the earth axis, where

$$[\psi] = \begin{bmatrix} \cos\psi & -\sin\psi & 0 \\ \sin\psi & \cos\psi & 0 \\ 0 & 0 & 1 \end{bmatrix} \tag{2}$$

$$[\theta] = \begin{bmatrix} \cos\theta & 0 & \sin\theta \\ 0 & 1 & 0 \\ -\sin\theta & 0 & \cos\theta \end{bmatrix} \tag{3}$$

$$[\phi] = \begin{bmatrix} 1 & 0 & 0 \\ 0 & \cos\phi & -\sin\phi \\ 0 & \sin\phi & \cos\phi \end{bmatrix} \tag{4}$$

$$[\alpha] = \begin{bmatrix} \cos\alpha & 0 & -\sin\alpha \\ 0 & 1 & 0 \\ \sin\alpha & 0 & \cos\alpha \end{bmatrix} \tag{5}$$

$$[\beta] = \begin{bmatrix} \cos\beta & -\sin\beta & 0 \\ \sin\beta & \cos\beta & 0 \\ 0 & 0 & 1 \end{bmatrix} \tag{6}$$

The Eq. (1) is the general matrix formula. During a typical AOA calibration flight test, we will assume the sideslip angle (β), and the bank angle (φ) are equal to zero. Then Eq. (1) simplified to the Eq. (7):

$$\begin{aligned} \sin(\alpha - \theta)V_t - V_{gD} = V_{wD} \\ \alpha = \sin^{-1}(\tfrac{V_{wD}+V_{gD}}{V_t}) + \theta \end{aligned} \tag{7}$$

Assuming that the vertical winds are zero usually is valid for a nonturbulent atmosphere. From Eq. (7), we can get the following:

$$\alpha = \theta - \gamma \tag{8}$$

$$\gamma = \sin^{-1}(\frac{\dot{h}}{V_t}) = \text{flight path angle} \tag{9}$$

$$\dot{h} = -V_{gD} = \text{rate of climb} \tag{10}$$

The Eq. (8) is the theory for attitude-flight-path angle method.

In level steady flight, $\gamma = 0$, so the attitude angle of an aircraft is equal to its angle of attack. The attitude angle is measured by inertial navigation system, and the accuracy is adequate enough.

Although this analysis requires minimum effort, it has a difficulty in the application involving the effect of vertical air currents, which is illustrated in the following example. Figure 6 is flight test data for AOA flight calibration of one aircraft. As we can see, the difference between AOA and attitude angle is not satisfied (the difference should be smaller than 0.35°), although the flight is very steady (as we can see from Fig. 7). Engineers have compared the attitude angle measured by inertial navigation system and by the additional device, it turns out the attitude angle has no abnormal phenomena. Furthermore the four AOA sensing device gives the approximate same results, and the aircraft's AOA calculation software works well too. Finally engineer check the lift coefficient curve with another flight (Fig. 8), in which the AOA is nearly the same as attitude angle. It turns out the lift coefficient curves match well, which proves there is no issue for the AOA signal. The biggest possibility of that phenomenon is that there is steady vertical wind during that flight, although lack of direct evidence. So each test condition should be repeated at least twice to ensure consistent results for the attitude angle method.

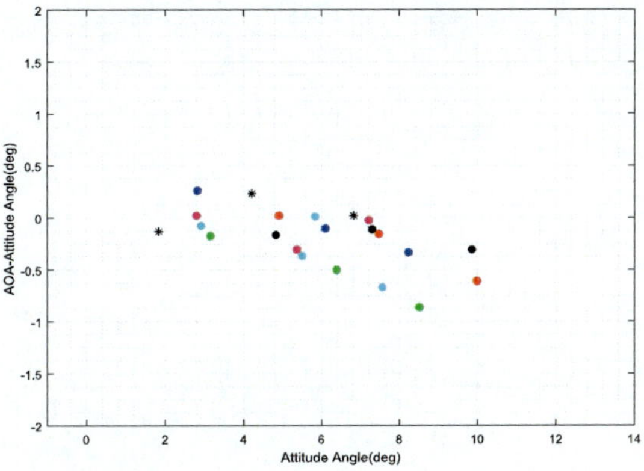

Fig. 6. Difference between AOA and attitude angle in one flight

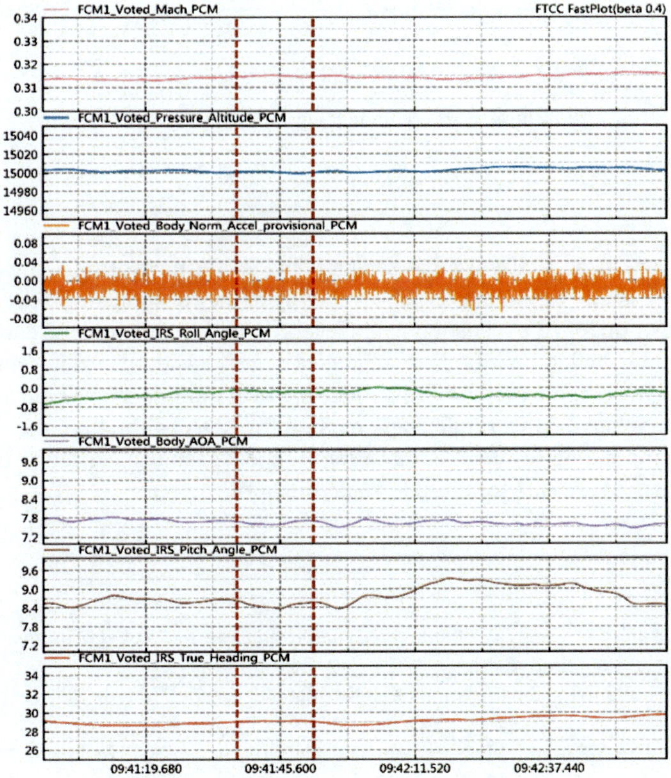

Fig. 7. Flight parameter at the test point

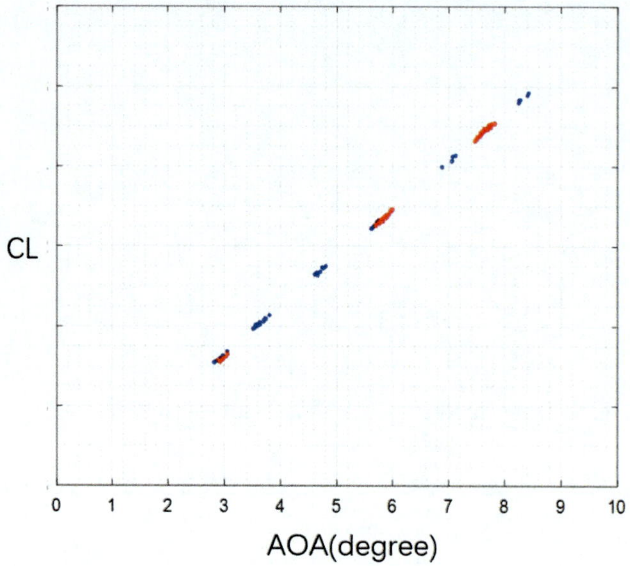

Fig. 8. Lift coefficient curve for two flights

4 Conclusion

The pivoted vane is generally used as sensing devices for measuring angle of attack for an aircraft, it typically be mounted on the forebody of the subsonic transport aircraft. The best install position for AOA sensor is near the largest width part of the fuselage. Once this position is established, the calibrations of angles of attack should be determined by flight test. Fuselage-nose-boom method and attitude angle or attitude-flight-path angle method are presented. When using attitude angle method for AOA calibration, the vertical wind could lead to the wrong results. So it's better to repeat each test condition at least twice to ensure consistent results for attitude angle method.

References

1. Yang, S.P., Sun, Y.F., Fang, Y., et al.: Aerodynamic analysis of AOA sensor location design on civil aircraft. Acta Aerodyn. Sinica **37**(2), 242–245 (2019)
2. Tang, H.: Installation and calibration of angle of attack sensor. Hongdu Sci. Technol. **3**, 16–20 (1999)
3. Haering Jr, E.A.: Airdata Measurement and Calibration. NASA Technical Memorandum-104316 (1995)
4. Gracey, W.: Summary of Methods of Measuring Angle of Attack on Aircraft. NASA, Washington DC (1958)
5. McFadden, N.M., Rathert Jr, G.A., Bray, R.S.: Flight Calibration of Angle-of-Attack and Sideslip Detectors on the Fuselage of a 35 Swept-Wing Fighter Airplane. NACA RM A52A04 (1952)

6. Seeley, L.W.: Flight Investigation of the Position Error Associated With Local Air Flow Detectors Installed on a Model F9F-2 Aircraft. NAVORD Rep. 2014(NOTES 646), U.S. Naval Ord. Test Station, Inyokern (China Lake, Calif.) (1953)
7. Tang, H.: Error correction for angle of attack flight testing. Hongdu Sci. Technol. **1**, 1–7 (1996)
8. McFadden, N.M., Holden, G.R., Ratchiff, J.W.: Instrumentation and Calibration Technique for Flight Calibration of Angle-of-Attack Systems on Aircraft. NACA RM A52123 (1952)
9. Olson, W.M.: Aircraft Performance Flight Testing. AFFTC-TIH-99-01, pp. 67–68 (2000)

Influence of Radius of Microphone Array on Virtual Rotating Array Beamforming for Rotating Noise Source Localization

Rachael Amoah, Cheng Wei Lee, and Wei Ma(✉)

School of Aeronautics and Astronautics, Shanghai Jiao Tong University, Shanghai, China
{rachael_amoah,chengwei-97,mawei}@sjtu.edu.cn

Abstract. The Virtual Rotating Array (VRA) beamforming algorithm is an efficient algorithm for localizing rotating noise sources, especially with ring arrays in the frequency domain. VRA has now been extended to work with arbitrary microphone arrays. Determining the proper array radius and its impact on VRA beamforming is still a major challenge in installing arbitrary microphone arrays for the better achievement of beamforming results. The purpose of this paper is to investigate the influence of arbitrary microphone arrays with different radii on virtual rotational beamforming. The arbitrary microphone configurations employed are the Underbrink, H0V5, and H0V3.13 arrays. The results show that at frequencies within $2.5 \geq f \geq 5$ kHz, there is usually a source intensity error when the array radius is half the size of the rotating source radius making other sources hard to identify. This may be due to array element spacing error or Rayleigh's criterion. Furthermore, localization results are significantly enhanced when the radius ratio (array radius/rotating source radius) is within 2–4 at frequencies(f) within $5 \geq f \geq 8$ kHz. Finally, employing deconvolution algorithms like CLEAN-SC for VRA beamforming greatly improves spatial resolution and source location precision.

Keywords: Microphone array radius · Rotating noise source localization · Arbitrary microphone array · Virtual rotating array beamforming · Spatial interpolation

1 Introduction

The problem of noise is a hot topic in the acoustic society and the community due to its negative effects. Efforts to limit noise and its effect have been a concern leading to the study of noise cancellation and source localization techniques (Collins et al. 1994). The most common approach to sound source localization in the acoustic society involves the use of microphone arrays.

Microphone arrays generally record sound pressure signals during an experiment followed by the identification of the noise characteristics using beamforming algorithms. Many acoustic researchers and engineers have studied and employed microphone arrays in localizing noise sources. Jae-Woong and Yang-Hann (1995) employed a line array in estimating the location and strength of spherical sound sources. Line arrays are classified

Z. Jing and X. Zhan (Eds.): ICASSE 2023, LNEE 1153, pp. 134–144, 2024.
https://doi.org/10.1007/978-981-97-0550-4_11

as one-dimensional (1-D) arrays as studied by Barsikow (1996) who designed a cross-array by combining two perpendicular line arrays of fifteen microphones. A significant increase in the use of 1-D arrays has evolved into various dimensions of microphone array technology. Array layouts now have standard array structural groups namely: the 1-D, 2-D, and 3-D planar arrays with a corresponding increase in the number of microphones. 2-D arrays include logarithmic spiral and square-shaped arrays (Rabinkin et al. 1997; Humphreys et al. 1998) while 3-D arrays include the spherical array (Elko and Nguyen Pong 1997). Scholars have examined the performance of various arrays among these groups indicating that each kind of array layout works with a suitable localization algorithm at different conditions. For example, Silverman and Patterson (1999), demonstrated that the best localization of sources that could be anywhere in a vast focal region is achieved with optimal sources and no reverberation using a fully encompassing array with randomly arranged microphones that are "uniformly" scattered. This was validated by an experiment conducted by Sachal et al. (2001). At larger frequencies, Underbrink (2002), also designed dobby helical arrays to estimate noise sources. This non-circular random microphone array used for sound source localization under various conditions will be referred to as an arbitrary microphone array in this paper.

The virtual rotating array (VRA) beamforming method is a powerful technique for identifying rotating sound sources, especially for ring microphone arrays. The VRA contains positions that have no real microphones installed hence sound pressures at these positions are linearly interpolated between the sound pressures measured by the neighboring real microphones (Herold and Sarradj 2015). Lowis (2007) is believed to propose the first approach to VRA using the modal decomposition method for in-duct measurements. It applies spatial Fourier transformation to a ring of microphones to achieve a "mode domain" data representation. Dougherty and Walker (2009); Pannert and Maier (2014) have also improved this approach to localize rotating sound sources. DAMAS and CLEAN-SC are the most used deconvolution algorithms for achieving high spatial resolution and dynamic range after beamforming (Ma and Liu 2017).

VRA beamforming faces a significant restriction, particularly in industrial experiments. The restriction is that VRA requires a ring array or several concentric rings to be centered around the rotating axis of the sound source (Zhang and Ma 2022). To suppress this restriction, Ma and Zhang (2020) presented the mode composition beamforming (MCB) technique, which can break this restriction even with an arbitrary microphone arrangement. However, comparing the ring array to an arbitrary configuration array, the side lobe level performance of the former is generally unsatisfactory as has been studied by Sarradj (2016). To properly reconstruct the pressure signals of the VRA microphone within a given precision, Jekosch and Sarradj (2020) successfully studied and applied two different interpolation methods using an arbitrary array in VRA beamforming. The first is a meshing method that creates a mesh using the Delauney-triangulation process. The second is a meshless procedure using radial basis function interpolation. The study demonstrated that both methods can accurately reconstruct the sound source positions and levels while compensating for the motion of rotating sources.

Since then, quite a few research works exist on the combination of arbitrary microphone array layouts with spatial interpolation methods for VRA beamforming technology. However, seven new spatial interpolation methods have recently been added to VRA beamforming by Yang et al. (2022). These methods were categorized under Delaunay-triangulation, radial basis functions, inverse distance weighting, and ordinary kriging. They were applied with arbitrary microphone configurations and simulated to study their performances and influence on the VRA beamforming technology. It was observed that the inverse distance weighting interpolation method (IDW) performs best in interpolating pressure time signals. In analyzing the performance of VRA beamforming technology with an arbitrary microphone array known as the "H0V5" array, Lee et al. (2022) also successfully employed the IDW interpolation method to interpolate microphone pressure time signals.

More importantly, the VRA beamforming result is greatly influenced by the array radius. With the use of a ring array, Zhou et al. (2019), studied the influence of changes in array radius on beamforming results and concluded that the resolution of the localized sources on the beamforming map is strongly related to the array radius. This paper aims to investigate the influence of arbitrary microphone arrays with different radii on virtual rotational beamforming. There are five sections in this paper: the second section states the problem and the important variables needed for the proper computation of VRA beamforming and deconvolution algorithm. Simulation applications are done in the third section, simulation results in the fourth, and the fifth concludes the paper.

2 Statement of the Problem

2.1 VRA Beamforming and Deconvolution Algorithm

A schematic diagram of VRA beamforming with an arbitrary microphone array for rotating source localization is shown in Fig. 1. The coordinate system's origin is located at the center of the microphone array. The center of the microphone array is aligned with the rotational axis of the sound source in the axial direction and rotates at an angular speed of Ω. There are M microphones in the microphone array and S grids in the scanning plane. The stationary microphones are positioned at $\mathrm{X} = [\mathrm{x}_1, \mathrm{x}_2, \mathrm{x}_3, \ldots \mathrm{x}_m, \ldots \mathrm{x}_M]^{\mathrm{T}}$. Y represents the scanning plane grid position vector and it is given by $\mathrm{Y} = \left[\mathrm{y}_1, \mathrm{y}_2, \mathrm{y}_3, \ldots \mathrm{y}_s, \ldots \mathrm{y}_S\right]^{\mathrm{T}}$. The VRA beamforming calculation represented here follows a similar pattern expressed by (Yang et al. 2022). It is given by:

$$\mathrm{b}(\mathrm{y}_{\Omega\mathrm{s}}, \omega) = \mathrm{h}^{\mathrm{H}}(\mathrm{y}_{\Omega\mathrm{s}}, \omega)\mathrm{C}_{\mathrm{v}}(\omega)\mathrm{h}(\mathrm{y}_{\Omega\mathrm{s}}, \omega), \mathrm{s} = 1, \ldots, \mathrm{S} \quad (1)$$

The denotation $(\cdot)^{\mathrm{H}}$, $(\cdot)^{\mathrm{T}}$, $(\cdot)_{\Omega}$ and $(\cdot)_{\mathrm{v}}$ implies conjugate transpose, non-conjugate transpose, parameter in rotating reference frame, and VRA microphones in rotating reference frame parameter respectively. $\mathrm{C}_{\mathrm{v}}(\omega)$ is the cross-spectral matrix (CSM) parameter. The "SV-C2" steering vector type proposed by Ma et al. (2020) will be used together with the CSM parameter in the VRA beamforming algorithm due to its high computational efficiency. The position of the sth rotating grid in the rotating reference frame is represented by $\mathrm{y}_{\Omega\mathrm{s}}$ and the angular velocity ω is mathematically represented as $\omega = 2\pi f$ where f is the frequency. Beamforming calculation in decibels (dB) can be done using the formula, $10\log_{10}[\frac{b(y_{\Omega s}, \omega)}{p_{ref}^2}]$ where p_{ref} represents the reference pressure of 2×10^{-5} Pa and $\mathrm{h}(\mathrm{y}_{\Omega\mathrm{s}}, \omega)$ in Eq. (1) is given by:

$$\mathrm{h}(\mathrm{y}_{\Omega\mathrm{s}}, \omega) = [h(\mathrm{x}_{\mathrm{v}1}, \mathrm{y}_{\Omega\mathrm{s}}\omega), \ldots, h(\mathrm{x}_{\mathrm{vm}}, \mathrm{y}_{\Omega\mathrm{s}}\omega), \ldots, h(\mathrm{x}_{\mathrm{v}M}, \mathrm{y}_{\Omega\mathrm{s}}\omega)]^{\mathrm{T}} \quad (2)$$

having x_{vm}, the mth virtual microphone's position in the rotating reference frame. The virtual microphone pressure $p_v(\omega)$ is expressed as:

$$p_v(\omega) = [p(x_{v1}, \omega), \ldots, p(x_{vm}, \omega), \ldots, p(x_{vM}, \omega)]^T \quad (3)$$

where the $p(x_{vm}, \omega)$ denotes the complex pressure spectrum at the mth virtual rotating microphone. This spectrum is obtained by performing fast Fourier transformation on the pressure time signals, $p(x_{vm}, t)$ at the mth virtual rotating microphone.

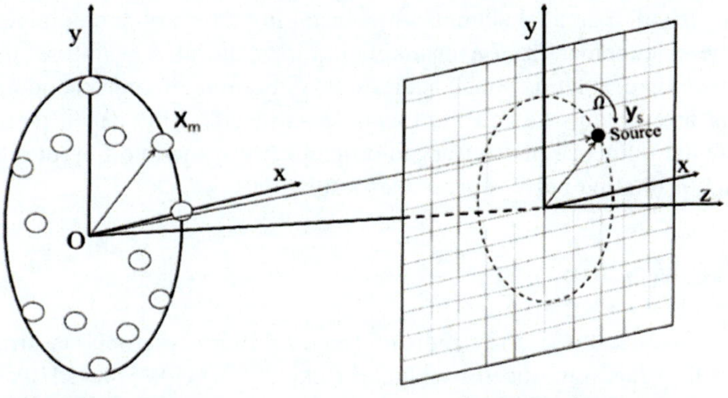

Fig. 1. VRA beamforming schematic diagram for rotating sources using arbitrary microphone array (Yang et al. 2022).

Conventional beamforming has an outcome of a dirty map, and all the sound sources are clear but with poor spatial resolution and side-lobe effects. However, with deconvolution algorithms, like the Deconvolution Approach for the Mapping of Acoustic Sources (DAMAS) and CLEAN based on source coherence deconvolution algorithm (CLEAN-SC), a clean map of noise source distributions can be obtained substantially to enhance spatial resolution and dynamic range with well-spread noise sources (Wang and Ma 2019a, 2019b). DAMAS is well known in the acoustic field for significantly improving spatial resolution and removing side-lobe effects however, it requires high computational efforts for most occasions and results may be poorer compared to CLEAN-SC (Ma and Liu 2017). CLEAN-SC is currently one of the most efficient algorithms because it enhances spatial resolution and dynamic range with high computational efficiency (Ma and Zhang 2020).

The main idea behind CLEAN-SC is that it determines a set of source powers to account for the initial, classical beamforming map. An iterative process is used to create a "clean" map from the initial, "dirty", map. By principle, the algorithm creates a model for the portion of the CSM connected to a partial source. Then, any sources coherent with the detected peak are accounted for in the model when computed from the "dirty" CSM and steering vectors. Finally, the beamforming map of the model CSM is subtracted from the "dirty" map and the model CSM from the dirty CSM. When there are no significant sources left in the dirty map, the iteration is finished (Sijtsma 2007).

This paper will employ the CLEAN-SC deconvolution algorithm after the application of VRA beamforming to achieve a clean map.

2.2 Influence of Radius of Microphone Array

The radius of microphone arrays generally affects the resolution size of the localized sound sources (Zhou et al. 2019). Thus, changing the array radius influences spatial resolution either positively or negatively depending on how close it is to the rotating source and other factors like frequency. Proper array radius and its influence on VRA beamforming is still a great challenge when installing arbitrary microphone arrays that can help achieve better localization results. To optimize the VRA beamforming technique for the localization of rotating sound sources, it is important to understand the influence that different array radius has on the localization results. This paper, therefore, aims to investigate the influence of arbitrary microphone arrays with different radii on VRA beamforming of rotating noise sources via simulations.

3 Simulation Setup

In this paper, we use three array radii of three arbitrary microphone arrays namely the Underbrink array, introduced by Underbrink (2002), H0V5, and H0V3.13 arrays introduced by Sarradj (2016) to investigate the influence of the radius of a microphone array on VRA beamforming of rotating noise sources via simulations. The simulation setup extends the subcase b of Herold (2017). It was conducted using the Acoular package in python (Herold and Sarradj 2017). Figure 2 shows the arrangement of the microphones with three-point sources rotating anti-clockwise. The critical parameters for simulation are displayed in Table 1. From the microphone array to the scanning plane is 0.5 m having z as the rotation axis. The sound source intensities of the three source points are 94 dB, 91 dB, and 88 dB respectively, the sound signal type emitted is a white noise signal and the evaluation time is 1 s. This paper will employ the inverse distance weighting interpolation method (IDW) since it is known to have the best sound source precision compared to other spatial interpolation methods (Yang et al. 2022).

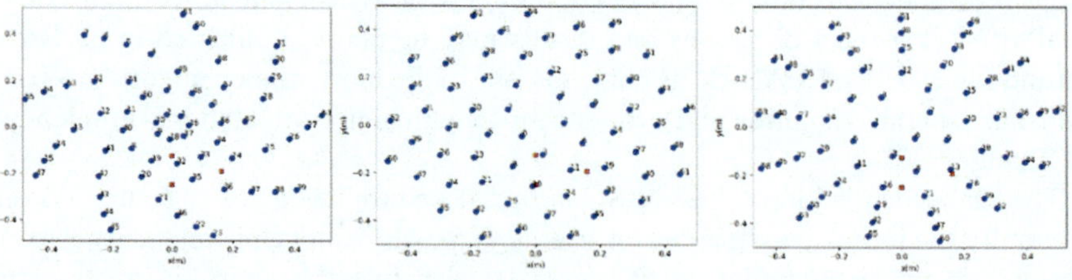

(a) (b) (c)

Fig. 2. An example of an arbitrary microphone placement with three rotating sources for VRA beamforming at an array radius of 0.5 m. (a) Underbrink (b) H0V5 array (c) H0V3.13 array. Blue points are microphones, and red points are point locations of rotating sources 1, 2, and 3 (Color figure online).

In this paper, the weighting power exponent, $\beta = 2.0$ will be used for the IDW interpolation method. Frequencies of 5 kHz and 2.5 kHz are employed as focus frequencies and an array radius variation range of 0.125–1 m is used for simulation study at a rotating source radius of 0.25 m. The influence of array radius on the localization performance will be observed and analyzed from VRA beamforming and CLEAN-SC maps.

Table 1. Critical simulation parameters

Microphone number	64
Sound rotating speed	1500 rpm
Source 1; 2; 3 (Initial position)	(0 m, – 0.25 m, 0.50 m); (0.161 m, – 0.192 m, 0.50 m); (0 m, – 0.125 m, 0.50 m)
Rotating source radius	0.25 m
Array radius	0.125 m, 0.55 m, 1 m
Array aperture	1 m
Evaluation time(tmax)	1 s
Size of Grid	3721 (61 × 61)
Resolution of grid	0.01 m
Block size (FFT)	1024
Window size (FFT)	Von Hann (50% overlap)
Number of iterations (CLEAN-SC)	1000

4 Simulation Results

Figures 3, 4, 5, 6, 7 and 8, show the VRA beamforming and corresponding CLEAN-SC maps of the three rotating point sources at different arbitrary microphone array radii, of $r = 0.125$ m, 0.55 m, and 1 m of the Underbrink, H0V5, and H0V3.13 microphone arrays. These are observed at low and high frequencies, of 2.5 kHz and 5 kHz respectively. As shown clearly in Figs. 3, 4, and 5, there is a source intensity error on the VRA beamforming map at $r = 0.125$ m when $f = 2.5$ kHz. It is hard to identify which source it is and this may be due to array element spacing error or Rayleigh's criterion (Zhou and Jiang 2019; Wang and Ma 2019b). Even with CLEAN-SC deconvolution, the position of the sources is still not perfectly accurate. At $r = 0.55$ m, the source intensity error drastically decreases and source 1 is identified as the point with the highest intensity for all arrays in Figs. 3, 4, and 5. Although source intensity error is reduced as the array radius is increased to $r = 0.55$ m and 1 m in Figs. 3, 4 and 5, it must be noted that there are still some ghost contributions on the map.

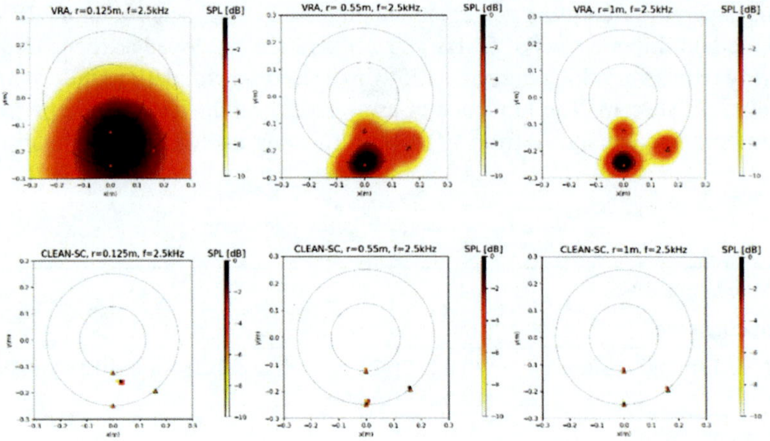

Fig. 3. VRA beamforming and CLEAN-SC map with Underbrink array at $f = 2.5$ kHz, $r = 0.125$ m, 0.55 m, 1 m.

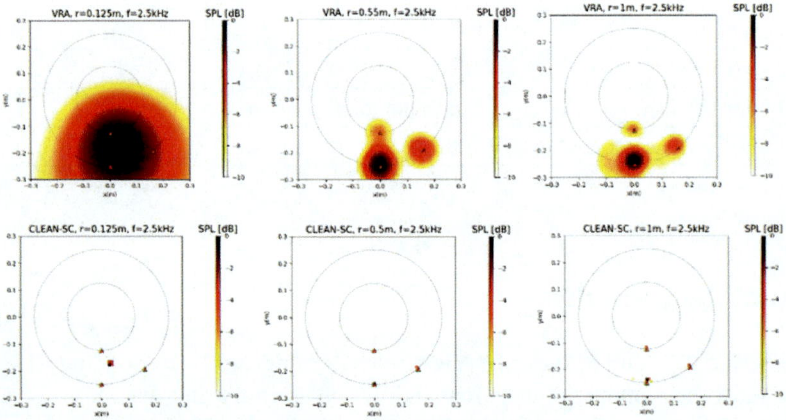

Fig. 4. VRA beamforming and CLEAN-SC map with H0V5 array at $f = 2.5$ kHz, $r = 0.125$ m, 0.55 m, 1 m.

Results at a high frequency of 5 kHz are displayed in Figs. 6, 7 and 8. As can be seen in these figures, the issue of source intensity error at $r = 0.125$ m is still prevalent as was observed in Figs. 3, 4 and 5 but with a minimized resolution size. Also, at $r = 0.55$ m in Figs. 6, 7 and 8, the VRA beamforming maps have a clearer source distribution, the source intensities are significantly improved, and the CLEAN-SC maps also pinpoint the sources accurately, thus improving spatial resolution. As the radius is further increased to $r = 1$ m, the sidelobe effect decreases and the source location is accurately identified using the CLEAN-SC map. Again, it can be found in Figs. 7, 8 that a similar occurrence

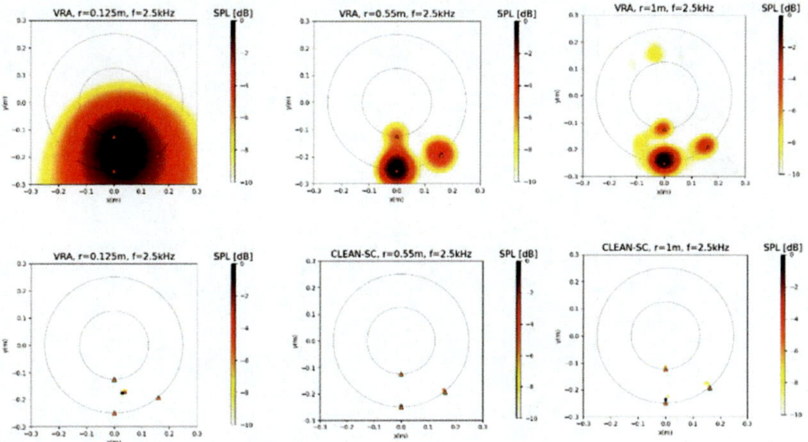

Fig. 5. VRA beamforming and CLEAN-SC map with H0V3.13 array at $f = 2.5$ kHz, $r = 0.125$ m, 0.55 m, 1 m.

of spatial resolution improvement and sidelobe reduction occurs as the radius is increased however, ghost contributions appear at $r = 1$ m. Overall, the sources seem to be easier to identify in the VRA beamforming maps at $r = 0.55$ m in Figs. 6, 7 and 8 compared to the other maps at $r = 0.125$ m and 1 m.

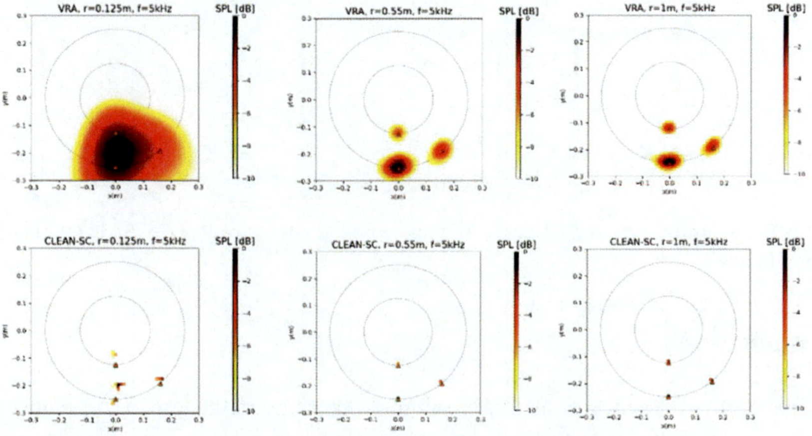

Fig. 6. VRA beamforming and CLEAN-SC map with Underbrink array at $f = 5$ kHz, $r = 0.125$ m, 0.55 m, 1 m.

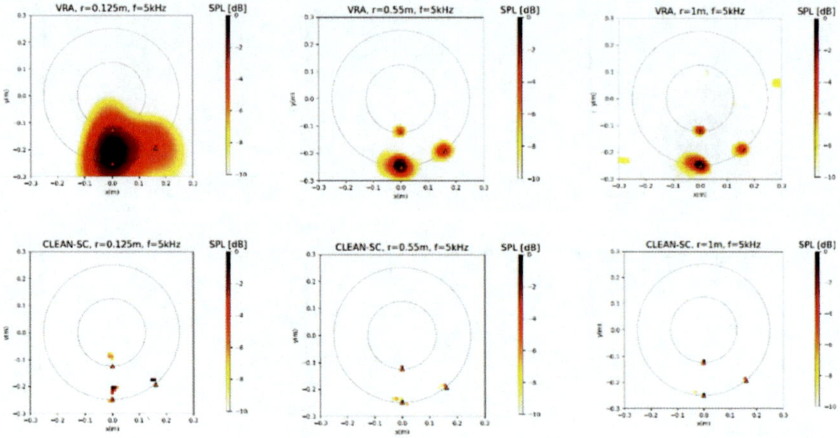

Fig. 7. VRA beamforming and CLEAN-SC map with H0V5 array at $f = 5$ kHz, $r = 0.125$ m, 0.55 m, 1 m.

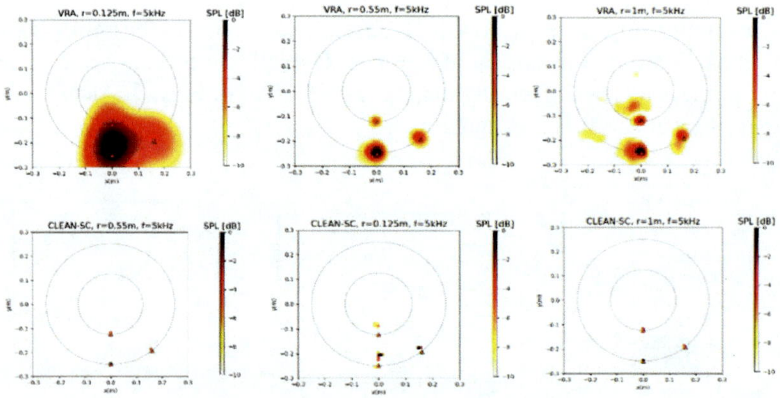

Fig. 8. VRA beamforming and CLEAN-SC map with H0V3.13 array at $f = 5$ kHz, $r = 0.125$ m, 0.55 m, 1 m.

5 Conclusions

In this paper, the influence of arbitrary microphone array radius on VRA beamforming for the localization of rotating noise sources is comprehensively studied. Three array radii ($r = 0.125$ m, 0.55 m, 1 m) of three arbitrary microphone arrays namely the Underbrink, H0V5, and H0V3.13 array are employed. The following conclusions were made.

1. The array radius greatly influences spatial resolution, especially at varied frequencies. At a high frequencies between $5 \geq f \geq 8$ kHz there is usually an enhanced special resolution, as the radius increases while at a low frequencies between $2.5 \geq f \geq 5$ kHz, the sources are hard to separate as the radius decreases.

2. At frequencies within 2.5 ≥ f ≥ 5 kHz, there is usually a source intensity error when the array radius is half the size of the rotating source radius making other sources hard to identify. This may be due to array element spacing error or Rayleigh's criterion as mentioned in Sect. 4.
3. CLEANSC deconvolution algorithm greatly enhances spatial resolution and source position precision.
4. Overall, localization results are significantly enhanced when the radius ratio (array radius/rotating source radius) is within 2–4 at frequencies(f) within 5 ≥ f ≥ 8 kHz.
5. Future works of VRA beamforming with arbitrary microphone arrays could include quantitative measurements of the dynamic range and spatial resolution for better map observation.

References

Barsikow, B.: Experiences with various configurations of microphone arrays used to locate sound sources on railway trains operate by the DB AG. J. Sound Vib. **193**, 283–293 (1996). https://doi.org/10.1006/jsvi.1996.0269

Collins, M.D., Makris, N.C., Fialkowski, L.T.: Noise cancellation and source localization. J. Acoust. Soc. Am. **96**, 1773–1776 (1994). https://doi.org/10.1121/1.410255

Dougherty, R.P., Walker, B.E.: Virtual rotating microphone imaging of broadband fan noise. In: 15th AIAA/CEAS Aeroacoustics Conference (30th AIAA Aeroacoustics Conference) (2009)

Elko, G.W., Nguyen Pong, A.-T.: A steerable and variable first-order differential microphone array. In: Acoustics, Speech, and Signal Processing, 1988, ICASSP 1988, 1988 International Conference, pp. 223–226 (1997)

Herold, G.: Microphone array benchmark b11:rotating point sources. Springer Verlag (2017)

Herold, G., Sarradj, E.: Microphone array method for the characterization of rotating sound sources in axial fans. Noise Control Eng. J. **63**, 546–551 (2015). https://doi.org/10.3397/1/376348

Herold, G., Sarradj, E.: Open-source software for the application of microphone array methods. Noise Vibr. Worldwide **48**, 44–51 (2017). https://doi.org/10.1177/0957456517699024

Humphreys, W.M., Brooks, T.F., Hunter, W.W., Meadows, K.R.: Design and use of microphone directional arrays for aeroacoustic measurements. In: 36th AIAA Aerospace Sciences Meeting and Exhibit. American Institute of Aeronautics and Astronautics Inc, AIAA (1998)

Jae-Woong, C., Yang-Hann, K.: Spherical beam-forming and MUSIC methods for the estimation of location and strength of spherical sound sources. Mech. Syst. Signal Process. **9**, 569–588 (1995). https://doi.org/10.1006/mssp.1995.0043

Jekosch, S., Sarradj, E.: An extension of the virtual rotating array method using arbitrary microphone configurations for the localization of rotating sound sources. Acoustics **2**, 330–342 (2020). https://doi.org/10.3390/acoustics2020019

Lee, C.W., Zhang, C., Ma, W.: Effects of radial misalignment on virtual rotating array beamforming for rotating sound source localization. Aerosp. Syst. (2022). https://doi.org/10.1007/s42401-022-00164-8

Lowis, C.R.: In-duct measurement techniques for the characterisation of broadband aeroengine noise. Ph.D. Dissertation, University of Southampton (2007)

Ma, W., Bao, H., Zhang, C., Liu, X.: Beamforming of phased microphone array for rotating sound source localization. J. Sound Vib. **467** (2020). https://doi.org/10.1016/j.jsv.2019.115064

Ma, W., Liu, X.: Improving the efficiency of DAMAS for sound source localization via wavelet compression computational grid. J. Sound Vib. **395**, 341–353 (2017). https://doi.org/10.1016/j.jsv.2017.02.005

Ma, W., Zhang, C.: A frequency-domain beamforming for rotating sound source identification. J. Acoust. Soc. Am. **148**, 1602–1613 (2020). https://doi.org/10.1121/10.0001939

Pannert, W., Maier, C.: Rotating beamforming - motion-compensation in the frequency domain and application of high-resolution beamforming algorithms. J. Sound Vib. **333**, 1899–1912 (2014). https://doi.org/10.1016/j.jsv.2013.11.031

Rabinkin, D.V., Renomeron, R.J., French, J.C., Flanagan, J.L.: Optimum microphone placement for array sound capture. J. Acoust. Soc. Am. **101**, 3114–3114 (1997). https://doi.org/10.1121/1.418950

Sachal, J.M., Silverman, H.E., Patterson III, W.R.: Large vs small aperture microphone arrays: performance over a large focal area. In: IEEE International Conference on Acoustics, Speech, and Signal Processing, pp. 3049–3052 (2001)

Sarradj, E.: A Generic approach to synthesize optimal array microphone arrangements. In: Berlin Beamforming Conference, BeBeC-2016-S4, Cottbus, Germany (2016)

Sijtsma, P.: CLEAN based on spatial source coherence. Int. J. Aeroacoust. **6**, 357–374 (2007). https://doi.org/10.1260/147547207783359459

Silverman, H.F., Patterson, W.R.: Visualizing the performance of large-aperture microphone arrays. In: International Conference on Acoustics, Speech, and Signal Processing (ICASSP), pp. 969–972 (1999)

Underbrink, J.R.: Aeroacoustic phased array testing in low speed wind tunnels. In: Muller, J.T. (eds.) Aeroacoustic Measurements. Experimental Fluid Mechanics, pp. 98–217. Springer, Heidelberg (2002). https://doi.org/10.1007/978-3-662-05058-3_3

Wang, J., Ma, W.: Comparison of deconvolution algorithms of phased microphone array for sound source localization in an airframe noise test. In: Jing, Z. (ed.) ICASSE 2019. LNEE, vol. 622, pp. 81–90. Springer, Singapore (2020). https://doi.org/10.1007/978-981-15-1773-0_7

Wang, X., Ma, W.: Arrangement optimization of phased microphone arrays based on deconvolution algorithm. In: International Conference of Aerospace Science Systems (2019b)

Yang, J., Zhang, C., Ma, W.: Spatial interpolation methods for virtual rotating array beamforming with arbitrary microphone configurations. Aerosp. Syst. **5**, 149–158 (2022). https://doi.org/10.1007/s42401-021-00117-7

Zhang, C., Ma, W.: Fast fourier transformation computation of modal transfer function for rotating source localization. AIAA J., 1–5 (2022). https://doi.org/10.2514/1.j061179

Zhou, S., Jiang, L.: Modern description of Rayleigh's criterion. Phys. Rev. A (Coll. Park) **99** (2019). https://doi.org/10.1103/PhysRevA.99.013808

Zhou, W., Yang, M., Ma, W., Bao, H.: Parameter investigation of ring microphone array for rotating sound source localization. In: Proceedings of Global Power and Propulsion Society (2019)

A Finite Element Method of Research on Transverse Mechanical Properties of Fiber-Reinforced Composites with Random Matrix Void Defects

Wei Zhang, Yin Yu, Yile Hu(✉), Haomin Li, and Zekai Zhao

Aerospace Structure Research Center, School of Aeronautics and Astronautics, Shanghai Jiao Tong University, Shanghai, China
yilehu@sjtu.edu.cn

Abstract. Matrix voids often appear in the process of manufacturing composite materials, and the existence of voids has a negative impact on the mechanical properties, especially the transverse mechanical properties, of composite materials. However, it will be costly to use experimental methods to determine the influence of matrix voids of different porosity on mechanical properties. This paper introduces a method of using finite element method to establish a randomly distributed matrix voids model, in order to analyze the change of mechanical properties of composite materials under different porosity conditions. Therefore, it provides a numerical calculation method with less computational cost and more efficient for studying the influence of matrix voids on the mechanical properties of the fiber-reinforced composite materials.

Keywords: Matrix void · Fiber-Reinforced Composites · FEM

1 Introduction

Fiber reinforced composites are widely used in the manufacture of modern aircraft structures. It has the characteristics of light weight, high specific strength, high specific stiffness, high fatigue resistance, high corrosion resistance and so on. Because these characteristics of composite materials meet the requirements of modern aircraft manufacturing, therefore, composite materials have become a widely used material in the aerospace field. The using of the materials has exceeded more than 50% of the total weight of the structure in some aircrafts. What's more, there are even some small and light aircraft manufactured by all composite materials [1].

Fiber-reinforced composite materials are generally made of reinforced fiber materials (such as carbon fiber, glass fiber, aramid fiber, etc.) and matrix materials (such as resin, rubber, ceramics, etc.) through some molding processes. It usually consists of three parts: fiber, matrix and interface. Carbon fiber composite materials (AFRP), glass fiber composite materials (GFRP), etc. are commonly used in the design and manufacture of aircraft.

Z. Jing and X. Zhan (Eds.): ICASSE 2023, LNEE 1153, pp. 145–158, 2024.
https://doi.org/10.1007/978-981-97-0550-4_12

It is obvious that there will be various manufacturing defects due to the limitations of the manufacture processes of the fiber reinforced composite materials. There are several different typical manufacturing defects at the mesoscale, including fiber wrinkle, matrix void, matrix crack defects and so on. The matrix void defect is one the most common manufacturing defects. In the manufacturing process of composite materials, the formation of void defect cannot be avoided, and the existence of it will inevitably affect the mechanical properties of composite materials, especially those dominated by the matrix [2, 3].

The matrix void can be divided into two types considering the different positions where they exist. One of the situations is between the fibers, which is caused by the fact that the resin is difficult to flow between the fibers due to the viscosity of the resin during the composite manufacturing process, and the other one is caused by the presence of air in the matrix that cannot be expelled [4]. These two types of void defects have been shown to negatively affect the mechanical properties of composites in previous experimental studies [5, 6]. The main parameters to characterize matrix void are porosity, void shape, void size, position, etc. Among them, porosity (the volume fraction of void) is the most widely used characterization parameter in calculation and analysis. Currently, we can use densitometry, ultrasonic testing, X-ray, CT and other techniques to obtain porosity [7–9].

Currently, there are three main research methods for the research of the influence of matrix void on the mechanical properties in composite materials: experimental method, analytical model solution and numerical simulation.

Several experiments have been set to study the effects of porosity, and the matrix void are made by different ways [10–18]. The effects of porosity have been studied experimentally by several people and have been made in a number of different ways. And it is shown in several sets of experimental results that when the porosity increases, the transverse tensile strength of the composite material will decrease significantly [12, 19]. Lei Y. showed in his research that the interlaminar shear strength, transverse tensile modulus, and shear modulus is greatly affected by porosity [20]. The experimental research results above show that the transverse strength of the composite is most sensitive to the porosity.

In some researches using analytical methods, some of them have proposed the concept of critical porosity, that is, below a certain porosity, the effect of void on the mechanical properties of composite materials can be ignored [11, 14, 15]. And some articles have proposed that porosity cannot be used as the single parameter to measure the influence of the mechanical properties. The influence of the shape, size, position, and distribution of the matrix void should also be considered [12, 14, 18].

Numerical simulation method is the most commonly used method at present. In the study of Vajari et al. on the influence of porosity on the transverse mechanical properties of unidirectional fiber reinforced composite materials, it was shown that when the porosity is in the range of 1–5%, the transverse strength will be greatly reduced with the increase of porosity. And this study also obtained composite failure trajectories under transverse tension/compression as well as out-of-plane shear [21, 22]. Based on the finite element method, Gonza'lez et al. studied the performance of the unit cell model of unidirectional fiber reinforced composites under transverse compressive loads. The results showed that the transverse strength of the unit cell is mainly controlled by the interface strength and the matrix yield strength under uniaxial compression. And two different fracture modes were analyzed. Li Bo et al. studied the influence of irregular void on the mechanical properties of transverse tension through the finite element mesoscopic model. The results showed that the shape of the void would affect the initial damage, crack growth and failure of the model [23]. Jiayue Wang studied the effect of micro-void on transverse stiffness and strength for fiber-reinforced composites subjected to load perpendicular to fiber using state-based peridynamic (PD) theory. It was found that the transverse modulus decreases with the void fraction in the range of 0.5–2%. A smaller void has a greater impact on the transverse modulus than a larger void under the same void content. The crack growth path for the multi-fiber model subjected to transverse tension was analyzed. The predictions show that cracks initiate at the interface between fiber and matrix, then deflect to the micro-voids nearby [24].

In this article, a micromechanical model is constructed to simulate the transverse mechanical properties of fiber reinforced composites with random matrix void. This article, using the finite element method with periodic boundary conditions simulate the reduction of the transverse tensile/compressive modulus of fiber reinforced composites with different porosity and void distribution under transverse tension and compression respectively.

2 Computational Micromechanical Model

2.1 Random Matrix Void Finite Element Model

In order to study the micromechanics of the fiber-reinforced composites, a representative volume element (RVE) of the microstructure should be constructed firstly. First, a 2D finite element model containing fibers and randomly distributed void needs to be established. In this paper, the shape of the void is assumed to be circle. In this model, we need these parameters: number of fibers, fiber volume fraction v_f, fiber radius r_f, maximum radius of void r_{vmax} and minimum radius of void r_{vmin}, and volume fraction of void v_f. The shape of the model is square, and the side length of the model can be calculated by the following formula.

$$d_{uc} = \sqrt{\frac{\pi n^2 r_f^2}{v_f}} \tag{1}$$

where n stands for the numbers of fiber of each roll. After obtaining the side length of the model, we need to create a square part with side length d_{uc} in ABAQUS, and use the Partition command to draw n circles representing fibers in the model. In this model, the fibers are evenly distributed.

The next step is to produce the random distributed void. The coordinates of the voids are randomly generated in the range of (0, d_{uc}), the radiuses of the voids are randomly generated in the range of (r_{vmin}, r_{vmax}), and the voids are evenly distributed randomly. In the process of generating random voids, we need to ensure that voids do not intersect each other, and voids do not intersect fibers. For the first case, the judgment conditions are as follows.

$$\sqrt{(x_i - x_j)^2 + (y_i - y_j)^2} > r_i + r_j \tag{2}$$

Where:

i – the new void
j – the exist voids
x – the coordinate along x axis
y – the coordinate along y axis
r – the radius of void.

For the second case, the judgement conditions are as follows:

$$\sqrt{(x_f - x_v)^2 + (y_f - y_v)^2} > r_f + 2r_v \tag{3}$$

Where:

f – fiber
v – void
x – the coordinate along x axis
y – the coordinate along y axis
r – the radius.

The reason for choosing $2r_v$ in formula (3) is to avoid the distance between pores and fibers being too small. It can be determined that the voids exist when the newly generated voids meet the two requirements above.

When the parameters of 2D model are set as: the fibers number of single-row is 3 ($n = 3$), the fiber volume fraction is 0.5 ($v_f = 0.5$), the fiber radius is 5.0 ($r_f = 5.0$), the maximum radius of void is 0.25 ($r_{vmax} = 0.25$), the minimum radius of void is 0.1 ($r_{vmin} = 0.1$), and the void volume fraction is 0.01 ($v_v = 0.01$), The corresponding multi-fiber with random void model is shown in the Fig. 1.

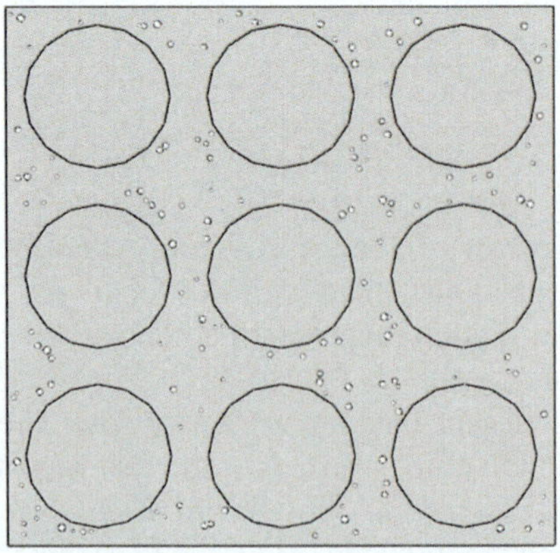

Fig. 1. Schematic diagram of multi-fiber with random void model

In the cycle of randomly generating voids, the condition for determining the stop of the cycle is that the actual void volume fraction is greater than or equal to the set void volume fraction, so it is likely that the actual void volume fraction is higher than the set value, so we need to recalculate the real void volume fraction of the model The real void volume fraction is calculated as:

$$v_{vreal} = \frac{\sum \pi r_i^2}{d_{uc}^2} \tag{4}$$

Where:

v_{vreal} – the real void volume fraction
r_i – the radius of each void
d_{uc} – the side length of the unit cell.

In this article, the 2D finite element model, a single fiber unit cell model, is used to simulate, as shown in the figure below (Fig. 2). And different porosity will be selected for analysis in the simulation.

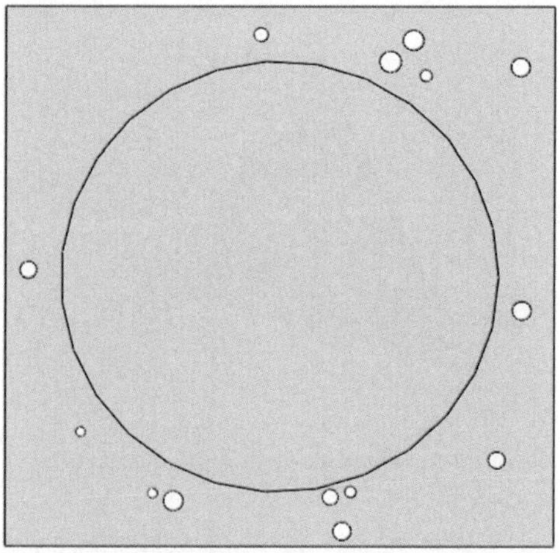

Fig. 2. Schematic diagram of singe fiber with random void model ($v_v = 0.0701923153628848$)

2.2 Material Properties

In this article, the unit cell model includes two parts: matrix part and fiber part. We assume the fiber is isotropic and homogeneous. The fiber's Young's modulus is $E_f = 40$ GPa, and Poisson's ratio is $\mu_f = 0.25$. The Young's modulus of the matrix is $E_m = 4$ GPa, and Poisson's ratio is $\mu_m = 0.35$.

2.3 Mesh

In this simulation, the plane strain state was adopted because the dimension along the fiber direction is much larger than that in the other two directions. The fibers and matrix are meshed with four-node bilinear plane strain-reduced integration elements (CPE4R) (as shown in Fig. 3) mainly and a few three-node triangle elements (CPE3) (as shown in Fig. 4).

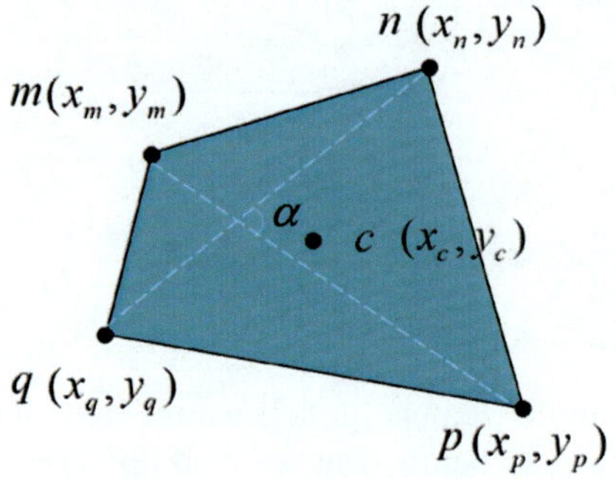

Fig. 3. Four-node bilinear plane strain-reduced integration elements

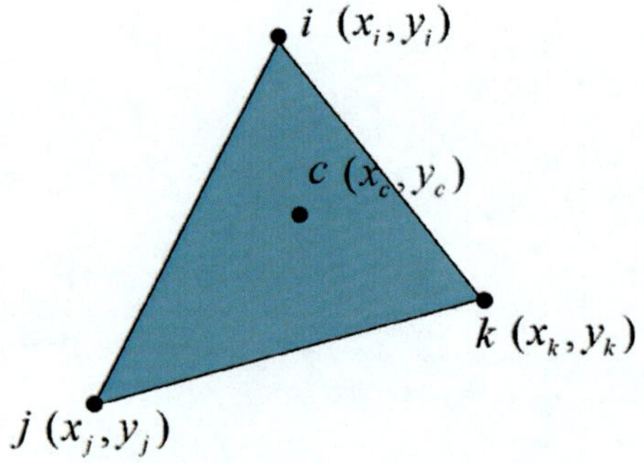

Fig. 4. Three-node triangle elements

The schematic diagram of the unit cell model meshing is shown in the figure below (Fig. 5).

3 Periodical Boundary Condition

Periodic boundary conditions (PBC) are needed to apply to the edges of the unit cell to ensure consistent deformation of opposite edges. As shown in the Fig. 6, the unit cell has four edges, which the left edge is parallel to the right edge, and the upper edge is parallel to the lower edge. Point C1–C4 are four corner nodes respectively.

Point i is any point on the left edge, and point j is the point corresponding point on the right edge. Equations for establishing periodic boundary conditions on left and right edges are expressed as Eq. (5):

$$\begin{aligned} u_i - u_{C1} &= u_j - u_{C2} \\ v_i - v_{C1} &= v_j - v_{C2} \end{aligned} \tag{5}$$

Fig. 5. Schematic diagram of the unit cell model meshing

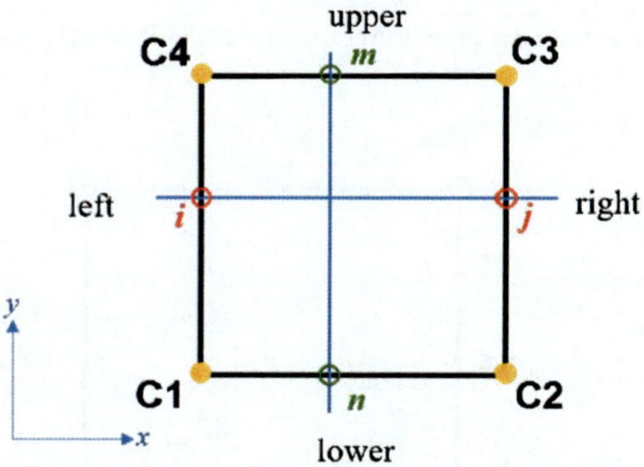

Fig. 6. Schematic diagram of periodical boundary conditions

where u and v represent the displacement along x axis and y axis of any node respectively.

Point m is any point on the upper edge, and point n is the point corresponding point on the lower edge. Equations for establishing periodic boundary conditions on left and right edges are expressed as Eq. (6):

$$\begin{aligned} u_m - u_{C4} &= u_n - u_{C1} \\ v_m - v_{C4} &= v_n - v_{C1} \end{aligned} \tag{6}$$

where u and v represent the displacement along x axis and y axis of any node respectively.

The body motions are prevented by constraining the vertex C1 and the Y-direction displacement of the vertex C2. As shown in Fig. 7, transverse tension and compression are then determined by applying the X-direction movement to the vertex C2 and C3 and simultaneously constraining the X-direction movement to the vertex C4.

Transverse shear is determined by applying X-direction displacement on the vertex C3 & C4, and simultaneously constraining the X-direction movement to the vertex C1 as shown in Fig. 8.

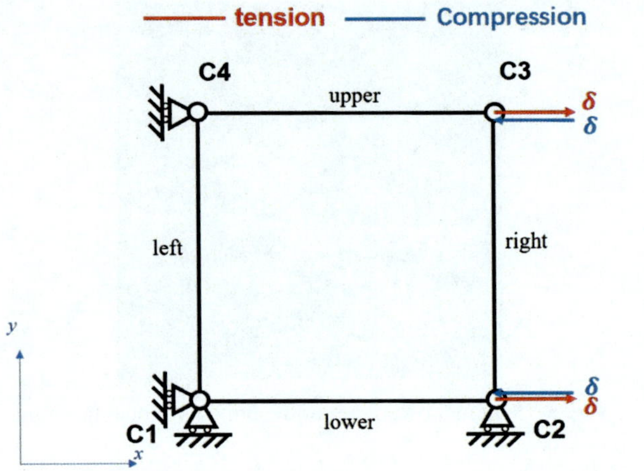

Fig. 7. The boundary condition under tensile or compressive load

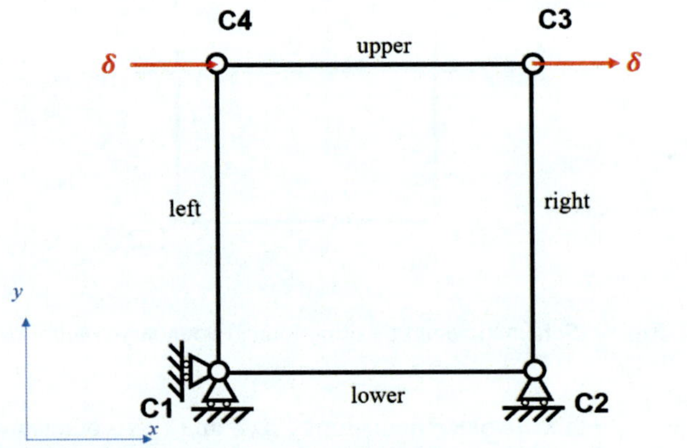

Fig. 8. The boundary condition under shear load

4 Results

In this article, the simulated model's parameters are: $r_f = 3$ mm, $v_f = 0.6$, $r_{vmax} = 0.25$ mm, $r_{vmin} = 0.1$ mm, and $v_{v1} = 0.5\%$. The models with preset porosity of 0.5% is simulated in transverse tension, compression and shear respectively. The actual porosity of the model with the preset porosity of 0.5% is 0.6911%.

4.1 Transverse Tension

The displacements of model with 0.6911% void volume fraction (preset porosity is 0.5%) under transverse tensile load are shown in Fig. 9 and Fig. 10. The displacement in the x direction gradually increases along the positive direction of the x axis, and the displacement in the y direction gradually decreases along the positive direction of the y axis, and the displacement of the matrix part is greater than that of the fiber part. The maximum displacement along x axis occurs on the right edge, and the minimum occurs on the left edge, which is consistent with the boundary conditions we set. Because there's no restriction on the upper edge and lower edge, so the displacement along y direction seems symmetric to the middle line of the model. Due to the presence of voids, the displacement distribution is not strictly symmetrical. This meets our preset periodic boundary conditions. The displacement distribution mode is similar to the results from Jiayue Wang's result.

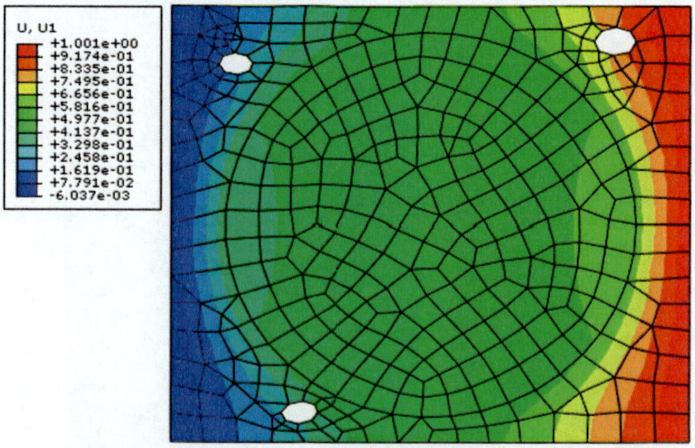

Fig. 9. The x-direction displacement under tension load ($v_v = 0.6911\%$)

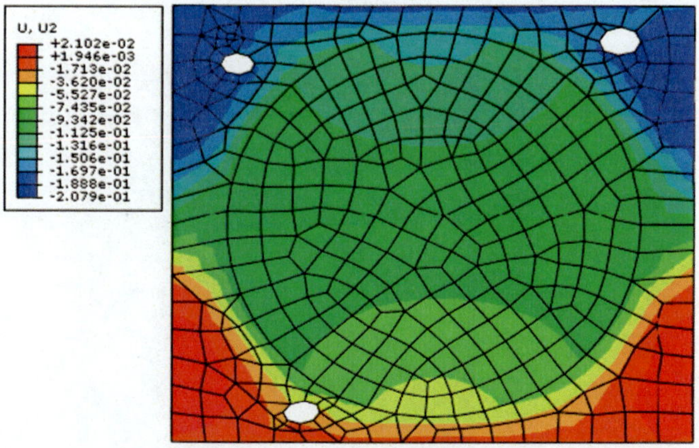

Fig. 10. The y-direction displacement under tension load ($v_v = 0.6911\%$)

4.2 Transverse Compression

The displacements of model with 0.6911% void volume fraction (preset porosity is 0.5%) under transverse compressive load are shown in Fig. 11 and Fig. 12. The displacement in the x direction gradually decreases along the positive direction of the x axis, and the displacement in the y direction gradually increases along the positive direction of the y axis, and the displacement of the matrix part is greater than that of the fiber part. The displacement distribution mode is similar to the results from Jiayue Wang's result [28].

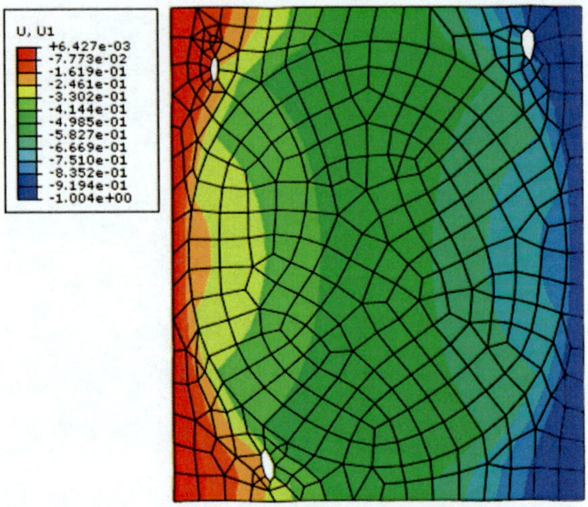

Fig. 11. The x-direction displacement under compression load ($v_V = 0.6911\%$)

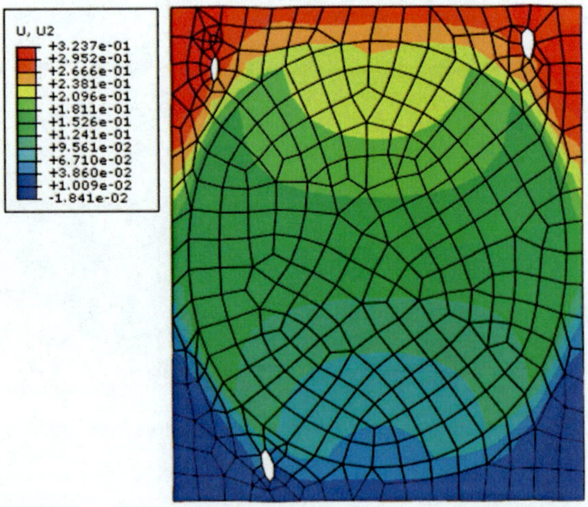

Fig. 12. The y-direction displacement under compression load ($v_V = 0.6911\%$)

The maximum displacement along x axis occurs on the left edge, and the minimum occurs on the right edge, which is consistent with the boundary conditions we set. Because there's no restriction on the upper edge and lower edge, so the displacement along y direction seems symmetric to the middle line of the model. Due to the presence of voids, the displacement distribution is not strictly symmetrical. This meets our preset periodic boundary conditions. The displacement distributions under compression load and tension load are opposite, which also proved that the simulation results are correct.

4.3 Transverse Shear

The displacements of model with 0.6911% void volume fraction (preset porosity is 0.5%) are shown in Fig. 13 and Fig. 14. The displacement in the x direction gradually decreases along the positive direction of the y axis, and the displacement in the y direction gradually increases along the positive direction of the x axis in general. The displacements seem symmetric about the middle line of the model, but it is not strictly symmetric.

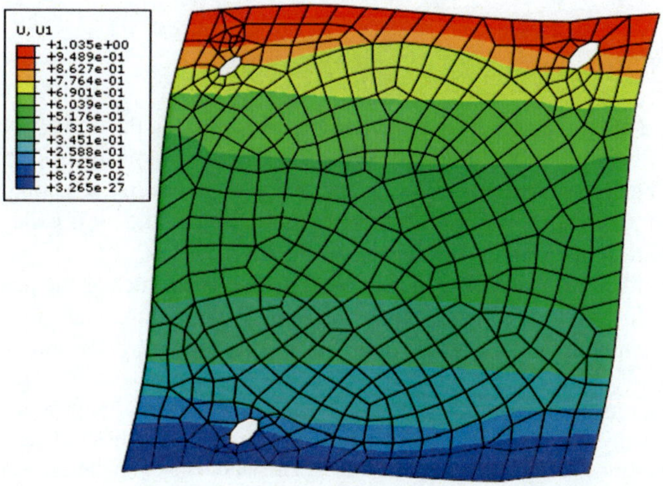

Fig. 13. The x-direction displacement under shear load ($v_v = 0.6911\%$)

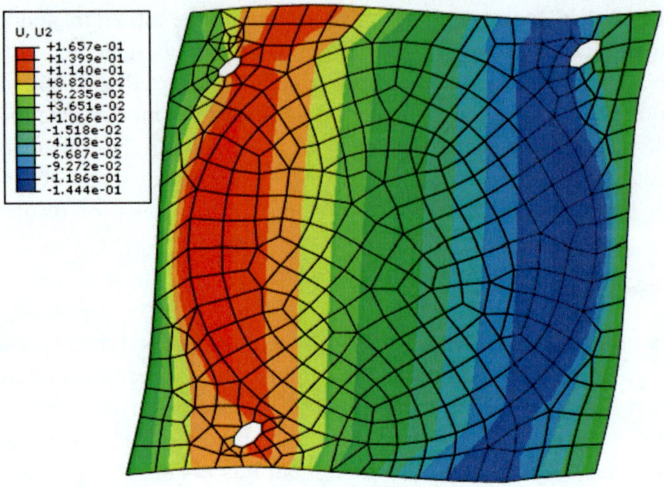

Fig. 14. The y-direction displacement under shear load ($v_y = 0.6911\%$)

5 Conclusion

In this article, a finite element modeling method for analyzing the transverse mechanical properties of fiber-reinforced composite materials with random matrix void defects is introduced. The simulation results are accurate under transverse tension, compression and shear load. Compared with peridynamics simulation results, it is can be considered that the simulation method has reference significance.

There are still some aspects to be improved in this work, such as the post-processing of the results, the initialization and expansion of damage, failure analysis, visualization, etc., so that the model can display the analysis results more directly and clearly.

Acknowledgements. This work is supported by School of Aeronautic and Astronautic, Shanghai Jiao Tong University. Thanks for the detailed instruction from my tutors Dr. Hu and Dr. Yu. Their insightful suggestions always light me in the lost. The authors also gratefully acknowledge the helpful comments and suggestions of the reviews, which have improved the presentation.

References

1. 孙盘滔.: 复合材料等同性试验. 科技尚品 **3**, 23(2016)
2. Mehdikhani, M., Gorbatikh, L., Verpoest, I., et al.: Voids in fiber-reinforced polymer composites: a review on their formation, characteristics, and effects on mechanical performance. J. Compos. Mater. **53**(12), 1579–1669 (2019)
3. Hcab, C., Slac, D., Jw, E., et al.: A focused review on the thermo-stamping process and simulation progresses of continuous fibre reinforced thermoplastic composites. Compos. B Eng. **224**, 109196 (2021)
4. Birt, E.A., Smith, R.A.: A review of NDE methods for porosity measurement in fibre-reinforced polymer composites. OR Insight **46**(11), 681–686 (2021)

5. Hernandez, S., Sket, F., Gonzalez, C., et al.: Optimization of curing cycle in carbon fiber-reinforced laminates: Void distribution and mechanical properties. Compos. Sci. Technol. **85**, 73–82 (2013)
6. Nikishkov, Y., Airoldi, L., Makeev, A.: Measurement of voids in composites by X-ray Computed Tomography. Compos. Sci. Technol. **89**, 89–97 (2013)
7. Standard Test Methods for Void Content of Reinforced Plastics. ASTM
8. Hamidi, Y.K., Aktas, L., Altan, M.C.: Three-dimensional features of void morphology in resin transfer molded composites. Compos. Sci. Technol. **65**(7–8), 1306–1320 (2005)
9. Patel, N., Rohatgi, V., Lee, L.J.: Micro scale flow behavior and void formation mechanism during impregnation through a unidirectional stitched fiberglass mat. Polym. Eng. Sci. **35**(10), 837–851 (2010)
10. Liu, L., Zhang, B.M., Wang, D.F., et al.: Effects of cure cycles on void con-tent and mechanical properties of composite laminates. Compos. Struct. **73**(3), 303–309 (2006)
11. Springer, G.S., Tang, J.M., Lee, W.I.: Effects of cure pressure on resin flow, voids, and mechanical properties. J. Compos. Mater. **21**(5), 421–440 (1987)
12. Olivier, P., Cottu, J.P., Ferret, B.: Effects of cure cycle pressure and voids on some mechanical properties of carbon/epoxy laminates. Composites **26**(7), 509–515 (1995)
13. Hernandez, S., Sket, F., Molina-Aldareguia, J.M., et al.: Effect of curing cycle on void distribution and interlaminar shear strength in polymer-matrix composites. Compos. Sci. Technol. **71**(10), 1331–1341 (2011)
14. Guo, Z.S., Liu, L., Zhang, B.M., et al.: Critical void content for thermoset composite laminates. J. Compos. Mater. **43**(17), 1775–1790 (2009)
15. Yoshida, H., Ogasa, T., Hayashi, R.: Statistical approach to the relationship between ILSS and void content of CFRP. Compos. Sci. Technol. **25**(1), 3–18 (1986)
16. Wisnom, M.R., Reynolds, T., Gwilliam, N.: Reduction in interlaminar shear strength by discrete and distributed voids. Compos. Sci. Technol. **56**(1), 93–101 (1996)
17. Kosmann, N., Karsten, J.M., Schuett, M., et al.: Determining the effect of voids in GFRP on the damage behaviour under compression loading using acoustic emission. Compos. Part B **70**, 184–188 (2015)
18. Carraro, P.A., Maragoni, L., Quaresimin, M.: Influence of manufacturing induced defects on damage initiation and propagation in carbon/epoxy NCF laminates. Adv. Manuf. Polymer Compos. Sci. **1**(1), 44–53 (2015)
19. Stamopoulos, A., Tserpes, K., Prucha, P., et al.: Evaluation of porosity effects on the mechanical properties of carbon fiber-reinforced plastic unidirectional laminates by X-ray computed tomography and mechanical testing. J. Compos. Mater. **50**(15), 2087–2098 (2016)
20. Lei, Y., Yan, Y., Liu, Y., et al.: Microscopic failure mechanisms of fiber-reinforced polymer composites under transverse tension and compression. Compos. Sci. Technol. **72**(15), 1818–1825 (2012)
21. Vajari, D.A., Gonzalez, C., Llorca, J., et al.: A numerical study of the influence of microvoids in the transverse mechanical response of unidirectional composites. Compos. Sci. Technol. **97**(16), 46–54 (2014)
22. Vajari, A.D.: A micromechanical study of porous composites under longitudinal shear and transverse normal loading. Compos. Struct. **125**, 266–276 (2015)
23. Li, B., Zhao, M.Y., Wan, X.P., et al.: Influence of irregular-void on transverse tensile mechanical properties of composites. Acta Materiae Compositae Sinica **36**(2), 356–361 (2019)
24. Jiayue, W., Yin, Y., Zhongcheng, M., et al.: Peridynamic meso-scale modeling for degradation in transverse mechanical properties of composites with micro-void defects. Acta Mech. Solida Sinica **35**(5), 813–823 (2022)
25. Taotao, Z., Ying, Y.: Micromechanical analysis of transverse damage of fibre-reinforced composites. Compos. Interfaces **23**(1), 75–88 (2016)

26. Mark, R.G., Ghodrat, K.: Finite element micromechanics for stiffness and strength of wavy fiber composites. J. Compos. Mater. **38**(4), 273–292 (2004)
27. Madenci, E., Barut, A., Phan, N.D.: Peridynamic unit cell homogenization. In: 58th AIAA/ASCE/AHS/ASC Structures, Structural Dynamics, and Materials Conference (2017)
28. Jiayue, W.: A Peridynamic analysis for effects of micro-void on transverse mechanical properties for unidirectional composites. Shanghai Jiao Tong University, Shanghai (China) (2022)

A Review of Some Key Issues in CFD-Based Throughflow Simulation

Zhihao Hu[1], Qitian Tao[1], and Xiaohua Liu[1,2](✉)

[1] Shanghai Jiao Tong University, Shanghai, China
{alanryen,sjtu-tqt,Xiaohua-Liu}@sjtu.edu.cn
[2] Key Laboratory (Fluid Machinery and Engineering Research Base) of Sichuan Province, Chengdu, China

Abstract. Throughflow simulation method is the backbone of compressor design and analysis. The CFD-based throughflow is a method developed in the recent decades. Comparisons between traditional throughflow method and CFD-based throughflow method are concerned in the present paper, and especially the development of the CFD-based method has been summarized. Some key issues of the CFD-based method are present. Modeling of the discontinuity model at leading and trailing edge, the flow losses and deviation angle are three key issues of the CFD-based method. Empirical flow losses and deviation angle models may not be accurate for advanced compressor throughflow simulation. New method of getting new empirical model is thus discussed. Four discontinuity models, the automatic incidence correlation model, the momentum flux discontinuity model, actuator disk model, and arbitrary discontinuity breakdown model, are presented, compared and the scope of application of each are discussed.

Keywords: Axial compressor · CFD-based Throughflow · Leading and Trailing Edge Discontinuity · Deviation Angle · Loss

1 Introduction

The numerical simulation method is in of the research methods which wildly used in many areas. The design and analysis of the multistage axial compressor involves numerical simulation. Since Wu (1952) proposed the general three-dimensional theory, the numerical simulation of the axial compressor flow field has been simplified the inviscid flow to a two kinds of relative stream surface iteration problem. As shown in the Fig. 1, the two kinds of stream surfaces are the S1 stream surfaces, blade-to-blade plane, and the S2 stream surfaces, meridian plane, respectively.

Throughflow calculation calculate the flow on the meridian surface developed based on the Wu's theory. Nowadays, three kinds of throughflow methods have developed and used, which are the stream function method, the streamline curvature method and CFD-based throughflow method. Throughflow simulation computer programs based on each method for either design and analysis problem came into being.

Z. Jing and X. Zhan (Eds.): ICASSE 2023, LNEE 1153, pp. 159–177, 2024.
https://doi.org/10.1007/978-981-97-0550-4_13

Even with the booming of computation ability and the development of the numerical method, the throughflow method is still the backbone of compressor design and analysis and wildly used in the axial multistage compressor design and analysis process (Denton and Dawes 1999). The aerodynamic design is a repetitive process of design and numerical analysis, it maybe dozens even hundreds round of test and adaptive before the compressor finally complete. Use the full three-dimensional (3D) simulation at the primary stage of design consumes a lot of time, which not only extend the design cycle, but also expends a plenty of computational resources. Therefore, the throughflow simulation is still a useful tool at the primary stage of compressor design (Jin 2011).

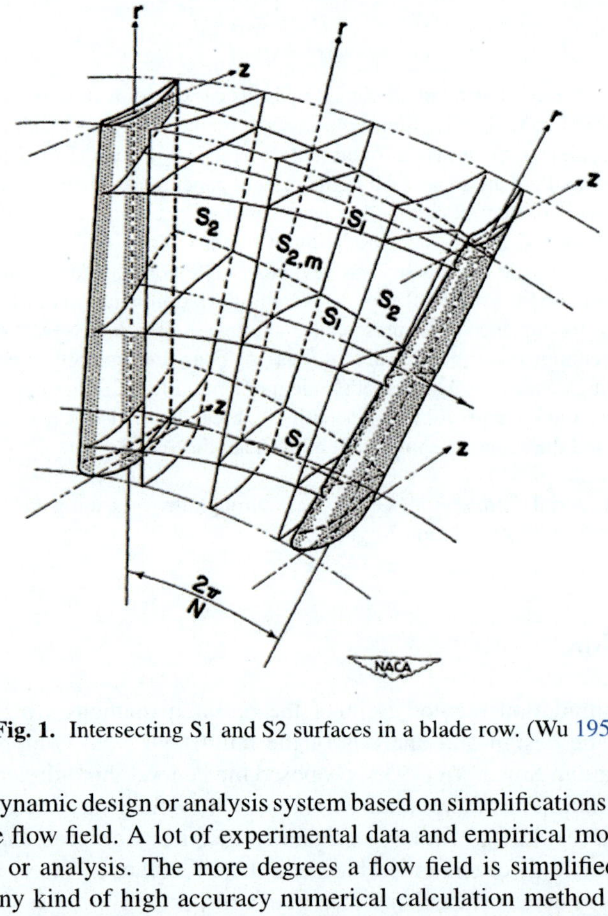

Fig. 1. Intersecting S1 and S2 surfaces in a blade row. (Wu 1952)

Any aero dynamic design or analysis system based on simplifications used hypothesis to simplify the flow field. A lot of experimental data and empirical models are applied for the design or analysis. The more degrees a flow field is simplified the more data are needed. Any kind of high accuracy numerical calculation method may fail owing to the mismatch between current problem with the previous data and empirical models. Meanwhile, some of the empirical models are based on the experimental results of low-load and low-speed cascade experiment. With the stage load and inlet Mach number of the compressors increasing, and the new blade shapes like lean and sweep are being used, the accuracy of the model is challenged. To get higher prediction accuracy of the flow-field in the throughflow, the feasibility of neural network method applied in the acquiring the up-to-date empirical models are discussed.

The CFD-based throughflow method using the circumferential averaged Reynolds averaged Navier-Stokes throughflow equations have a more complete set of control equation comparing to the traditional streamline curvature throughflow method. The more complete the equation a throughflow method uses, the less empirical inputs it demands. Meanwhile, for transonic flow simulation, the CFD-based throughflow is capable of maintaining the stability of the control system. In consequence, the CFD-based throughflow method is more and more welcomed among the compressor designers.

Losses and deviation angle have long been the focus among the throughflow method. Since CFD-based throughflow method uses a different set of equations compared with the classical streamline curvature method, the derivation and models of the losses and deviation angle in CFD-based throughflow method should be introduce.

2 Three Throughflow Methods

2.1 Stream Function Method

Marsh (1966) firstly conducted the stream function throughflow method use a matrix method to solve the axisymmetric function. The stream function method transformed the differential form of stream function into the difference form by the finite difference approximation. The result showed the matrix method can provide a good prediction for the axial velocity profile. However, the stream function method in not wildly used for turbomachinery for its incapable to solve the transonic function and it cannot tell form the subsonic solution and supersonic solution (Denton and Dawes 1999). Marsh also confirmed use the matrix to solve the stream function was limited for it requires a high-speed computer with relatively large storage to conduct the analysis. Both factors setback the further the utility of the matrix method on solving the stream function.

2.2 Streamline Curvature Method

Streamline curvature throughflow method is a very popular throughflow method. The streamline curvature method introduces the concept of streamline curvature, which assume the location of the initial streamline, iterate the radial equilibrium equations and the equations of conservation of mass and energy. Such method has advantages of clear physical concept, simple formulars, and easily to program. In consequence, the streamline curvature method is wildly used in the engineering field. Novak (1967) is the pioneer to develop the throughflow simulation program using streamline curvature method. Several simplify assumption has made in streamline curvature method. (1) flow is adiabatic. (2) flow is steady. (3) the viscous forces are neglected in the momentum equation, the loss is presented by the increase of the entropy. The entropy increases are calculated by the total pressure recovery coefficient and efficiency. (4) the mass flow coefficients are introduced owing to the influence of viscous. (5) the fluid is assumed as the prefect gas. The meridional distribution of streamline is shown in Fig. 2. However, the streamline curvature has a major setback. When calculate supersonic flow field, the channel may be partially or completely choked. The maximum flow rate is only decided by stagnation temperature, stagnation pressure and transaction area. When the mass flow rate does not match such relation, the simulation will be divergent.

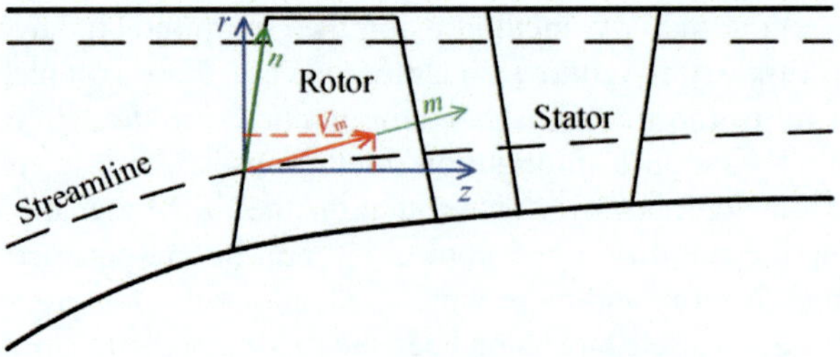

Fig. 2. Meridional distribution of the streamline. (Fei 2021)

Owing to the assumptions of the stream function methods and streamline curvature method, a lot of empirical models are applied. To lessen the dependent of the empirical models, throughflow methods with high order of dimensions of control equations should be investigated.

2.3 CFD-Based Throughflow Method

With the development of computer and numerical method, the CFD-based throughflow simulation program came into being. The CFD-based throughflow method, also called time-marching throughflow. This kind of model is based on Euler equation or Navier-Stokes(N-S) equations and solved by time-marching finite volume method. The CFD-based throughflow method has the following capabilities. (1) this method can predict the mass flow rate at the choke state for the mass flow is calculated form the flow field. (2) it can capture some kind of shock wave. (3) the method can be used for subsonic, transonic, and supersonic simulation simultaneously. (4) Owing to the time-marching method, the throughflow simulation inherent unsteady (Jin 2011).

The first CFD-based throughflow program was conducted by Spurr (1980). Spurr use the time-marching throughflow program to solve the circumferential averaged Euler equation and iterative with a blade-to-blade simulation program to perform simulation on a transonic nozzle. His result matches well with that from the full three-dimensional Euler method. Dawes (1992) modified a three-dimensional Navier-Stokes solver to make it possible to solve blade rows in multistage compressors. Dawes also explored the feasibility of combination of full three-dimensional solver and throughflow solver. Yao and Hirsh (1995) added body force, friction force and introduced losses and deviation angle data for each blade subroutines to a three-dimensional Euler/N-S solver and extend the model to design problem. And they discussed the effect of flow angles and blockage distribution on the blade rows. Body-force term represent blade force and viscous losses in the Euler solver for design and analysis, which is capable of solving subsonic, transonic and supersonic flow with a same set of code as well as capture shock (Damle 1996). Baralon et al. (1998) modeled the profile losses, deviation angle, end-wall skin friction in Euler throughflow model and imported the spanwise mixing model developed by Gallimore (1986). Their model treats the axisymmetric shock as normal blade passage shock through their theoretical and numerical study. They also investigated the method of shock wave capture and proposed a novel method to calculate the cascade blockage factor. They use a three-stage transonic fan to validate the models, the validation result showed the solver is capable of providing a meridional picture of transonic flow for different operating points. Pacciani et al. (2016) develop an axisymmetric Euler solver to naturally fit the secondary and tip leakage effects into throughflow solvers.

Throughflow simulation models based on the circumferential averaged Euler equation neglect the shear stresses. Losses are presented by total pressure drops and entropy changes (Fay et al. 1999). To better represent the actual flow field, more through elements of the flow field should be take into consideration. In consequence, the time-marching Navier-Stokes throughflow method are researched.

Fay et al. (1999) developed a method numerically solve the viscous equations of the meridional flow. They also model the secondary flow due to vortices lead by the tip clearance. And by radial mixing approach, the incoming shear layer vorticity are added. Their model is capture to represent the important feature of viscous flow inside the cascade. Sturmayr and Hirsch (1999) extend a multigrid N-S solver to include throughflow model, investigated shock and associated losses. In their solver, the blades are described by distributed profile blade thickness, deviation angle, and losses. Their model, capture quasi-normal shocks in analysis mode and axisymmetric shock in design mode. Simon et al. (2007) detailed develop a N-S throughflow solver based on the circumferential averaged N-S equation. Their solver use circumferential averaged Reynolds averaged N-S equations. In their throughflow model both Reynolds stress term from Reynolds average and circumferential unevenness term from circumferential average are included. They also discussed the influence of blade force, circumferential stress, and circumferential unevenness term to the flow field. They found the inviscid blade force have the largest impact, the circumferential unevenness secondary to it, and the viscous blade force take the least effect. Taddeo and Larocca (2008) continuing conduct the research on the time-marching throughflow. They applied the distributed loss model and actuator model in the time-marching throughflow. The distributed loss model acts as a compact blade force that turns the flow without producing the unphysical entropy. They also explore the feasibility of combining the time-marching throughflow and empirical models. They found the combination can not only increase the numerical accuracy, but also apply the existing compressor design experiences.

In China, Ji et al. (1999) are the pioneer in researching the CFD-based throughflow. They explored the feasibility of time-marching N-S throughflow in a high load transonic turbomachinery, and verified by design and analysis of an axial supersonic inflow shock rotor. They also announced, the time-marching method inherited unsteady term which can be used for the exploring of the flow stability. Jin (2011) developed a time-marching throughflow solver, which can do both time-marching Euler simulation and N-S simulation. The inviscid blade force is associated with the circumferential pressure, and the inviscid blade force is associated with viscous stress. Jin also declared that at the off-design result from the circumferential averaged N-S simulation differ a lot from the full three-dimensional results owing to the loss and deviation applied for the off-design points have not been fully investigated. Wan et al. (2013) investigated high loaded transonic single-stage fan a four-stage fan flow field performance by both Euler and N-S throughflow solver. By comparing with three-dimensional simulation and experimental results, the throughflow models can provides a credible performance characteristics and better prediction in end-wall regions. Wu (2019) based on the time-marching N-S throughflow, developed the incident, deviation angle, losses and spanwise distribution model at the design point for analysis cases and design point deviation angle and losses model at off-design points. Yang et al. (2019) develop an off-design performance analysis solver, employ novel inviscid blade force model to achieve desired flow deflection, use cubic spline interpolation to solve the discontinuity problems at the leading and trailing edge of the blade rows, and integrated empirical loss model to simulate real three-dimensional viscous force effects. Liu et al. (2021) uses a liner cascade based on N-S throughflow and NUMECA to modify the inviscid blade force, which previously based on the assumption that flow thorough the averaged stream surface with no entropy increasing. Their model decreases 4 times of the prediction error of the adiabatic efficiency than the previous method. Tao et al. (2022) use a distributed inviscid blade force and viscid blade force to reproduce the flow deflection and loss and use uniform inlet 3-D simulation results to interpolate the deviation angle and loss parameters. Then they computed a multistage fan and booster with inlet radial and circumferential distortion.

The time-marching throughflow has developed for about 30 years, and has been widely used in the primary design and analysis phase of the compressor. Several issues have not yet been solved, which will be discussed in the next section.

2.4 Several Issues of CFD-Based Throughflow Method

As conclude by Yang et al. (2017), there are several issues should be further investigated to set up the high accuracy time-marching throughflow method.

Firstly, standard radial blockage factor calculation should be set up to avoid artificial adjustment for getting the accuracy geometry description of the flow field to reflect the physical blockage phenomenon more accurately. At present, the blockage factor has two kinds. One is the circumferential blockage factor, directly derived from the circumferential averaged N-S equation, has a significant influence on the simulation of the mass flow rate and efficient at the choke state. The other is the normal blockage factor, which is based on the geometry of the blade cascade, which can be applied for the mean flow surface throughflow.

Secondly, boundary layer thickness model should be added to get a much accurate reflection of the blockage. Meanwhile, the relation between the boundary layer thickness and the blade force should be further investigated.

Thirdly, for transonic/supersonic flow, the capture of the shock wave should be further investigated, the incapability of capturing shock wave other than the normal shock wave in both design and analysis, which will not being able to simulate all kinds of shock in the flow field.

Fourthly, the flow discontinuity at the leading edge and trailing edge should be further considered. At the leading and trailing edge, instantaneously turn of flow happened. Sever methods have been proposed and discussed. This issue will be fully discussed in next chapter.

3 Leading and Trailing Edge Discontinuity Models

The incidence at the leading edge and deviation at the trailing edge is a key issue of the CFD-based throughflow simulation. Owing to the axisymmetric hypothesis, the CFD-based throughflow simulation cannot sense the blade at the off-blade area to get the pre-rotational velocity. However, once enter the blade region the flow must tangent to the stream surface, the circumferential velocity is constrained by the averaged blade flow surface angle, hence velocity discontinuity at the grid interface of blade leading edge happens. If the blade force term is used to simulate the velocity transition of the leading edge, the sudden change of the volume force will occur, even if the mesh is continuously densified, it will not help. In practice, it is found that as long as the incidence of the incoming flow at the leading edge is greater than 2°, there will be obvious numerical loss. However, the entropy increase corresponding to this kind of loss is non-physical. For the trailing edge of the blade, similar discontinuities will also occur.

3.1 Automatic Incidence Correction Model

Baralon proposed the automatic incidence correction technique (Baralon et al. 1998). In this technique, the blade angles of the first 20% of the chord are modified linearly, parabolically in (x, rθ)coordinates, so that as to adapt progressively to the flow angle upstream of the leading edge. This approach can be defined conceptually as introducing a length scale into the singularity problem. Indeed, when there is very little incidence, the blade force applied to the cell just downstream of the leading edge do not cause any oscillation thereby no entropy. Hence, it appeared natural to solve the singularity approach by applying a progressive change in blade angles, in volume forces. The incidence correction is applied periodically during the time-marching of the computation. The computed flow then adapts itself smoothly to the changes caused by the incidence correction.

The blade shape modifications may have a local effect on the blade loading, but the uncertainty in blade loading is large anyway because the mean line constraint of the throughflow approach is already only an approximation of the real mean flow. Furthermore, the turning that is of interest for performing the work is the turning that is of the flow itself and not of the blade.

Nevertheless, even if the automatic incidence correction is very efficient for subsonic flows, this approach may cause problems for supersonic flows in terms of stream tube are variation.

In case of significant incidence, the modification of the blade geometry at the leading edge may trigger a strong normal shock which is naturally followed by a second one for reasons of stream tube variation. This behavior could be associated with the real blade-to-blade flow of typical transonic rotor blades where, for high mass flows, a first bow shock occurs at the leading edge followed by a weaker normal shock. However, the blade-to-blade bow shock is usually an oblique shock. Therefore, it is believed that the two strong normal shocks obtained in the throughflow solution may correspond to an overproduction of shock losses in this case. It should be emphasized that the problem of successive shocks is depending on the upstream Mach number, the blade geometry and the local incidence. It may not arise in a different configuration.

3.2 Momentum Flux Discontinuity Model

In order to exert a full control on the leading-edge flow without modifying the blade shape, an approach based on a discontinuity in momentum is proposed (Baralon et al. 1998).

The conservation of mass and energy are respected but additional source terms are introduced in the momentum equations in order to control both the flow angle downstream of the discontinuity and generation of entropy.

The technique can be divided in two stages: first, we must derive the target forces that will be included in the momentum fluxes to obtain the desired conditions at the discontinuity. Afterwards, we must derive at the flux calculation level the flow state downstream of the leading edge with the forces included.

This technique has been validated on subsonic and supersonic applications using a quasi-one-dimensional code with blade modelling included. Mach number as high as 3 and incidence angles up to 30° were computed successfully with this 1D code. Computed total pressure and leading-edge flow angle conditions were exactly equal to those specified.

This method solves the leading and trailing edge problems through the discontinuous relationship, but there are still problems in practical application, that is, the iterative solution of flow parameters is difficult in mathematics, especially when the Mach number is close to 1, so it cannot be applied to transonic flow, such as transonic rotor. In addition, this method also requires manual input of additional parameters, namely total pressure loss.

3.3 Actuator Disk Model

An isentropic flow through a linear cascade can be described using 1D Euler equations with a distribution of inviscid blade force that is responsible for turning the flow (Taddei and Larocca 2014a, 2014b). A further source term includes blade blockage effects through a prescribed blockage factor, $h(x) = 1 - \sigma_x \Delta z(x)/(x_{TE}\text{-}x_{LE})$. The equations have the divergence form

$$\frac{\partial W}{\partial t} + \frac{\partial F_c}{\partial x} = Q_i + Q_h \tag{1}$$

where

$$W = \begin{Bmatrix} \frac{\rho}{\rho V_x} \\ \frac{\rho V_z}{\rho E^0} \end{Bmatrix} F_c = \begin{Bmatrix} \frac{\rho V_x}{p+\rho V_x^2} \\ \frac{\rho V_x V_z}{\rho V_x H^0} \end{Bmatrix} Q_i = \begin{Bmatrix} \frac{0}{f_{ix}} \\ \frac{f_{iz}}{\vec{f_i} \cdot \vec{v}} \end{Bmatrix} Q_h = -\frac{\rho V_x}{h}\frac{\partial h}{\partial x}\begin{Bmatrix} \frac{1}{V_x} \\ \frac{V_z}{H^0} \end{Bmatrix}$$

A given streamline $z(x)$ is tangent to the relative flow velocity and orthogonal to the inviscid blade force vector

$$\frac{\partial z}{\partial x} = \frac{V_z - V}{V_x} = -\frac{f_{ix}}{f_{iz}} \tag{2}$$

Equations (2) close the system in Eq. (1) together with the usual thermodynamic relations. Outside of the cascade, since both the source terms vanish, a uniform flow solution satisfies Eq. (1) and the flow angle is only determined by the upstream conditions. Inside of the cascade, if streamline $z(x)$ is assumed to coincide with the blade camber line, the relative flow angle is constrained by the blade angle. Whenever the incoming flow angle does not match the LE angle, a discontinuity occurs at the LE and unphysical production of entropy will appear in the numerical solution of Eq. (1). Furthermore, the flow leaves the cascade at the TE blade angle and no deviation can be introduced. The most common method to overcome the drawback follows a distributed approach. A modified streamline geometry is assumed, which (i) fits the direction of the incoming flow and (ii) gives the outgoing flow the desired deviation from the TE blade angle.

The main idea underlying the proposed model is to concentrate all the incidence and deviation at the LE and TE, respectively, with no production of unphysical entropy. This can be done by treating the edges as ADs. Due to hyperbolic nature of the system in Eq. (1), the most physically consistent approach to its time-marching solution is to recast it in the form of a conservation law and integrate this form using upwind finite-volume schemes. For the purpose of the present work, a flux difference splitting technique is adopted. The numerical flux vector at a generic interface between two contiguous cells is evaluated using an approximate Riemann solver. Let $i + 1/2$ be the interface at the LE and TE. A modified Riemann solver is adopted, which places a discontinuity exactly at the interface (Fig. 3). The relations of a quasi-steady compressible AD are assumed to be valid across the discontinuity.

$$
\begin{aligned}
&\rho_1 V_{x1} = \rho_2 V_{x2} \\
&\frac{a_1^2}{\gamma - 1} + \frac{V_{x1}^2 + (V_{z1} - v)^2}{2} = \frac{a_2^2}{\gamma - 1} + \frac{V_{x2}^2 + (V_{z2} - v)^2}{2} \\
&S_1 = S_2 \\
&V_{z2} = v + V_{x2} \tan\beta
\end{aligned}
\tag{3}
$$

where subscripts 1 and 2 refer to the flow states upstream and downstream of the AD, respectively. The first, second, and third equations of the system in Eq. (3) prescribe conservation of mass flow rate, relative total enthalpy, and entropy between states 1 and 2 (note that $h = 1$ at the LEs and TEs). In particular, the third equation avoids production of entropy across the interface. The fourth equation forces the relative flow in state 2 to take a specified β angle. At the LE, this angle will be the LE blade angle. At the TE, it will be the TE blade angle plus the desired deviation. Equation (3) is combined with the eight compatibility equations of the approximate Riemann solver across waves $V_x - a$ and $V_x + a$ and contact surface V_x. In particular velocity V_z is assumed to be simply conveyed, together with entropy, along contact surfaces. If U stands for the primitive variable vector $\{a\ V_x\ V_z\ S\}$ T, one obtains a nonlinear system that provides U_1 and U_2 for a given β angle. As in the conventional Riemann problem first- or higher-order approximate solutions can be performed. Second-order accuracy is achieved, in the spirit of ENO approaches, through definition of a linear distribution of the primitive variables over the cells and use of slope limiters. Equation (3) involves the assumption of an unchoked flow across the AD. However, Eq. (3). Can be modified to treat choked flows. Due to its importance, this topic will be addressed in a separate contribution.

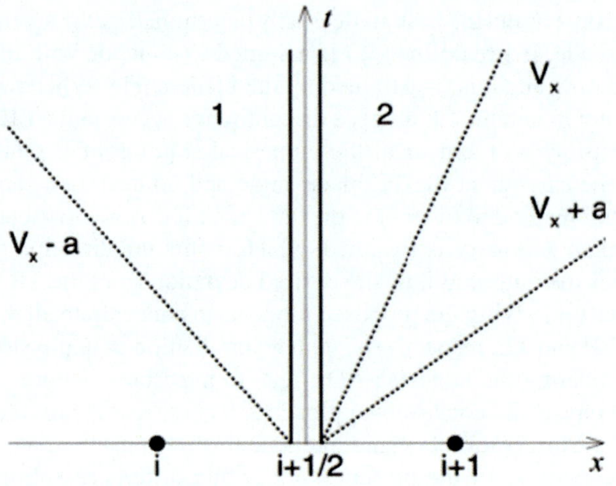

Fig. 3. Representation of an AD at a LE or TE (Taddei and Larocca 2014a, 2014b).

3.4 Arbitrary Discontinuity Breakdown Model

The equations describing gas flows in bladed subdomains and in axisymmetric regions are different. What is more, small perturbances propagate in these subdomains with different velocities (Nigmatullin et al. 1994). Indeed, in bladed regions local acoustic perturbances propagate in the direction perpendicular to grid lines $\xi = const$ with the velocities $\frac{U_\xi}{C_\xi} \pm a$.

It is evident that, first, these boundaries are in common case the discontinuity surfaces and secondly, it is necessary to set enough boundary conditions at these surfaces which depends on number of coming and leaving perturbances. In order words it is necessary to consider an arbitrary discontinuity breakdown problem at these boundaries.

Let us first consider the boundary corresponding to leading edges. A treatment of the boundary conditions means that the usual arbitrary discontinuity breakdown procedure must be placed by the special one described below.

If it is necessary the corresponding right or left limit is used. The parameters at the bladed side will be marked by index 1 and parameters at the axisymmetric subdomain will be marked by index 2. Large left and right parameter (at both sides of the surface) will have the indices L and R.

It's convenient to consider leading triangle drawn at Fig. 5. This triangle may be corresponded to some real triangle in blade-to-blade space. Similar to the trailing edge boundary. Similar to leading edge case one may define trailing edge triangle.

The leading and trailing edge discontinuity processing model is equivalent to modifying the intermediate wave in the general Riemannian problem. The intermediate wave here is no longer a contact discontinuity, but an "intermediate" wave that considers the flow transition and flow picture between L and R, and has the property of entropy conservation under normal conditions. At the same time, the treatment methods for various special cases such as the presence of detached shock wave, leading edge blockage, trailing edge blockage and backflow at the leading edge, so that the leading and trailing edge discontinuity problems can be well handled under any flow state (Fig. 4).

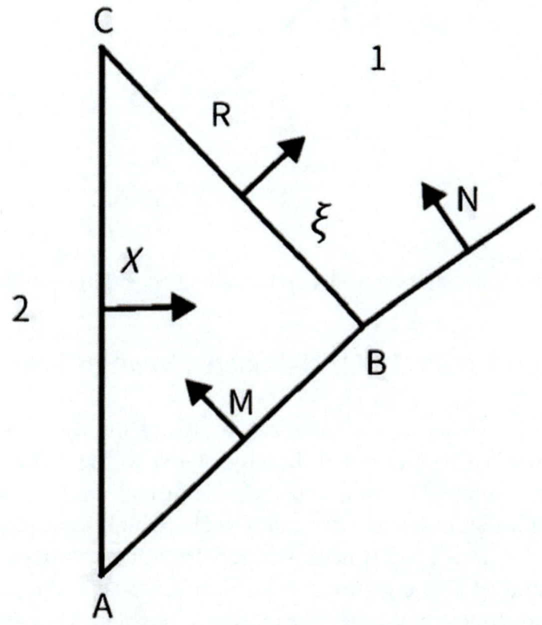

Fig. 4. Arbitrary discontinuity model at the leading edge (Nigmatullin et al. 1994).

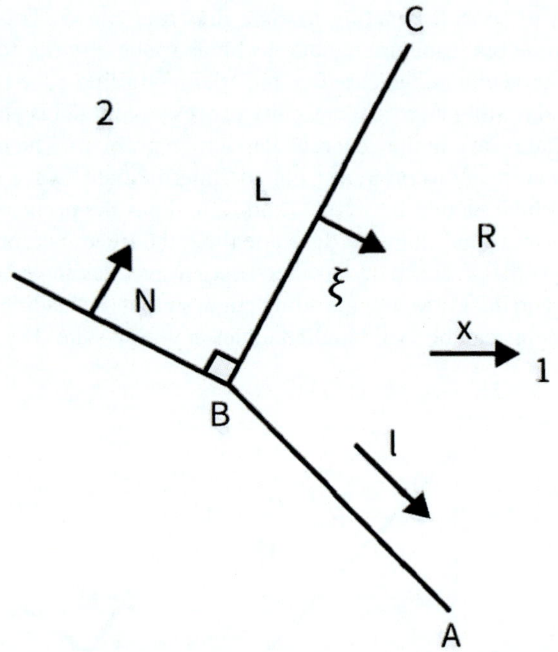

Fig. 5. Arbitrary discontinuity model at the trailing edge (Nigmatullin et al. 1994).

4 Deviation and Losses in CFD-Based Throughflow

Unlike the full three-dimensional simulation, by using throughflow method, the loss and flow deflection can't be simulated directly owing to the reduction of dimension. In consequence, the researchers usually use the empirical model based on the cascade experiment to modify the losses and deviation angle, which represents the deflection.

Most currently developed N-S time-marching throughflow programs are based on the circumferential averaged N-S equations. In the blade passage, the material appearance of the blades does not appear anymore. The effects of the blades are replaced by the force distributions. Usually, flow deflations are modelled by inviscid blade force distribution. While the generated losses are modelled by the distribution of viscous forces. The generated loss is a combination of the blade walls shear stresses and the circumferential stresses (Simon 2007).

4.1 Modeling of Deviation Angle and Losses

At the outlet of blade, the deviation angle is defined as the difference between the flow angle and blade angle. For blade, who has a fixed geometry, the deviation angle represents the flow deflection in the blade region, which represents the work done by the compressor. A good prediction of the deviation angle will lead to much accurate results. As far as the knowledge of author, there are mainly two methods to modify the deviation angle in the time-marching throughflow. One is use inviscid blade force to represent the flow deflation in the blade region (Simon 2007), the other is use an actuator disk model (Taddei and Larocca 2014a, 2014b).

As presented by Simon, in the blade section, several prescriptions have been made to model the deviation angle. Firstly, an orthogonal condition between the blade force and the flow direction are imposed. The flow deflection is generated by the inviscid blade force, which deflects the flow while generating no losses. In the reference frame of the blade, the blade force produces no work. However, a force generates no work and has a component aligned to the flow direction will generate entropy. Therefore, the blade forces are prescribed to be orthogonal to flow. Secondly, as shown in Fig. 6, a relation between the radial component and the circumferential components of the blade force are prescribed by the angle between the normal to the camber surface and the axial direction. Thirdly, the prescribed intensity of the blade force can be obtained by analysis formulation, which take the advantages in the usage of only blade geometry and the robustness in magnitude the blade force than the design formulation and analysis formulation from the circumferential momentum.

Taddei and Larocca (2014a, 2014b) apply the compressible actuator disk equations to modify the evaluation of the deviation angle. In the blade region, by the assumption of the throughflow, the relative flow angle constrains to the meridional streamline direction surface geometrical angle, while in the passage region, no blade force exists. This arises discontinuity of the flow at the leading and trailing edge of the blade. The actuator disk applied at both the leading edge and the trailing edge of the blade instantaneously turns the flow while not producing unphysical entropy. This method avoids unphysical incidence owing to the discontinuity at the leading edge and provides desired deviation angle at the trailing edge, spares handmade modification of the throughflow surface, and allows coping with strong incidence gradient by not specially treating between the inviscid and viscous meridional flows.

Wu (2019) propose a deviation angle model for off-design states, analytically solved the implicit function, which predicts the deviation angle. This model includes no empirical coefficient, and the model explained the relation between the incidence angle and deviation angle, inconsequence, the deviation angle at all spanwise and all working condition can be applied. However, this model is limited to the rotors and stators included in the airfoil database, whose data is already acquired by the experimental.

The generated loss inside the blade passage can be modelled by the distribution of viscous forces. Both the shear stresses acting on the blade wall as well as the circumferential stresses contributes to the loss. The flow losses model and the deviation angle model are respectively the empirical model used for predicting the total pressure loss and outflow direction in the design of compressor. Denton defined the loss in term of entropy increase (Denton 1993). Denton also classified the loss in the compressor into four kinds, blade profile loss due to diffusion on the surface and thickness at trailing edge, end-wall loss due to boundary layer and clearance, leakage loss, and shock loss. While the end-wall loss and leakage loss can be predicted by the loss spanwise distribution model. The shock loss can be solved directly by the time-marching throughflow simulation. In consequence, the profile loss model, which mainly generated form the viscosity friction and flow separation is the only loss model needed to be discussed in throughflow simulation (Wu 2019).

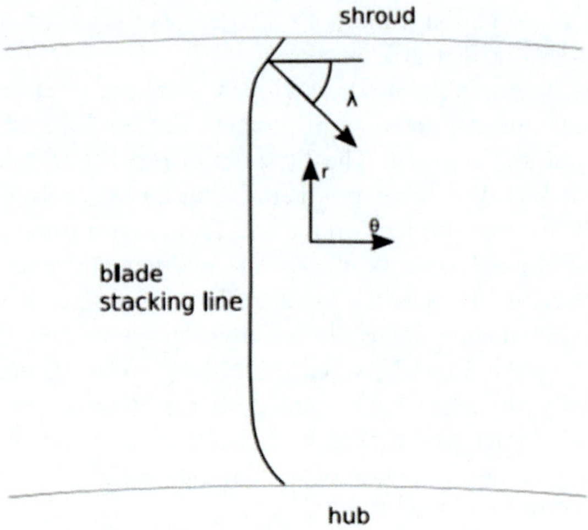

Fig. 6. Definition of the lean angle: camber surface inclination (axial view) (Simon 2007)

Lieblein (1953, 1957) conducted some basic work to predict the profile losses. With the experimental data form low-speed cascade. Lieblein derived the correlation between the momentum thickness and blade diffusion factor, D, or the equivalent diffusion factor, D_{eq}. Lieblein's model is a milestone since then, even though it only valid for low-speed traditional cascade. Koch and Smith (Koch et al. 1976) constructed a more comprehensives losses model based on a large number of cascade, single rotor and compressor experimental result. Their model takes the equivalent diffusion factor, D_{eq}, as a variable and considers the actual flow conditions such as Mach number, Reynolds number, airfoil surface finish, etc. The classical deviation angle model includes the Cater mode (Cater 1950) and Lieblein model (Lieblein 1957). The Cater model is based on the result of the C-4 subsonic airfoil, while the Lieblein model refers the experimental result of the NACA 65 cascade.

By hypothesis two-dimensional flow in the blade-to-blade surface two-dimensional profile losses are generated. The empirical correlations of profiles losses and the deviation angle are computed through design value plus an off-design value.

Deviation angle and losses spanwise distribution model described the effect of the three-dimensional flow phenomenon like end-wall viscosity, leakage flow and secondary flow to the losses and deviation angle Petrovic et al. (2010). Summarized a simplified deviation angle and loss distribution model, which can be applied for reasonably describe the span wise distribution while a leakage of the experimental data.

4.2 Novel Method to Develop Empirical Model

The empirical models discussed above match quit well for classical blade families. However, geometries of modern compressors are generally designed with customized blade, the previous profile loss model may fail to match the up-to-date blade. In consequence, large database with abundant airfoil geometry and flow condition should be used for design and analysis in the throughflow methods. A novel method based the neural network to get the empirical model of deviation angles and losses will be discussed in this section.

A novel approach to get a more compete empirical model is by neural network (Schmitz et al. 2011; Li et al. 2021). The basic idea of this method can be summarized in 3 steps. Firstly, set up a database, the database includes the geometry of airfoil and flow field paraments, both geometry and flow field data form experimental and numerical simulation can be used. Secondly, surrogate models are trained by neural network to obtain the correlation of between loss or deviation angle with geometry and flow field data to get new empirical model. Thirdly, the new empirical models are combined with throughflow simulation models to get the prediction of the flow paraments.

Schmitz et al. (2011) use a database includes 106 randomly created airfoils, and for each airfoil both geometry paraments like the metal angel, maximum profile thickness, leading edge paraments, curvature parameters, chord length, and flow field paraments like relative Mach number, relative inflow angle and Reynolds number are included. Figure 7, show the distribution of the airfoil losses over the Mach number and the inflow angle. They develop a novel method to get loss and deviation angel prediction model for compressor. They pointed out that, the accuracy can be increased globally or just for small areas by adding more sample points to the training. They also confirm that secondary effect flow such as the end-wall boundary layer and spanwise mixing should be further investigated.

Li et al. (2021) use genetic algorithm-back propagation neural network (GA-BGNN). Loss prediction surrogate model at design and off-design condition of the compressor blade span was developed by GA-BGNN. They established a database, which contains several sets of blade element geometry and blade performance data. Considering different working condition of rotor and stator, different database and surrogate model are used. They also take into consideration of the mechanism of different losses generated between the rotor and stator. The prediction results from the surrogate model are compared with the result from traditional empirical results and experimental data, shows the surrogate model matches much better than the traditional model.

With the help of the neural network, a database contains a hug amounts data of the airfoil can be applied for the construction of the empirical model of loss and deviation angle with higher accuracy, which will better reflection of the flow in the blade region and improve the performance prediction accuracy.

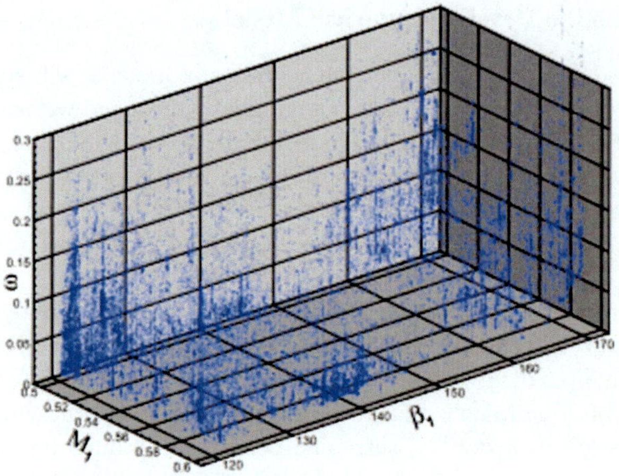

Fig. 7. Distribution of the airfoil losses over the Mach number and the inflow angle (Schmitz et al. 2011)

5 Conclusion

This paper detailed reviewed the development and several issues of the CFD-based throughflow simulation method. The discontinuity at the leading and trailing edge, modeling of loss and deviation are major concerns of the CFD-based throughflow. Several conclusions have been made.

1. Time-marching throughflow method has advantages over the streamline curvature and stream function method in including capable of predicting mass flow rate at the choke state, capturing shock wave, able to perform transonic simulation and inherent unsteady.
2. Treatment of blockage factor, shock waves capture, discontinuity at the leading and trailing edge of the blade rows are some key issues of the accuracy of the time-marching throughflow simulation. These issues should be further investigated and common sense may be discussed among researchers.
3. The discontinuity models at the leading and trailing edge of the blade rows have long been concerned in the throughflow simulation. Automatic incidence correlation model is limited by the incidence, which cannot be too lager, and limited in the subsonic condition. Momentum flux discontinuity model have problem in transonic flow and require manual input of additional total pressure loss parameters. The actuator disk model cannot be applied to blocking conditions and describe the shock wave. Arbitrary discontinuity breakdown model is an analytical solution to the leading and trailing edge discontinuity problem under the physical flow picture, and can be applied to the calculation of various working conditions.
4. Modeling the losses and deviation angle has long been concerned in throughflow simulation. Previous researchers use the cascade experimental data to get the empirical model of Loss and deviation angle. However, with the development of the airfoil and compressor design methodology, those empirical models are not suitable for the modern throughflow simulation. Neural network method with a large airfoil and

flow parameter are maybe a novel approach to get empirical models apply for model throughflow simulation.

Acknowledgements. This work is partially supported by National Nature Science Foundation of China (No. 51976116), Natural Science Fund of Shanghai (No. 19ZR1425900), and the Open Research Subject of Key Laboratory (Fluid Machinery and Engineering Research Base) of Shanghai Province (No. Szjj2019-022), which are greatly acknowledged. The authors also gratefully acknowledge the helpful comments and suggestions of the reviewers, which have improved the presentation.

References

Wu, C.: A General Theory of Three-Dimensional Flow in Subsonic and Supersonic Turbomachines of Axial-, Radial-, and Mixed-Flow Types. NACA TN 2604, pp. 1–90 (1952)

Denton, J.D., Dawes, W.N.: Computational fluid dynamics for turbomachinery design. Proc. Inst. Mech. Eng. C J. Mech. Eng. Sci. **213**(2), 107–124 (1999). https://doi.org/10.1243/0954406991522211

Jin, H.: Application of circumferentional average method in multistage axial fan/compressor design and analysis. Dissertation, Beihang University (2011)

Harry, M.: A Digital Computer Program for the Through-Flow Fluid Mechanics in an Arbitrary Turbomachine Using a Matrix Method. London (1966)

Novak, R.A.: Streamline curvature computing procedures for fluid-flow problems. J. Eng. Power **89**(4), 478–490 (1967)

Spurr, A.: The prediction of 3D transonic flow in turbomachinery using a combined throughflow and blade-to-blade time marching method. Int. J. Heat Fluid Flow **2**(4), 189–199 (1980)

Dawes, W.N.: Toward improved throughflow capability: the use of three-dimensional viscous flow solvers in a multistage environment. J. Turbomach. **114**(1), 8–17 (1992)

Yao, Z., Hirsch, C.: Throughflow model using 3D Euler or Navier-Stokes solver. In: Turbomachinery - Fluid Dynamic and Thermodynamic Aspects, Germany (1995)

Damle, S.V.: Throughflow method for turbomachines using Euler solvers. In: 34th Aerospace Sciences Meeting and Exhibit. (1996)

Damle, S.V., Dang, T.Q., Reddy, D.R.: Throughflow method for turbomachines applicable for all flow regimes. J. Turbomach. **119**(2), 256–262 (1997)

Baralon, S., Eriksson, L.E., Hall, U.: Validation of a throughflow time-marching finite-volume solver for transonic compressors. In: ASME 1998 International Gas Turbine and Aeroengine Congress and Exhibition (1998)

Baralon, S., Eriksson, L.E., Hall, U.: Evaluation of higher-order terms in the throughflow approximation using 3D Navier-Stokes computations of a transonic compressor rotor. In: ASME 1999 International Gas Turbine and Aeroengine Congress and Exhibition (1999)

Gallimore, S.J.: Spanwise mixing in multistage axial flow compressors: Part II—throughflow calculations including mixing. J. Turbomach. **108**(1), 10–16 (1986)

Gallimore, S.J., Cumpsty, N.A.: Spanwise mixing in multistage axial flow compressors: Part I—experimental investigation. J. Turbomach. **108**(1), 2–9 (1986)

Pacciani, R., Filippo, R., Michele, M., Andrea, A., Strfano, C., Federico, D.: A CFD-based throughflow method with an explicit body force model and an adaptive formulation for the s2 stream surface. Proc. Inst. Mech. Eng. A J. Power Energy **230**(1), 16–28 (2016). https://doi.org/10.1177/0957650915607091

Fay, G., Lawerenz, M., Przewozny, H.: Calculation of pitch averaged viscous flow in an annular compressor cascade. Comput. Methods Exp. Meas. IX **21**, 185–194 (1999)
Sturmayr, A., Hirsch, C.: Throughflow model for design and analysis integrated in a three-dimensional Navier-Stokes solver. Proc. Inst. Mech. Eng. A J. Power Energy **213**(A4), 263–273 (1999)
Simon, J.F., Thomas, J.P., Leonard, O.: On the role of the deterministic and circumferential stresses in throughflow calculations. J. Turbomach. Trans. ASME **131**(3), 1–12 (2009)
Simon, J.F., Leonard, O.: Modeling of 3-D losses and deviations in a throughflow analysis tool. J. Therm. Sci. **16**(3), 208–214 (2007)
Simon, J.F.: Contribution to throughflow modelling for axial flow turbomachines. Dissertation. Ingenieur Civil Electro-Mecanicien (2007)
Taddei, S.R., Larocca, F.: Axisymmetric design of axial turbomachines: an inverse method introducing profile losses. Proc. Inst. Mech. Eng. A J. Power Energy **222**(A6), 613–621 (2008)
Taddei, S.R., Larocca, F.: CFD-based analysis of multistage throughflow surfaces with incidence. Mech. Res. Commun. **47**, 6–10 (2013)
Taddei, S.R., Larocca, F.: An actuator disk model of incidence and deviation for RANS-based throughflow analysis. J. Turbomach. **136**(2), 1–9 (2014)
Ji, L., Meng, Q., Zhou, S.: Time-marching method for throughflow computation of turbomachinery. J. Aerosp. Power **14**, 23–26 (1999)
Wan, K., Jin, H., Jin, D., Gui, X.: Influence of Non-Axisymmetric Terms on Circumferentially Averaged Method in Fan/Compressor (2013). https://doi.org/10.1007/s11630-013-0586-1
Wu, D.: Throughflow moldelling and validation of multistage axial flow compressor. Dissertation, Shanghai Jiao Tong University (2019)
Yang, C., Wu, H., Yang, J., Ferlauto, M.: Time-marching throughflow analysis of multistage axial compressors based on a novel inviscid blade force model. Proc. Inst. Mech. Eng. G J. Aerosp. Eng. **233**(14), 5239–5252 (2019). https://doi.org/10.1177/0954410019840588
Liu, X., Wan, K., Jin, D., Gui, X.: Development of a Throughflow-Based Simulation Tool for Preliminary Compressor Design Considering Blade Geometry in Gas Turbine Engine (2021). https://doi.org/10.3390/app11010422
Yang, J., Wang, C., Wang, D., Shao, F., Yang, C., Wu, H.: Time marching based throughflow method: current status and future development. Chin. J. Aeronaut. (**9**), 520996–520996 (2017). https://doi.org/10.7527/s10000-6893.2017.620996
Tao, Q.: An Aerodynamic performance calculation method of fan and booster with inlet distortion. Dissertation, Shanghai Jiao Tong University (2021)
Tao, Q., Jin, H., Liu, X., Zhu, Z.: A fast computational method of the aerodynamic performance of fan and booster with inlet distortion (2022). https://doi.org/10.1177/09544062221115346
Barlon, S., Eriksson, L.E., Hall, U.: Validation of a throughflow time-marching finite-volume solver for transonic compressor. ASME. 98-GT-47 (1998)
Simon, J.F.: Contribution to throughflow modelling for axial flow turbomachines. Dissertation. University of Liege (2007)
Persico, G., Rebays: A penalty formulation for the throughflow modeling of turbomachinery. Comput. Fluids **60**(10), 86–98 (2012). https://doi.org/10.1016/j.compfluid.2012.03.001
Taddei, S.R., Larocca, F.: CFD-based analysis of multistage throughflow surfaces with incidence. Mech. Res. Commun. **47**(47), 6 (2013). https://doi.org/10.1016/j.mechrescom.2012.10.005
Taddei, S.R., Larocca, F.: An actuator disk model of incidence and deviation for RANS-based throughflow analysis. J. Turbomach. **136**(2), 021001 (2014). https://doi.org/10.1115/1.4025155
Nigmatullin, R.Z., Ivanov, M.J.: The mathematical models of flow passage for gas turbine engines and their components (1994)
Denton, J.D.: Loss mechanisms in turbomachines. J. Turbomach. **115**(4), 621–656 (1993)

Lieblein, S.: Analysis of experimental low speed loss and stall characteristics of two dimensional compressor blade cascades (1957)

Lieblein, S., Schwenk, F.C., Broderick, R.L.: Diffusion factor for estimating losses and limiting blade loadings in axial-flow-compressor blade elements (1953)

Koch, C.C., Smith, L.H.: Loss sources and magnitudes in axial-flow compressors. J. Eng. Power **98**(3), 411–424 (1976)

Cater, A.D.S.: The low-speed performance of related aerofoils in cascade, London (1950)

Petrovic, M.V., Wiedermann, A., Manjac, M.B.: Development and validation of a new universal through flow method for axial compressors. Proc. Inst. Mech. Eng. A J. Power Energy **224**(6), 869–880 (2010). https://doi.org/10.1243/09576509JPE991

Schmitz, A., Aulich, M., Nicke, E.: Novel approach for loss and flow-turning prediction using optimized surrogate models in two-dimensional compressor design. In: ASME 2011 Turbo Expo: Turbine Technical Conference and Exposition (2011). https://doi.org/10.1115/GT2011-45086

Li, J., Teng, J., Zhu, M., Qiang, X.: Loss prediction of axial compressors using genetic algorithm–back propagation neural network in throughflow method. Proc. Inst. Mech. Eng. G J. Aerosp. Eng. **0**(0), 1–13 (2021). https://doi.org/10.1177/09544100211041490

Influence of Multiple Parameters on the Efficiency of a Single Nozzle in a Heavy Gas Turbine Combustor

Lei Hu[1,2], Fang Chen[1](✉), Yu Meng[1], Zhenkun Sang[1], and Yuhuang Chen[1]

[1] Shanghai Jiao Tong University, Shanghai, China
{ray-0707,fangchen}@sjtu.edu.cn

[2] China United Gas Turbine Technology, Ltd., Shanghai, China

Abstract. In order to extend the safe, stable and efficient operation range of lean burn combustion technology, an experimental investigation was conducted to get the combustion and emission performance of a singular nozzle including peripheral nozzle and double swirl nozzle. The investigation involves variations of operational parameters such as inlet air pressure, temperature, flow rate, and equivalence ratio. The results demonstrated that all tested nozzles achieved a combustion efficiency exceeding 99.8%, with the majority surpassing 99.9%. At 5 atmospheres, combustion efficiency was greater than 99.99%, which meeting standard requirements. Under partial load conditions, employing a control strategy of a constant equivalent ratio yielded positive benefits for the exit gas temperature but negative effects on emissions. The Peripheral nozzle end outperformed reference end scheme. NOx emissions are expected to increase as intake air temperatures increase.

Keywords: nozzle · combustion · emission · ignition · flameout

1 Introduction

Due to the intrinsic intermittency of wind and solar energy, the power grid based on new energy sources is facing the demand of frequent peak shaving, which requires the heavy-duty gas turbine to have high-efficiency and wide-load operation capability. Otherwise, it is a difficult problem for lean premixed combustion technology to have excellent combustion efficiency and low pollutant emission under low load. The combustion chamber is one of the three core components of a gas turbine, which directly affects the combustion efficiency and emission. The working environment of high temperature, high pressure and high combustion heat intensity put forward more stringent requirements for the design, processing and post-maintenance of the combustion chamber. Therefore, the design and test technology are the core technologies that must be mastered in the development of gas turbines [1]. The fuel nozzle is the key structure of the combustion chamber. Studying the difference in performance of different nozzles is essential to combustion chamber design. Ji [2] evaluated the influence of burner outlet length on pressure loss, velocity field, temperature field, fuel concentration distribution, flame

Z. Jing and X. Zhan (Eds.): ICASSE 2023, LNEE 1153, pp. 178–191, 2024.
https://doi.org/10.1007/978-981-97-0550-4_14

length and combustion efficiency. Yang et al. [1] analyzed the flow characteristic of the combustor fuel nozzle. Ji et al. [3] investigated the fuel supplying performances, spray cone angle changing characters, droplet size and its rules for a group of the injector. Zhao et al. [4] studied the effects of equivalence ratio, pilot fuel flowrate, and pressure on emission characteristics of premixing fuel nozzle in gas turbine combustor. Ramraj H. Sundararaj et al. conducted experiments to study the effect of reducing nozzle area on the performance of turbojet engines. [5] Yong-Ik Hyun et al. analyzed the influence of tip clearance on the turning angle and the pressure loss of turbine nozzles. [6] Wang Bin et al. designed the nozzle structure and analyzed the fuel flow field in the nozzle. [7] Wang Weilong et al. researched and analyzed the nozzle Wear, Ablation and Charcoal Accumulatio. [8] Bing-bing Sun et al. explored the impact of gas temperature on the nozzle damping characteristics [9] Zhixia He et al. analyzed the effects of the nozzle sac volume, orifice inlet curvature, orifice inclination angle, injector needle lift and needle eccentricity on the cavitating flow inside the nozzle. [10] G. Costa et al. studied on the influence of the nozzle throat shape in relation to the uniformity and density of the generated plasma profile. [11] Yuanzhi Tang et al. studied the influence of different nozzle orifice diameters on the trends of spray, ignition and flame diffusion under the flame-wall impinging combustion. [12] But there is still a lack of research on combustion and emission performance of a single nozzle under different operating parameters for a single nozzle, which directly reflects the performance of chamber at part loads.

Therefore, an experimental study was conducted on a combustion test-bed to investigate these factors, including the inlet air pressure, temperature, flow (or velocity), and combustion equivalence ratio.

2 Experimental Facility

To evaluate the performance of nozzles, including peripheral and central nozzles, an experiment was conducted on a heavy-duty gas turbine test bench under partial load conditions. Figure 1 shows the schematic diagram of the experimental system. The compressed air provided by the gas source was first passed through a flow orifice meter, and then heated by a natural gas heating furnace before entering the test piece. The air entered the combustion chamber through the inlet end and the gas-collecting chamber, and then the high-temperature gas was discharged into the atmosphere after being cooled by water.

The test experiment was conducted on a heavy-duty gas turbine test bench, with the primary design of the test section, illustrated in Fig. 2, consisting of a flame tube and a 400 MW nozzle. The gas was transported to the nozzle through an external pipe before being injected into the combustion chamber, where it combines with air to initiate combustion. The schematic of a single nozzle with a swirl angle of 42°. is depicted in Fig. 2.

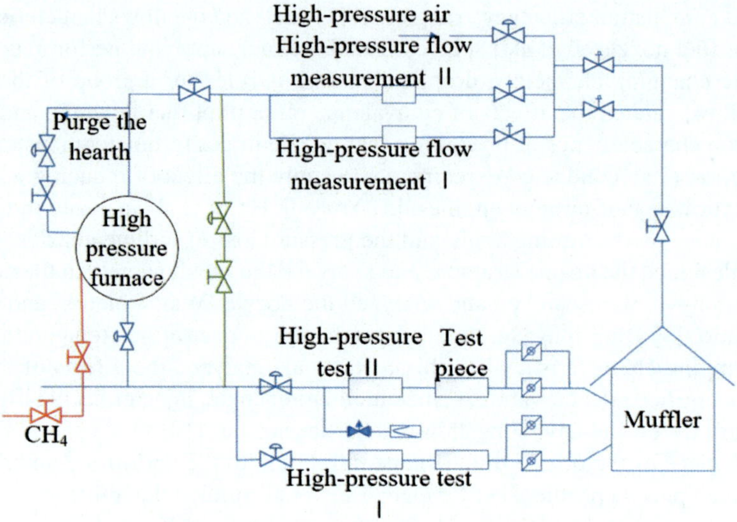

Fig. 1. The experimental system

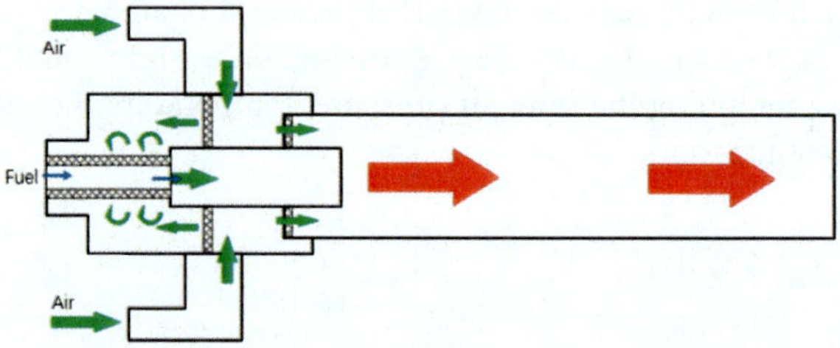

Fig. 2. Schematic diagram of nozzle structure

The test was conducted for the 400 MW peripheral nozzle, the working pressure is 1 atm, 5 atm. The related experimental conditions are shown in Table 1. The outlet of gas temperature was measured by a 7-point B-type thermocouple with an accuracy of ±1.5%. The velocity was calculated by the flow rate which was measured by the flow orifice meters (flow range 0.5–1.8 kg/s and 0.5–5.5 kg/s respectively, accuracy ±1% of full scale). The nitrogen oxide analyzer uses a CAI Model 600 HCLD (measurement range 0–3000 ppm, repeatability ±0.03% of selected scale span) which was based on the principle of HCLD chemiluminescence. The CO analyzer adopts Siemens ULTRAMAT6E based on NDIR principle (measurement range 0–1000 ppm, repeatability ±1% of selected scale).

Table 1. Experimental conditions

Nozzles	Pressure	Swirl angle	Equivalent ratio	Relative load, %	Average velocity, m/s
Peripheral nozzle	1atm	42°	0.55–0.75	40–65	45
				50–80	55
				60–100	65
Double swirl nozzle	1atm			40–65	45
	5atm			50–80	55
				60–100	65

3 Result Analysis

3.1 Performance Analysis of Peripheral Nozzle

3.1.1 The Relationship Between Combustion Efficiency and Equivalence Ratio

The impact of nozzle gas equivalent ratio on combustion efficiency under the condition of air temperature is 550 K and flow velocities are set at 45 m/s, 55 m/s, and 65 m/s respectively. Figure 3 shows the relationship between the combustion efficiency and equivalent ratio. The result shows that under a various premix channel speed and equivalence ratio conditions, the combustion efficiency exceeded 99.9%, and the equivalence ratio is not affected significantly by the air. As the equivalence ratio increase, the efficiency of the combustion declines slightly, corroborating with theoretical analysis. Moreover, at higher equivalence ratio, the flame temperature increases, and the generation of CO also rises. The corresponding results is shown in Fig. 3.

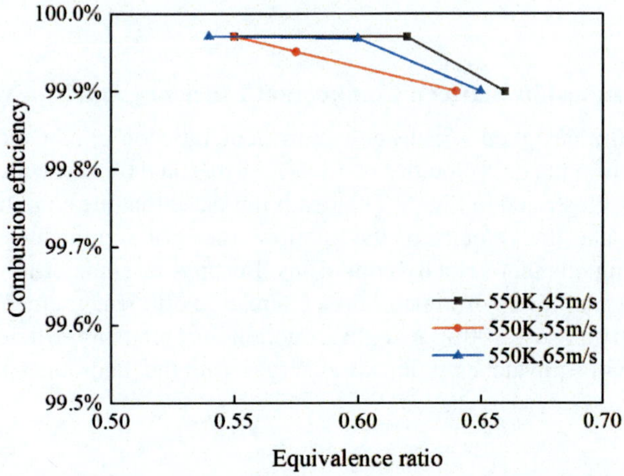

Fig. 3. Relationship between combustion efficiency and equivalent ratio

3.1.2 The Relationship Between Pollutant Emission and Equivalence Ratio

At a specific air velocity, the increase in equivalence ratio will increase the emissions of NOx, UHC and CO. The NOx emission concentration is below 20 ppm, the UHC emission concentration is less than 10 ppm, and the CO emission concentration is less than 70 ppm, as shown in Fig. 4. A high NOx emission concentration was observed under conditions of a large equivalence ratio and low premix channel velocity (45 m/s), which can be attributed to the prolonged residence time in a hot state and high flame temperature. The low combustion pressure and the cooling effect of the water-cooled wall of the flame tube primarily cause the high emission of CO.

3.1.3 The Relationship Between the Outlet Temperature of the Combustor and the Equivalence Ratio

The maximum and average values of the outlet temperature of the combustor are presented in Fig. 5. The figure illustrates that an increase in the velocity of the premixing channel results in a higher outlet temperature of the combustor. The difference between the maximum value and the average value of the combustion chamber outlet temperature is approximately 40K, indicating effective premixing of fuel-air.

3.2 Performance Analysis of Peripheral Nozzle

Initially, the ignition and normal pressure performance tests were conducted to determine the optimal nozzle combination scheme. A comparison between the newly developed scheme and the baseline scheme (050 nozzle + 055 end) was carried out on-site, demonstrating the superior comprehensive performance of the former. Notably, the NOx reduction ranged from approximately 20% to 70%, with an average reduction of 42%. Subsequently, a performance test was conducted at low pressure (5 atmospheres), and the results were presented in conjunction with those obtained during the normal pressure test.

3.2.1 The Relationship Between Combustion Efficiency and Equivalence Ratio

The impact of the designed nozzle gas equivalent ratio on combustion efficiency at different premixing channel velocities of 45 m/s, 55 m/s, and 65 m/s, with air temperature fixed at 550 K is illustrated in Fig. 6. The result indicates that the combustion efficiency exceeds 99.8%, and the velocity of the air flow does not significantly affect it under various premixing channel velocity conditions. Furthermore, under all test conditions, including at 5 atm of air environment, the combustion efficiency remains above 99.8% and is consistently close to 100%. A slight reduction in combustion efficiency is observed with an increase in equivalence ratio, which aligns with the theoretical analysis (Fig. 7).

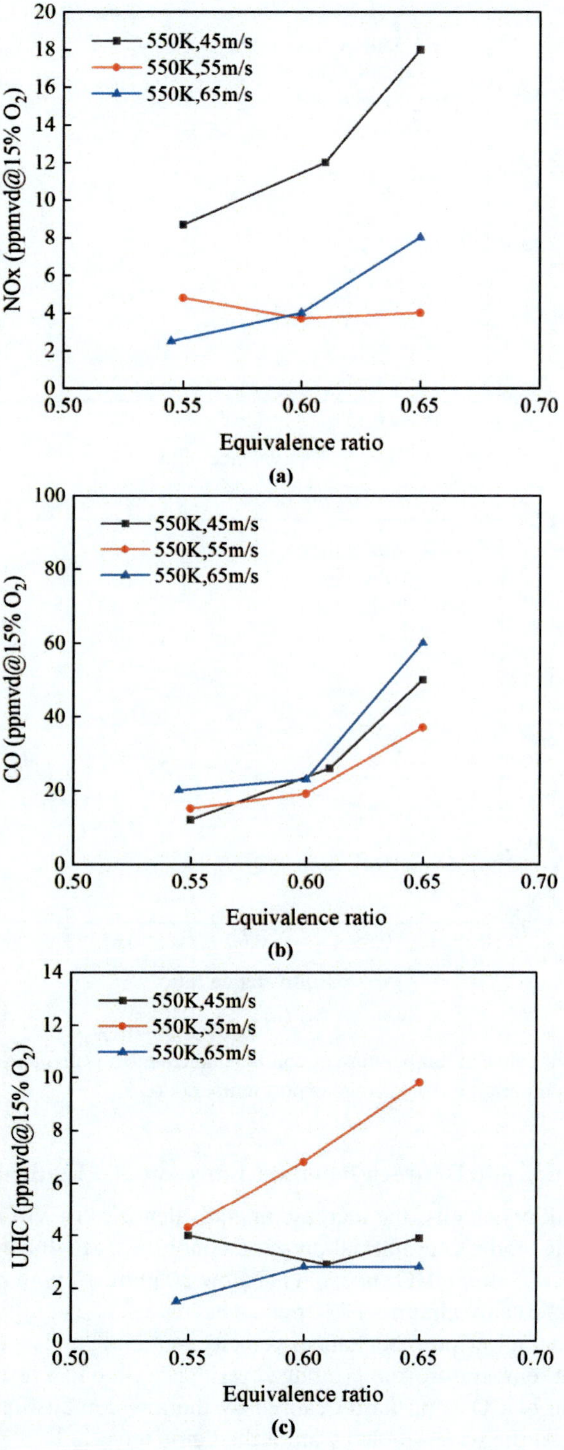

Fig. 4. Relationship between pollutant emission and equivalent ratio (a) NO_X (b) CO (c) UHC

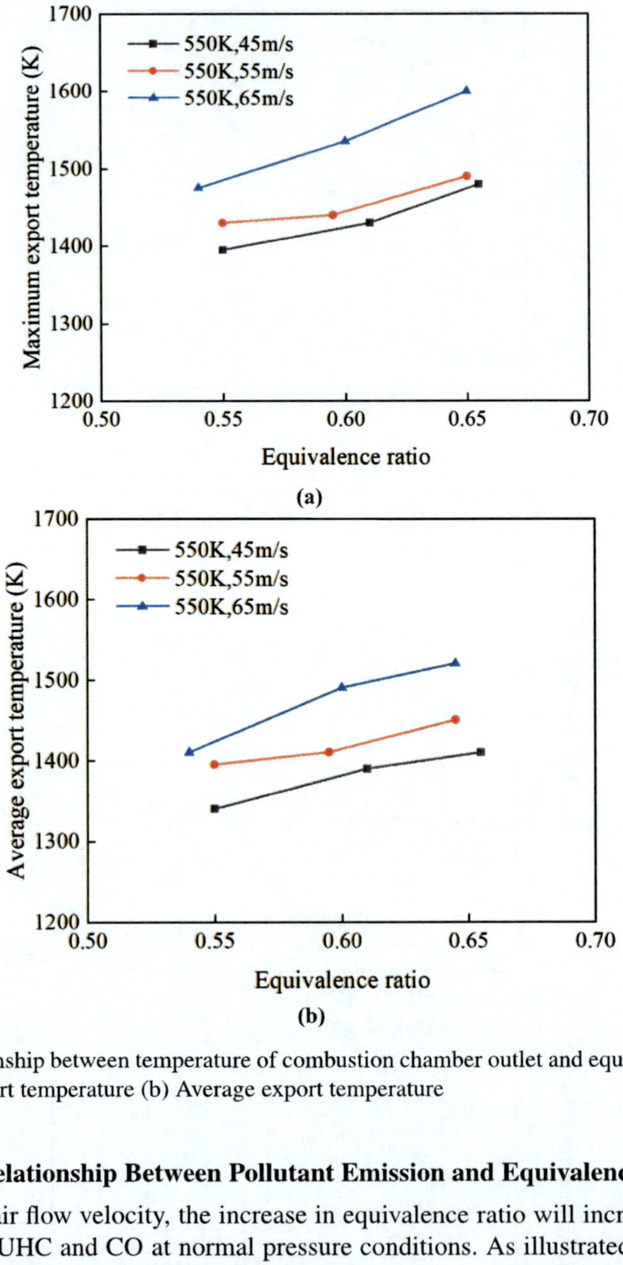

Fig. 5. Relationship between temperature of combustion chamber outlet and equivalent Ratio (a) Maximum export temperature (b) Average export temperature

3.2.2 The Relationship Between Pollutant Emission and Equivalence Ratio

At a specific air flow velocity, the increase in equivalence ratio will increase the emission of NOx, UHC and CO at normal pressure conditions. As illustrated in Fig. 8, the concentration of NOx and UHC emission is below 20 ppm, 10 ppm respectively, while CO emission reaches a maximum of 130 ppm. The highest NOx emission concentration was observed under a low premix channel velocity (45 m/s) and large equivalence ratio due to high flame temperature and prolonged residence time in a hot state. Conversely, the high emission of CO is primarily caused by the low combustion pressure and the quenching effect of the water-cooled wall of the flame tube.

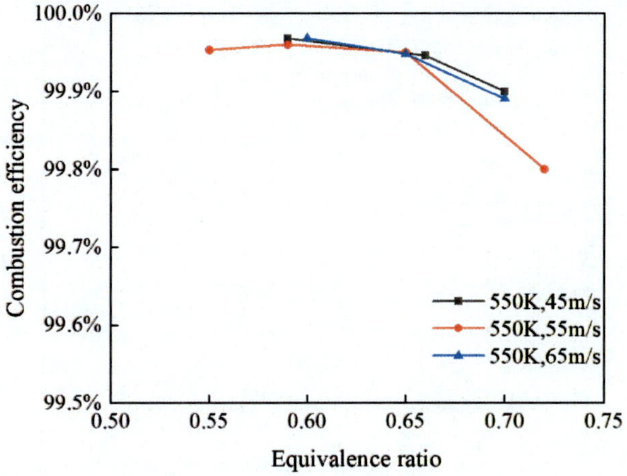

Fig. 6. Relationship between combustion efficiency and equivalent ratio under atmospheric pressure

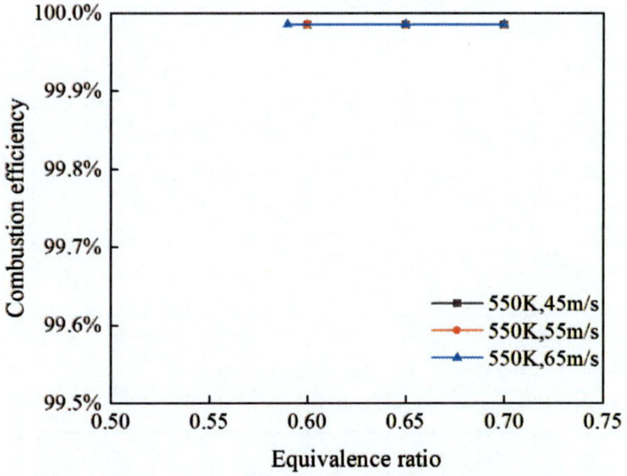

Fig. 7. Relationship between combustion efficiency and equivalent ratio under 5 atmospheric pressures

Figure 9 illustrates the emission characteristics of pollutants from nozzles operating at a pressure of 5 atm. The emission concentrations of NOx, UHC, and CO are compared with those observed under normal pressure conditions. The results show that as the load decreases, the NOx emissions increase under the same equivalent ratio condition, especially at low load (45 m/s) where the NOx emissions increase sharply. Under the condition of low load (45 m/s), it is not suitable to adopt the control strategy of same equivalent ratio. Moreover, an increase in the equivalence ratio and a decrease in premix channel velocity (45 m/s) lead to an increase in NOx emission concentration and a decrease in UHC and CO emissions concentration. Furthermore, it was observed that the

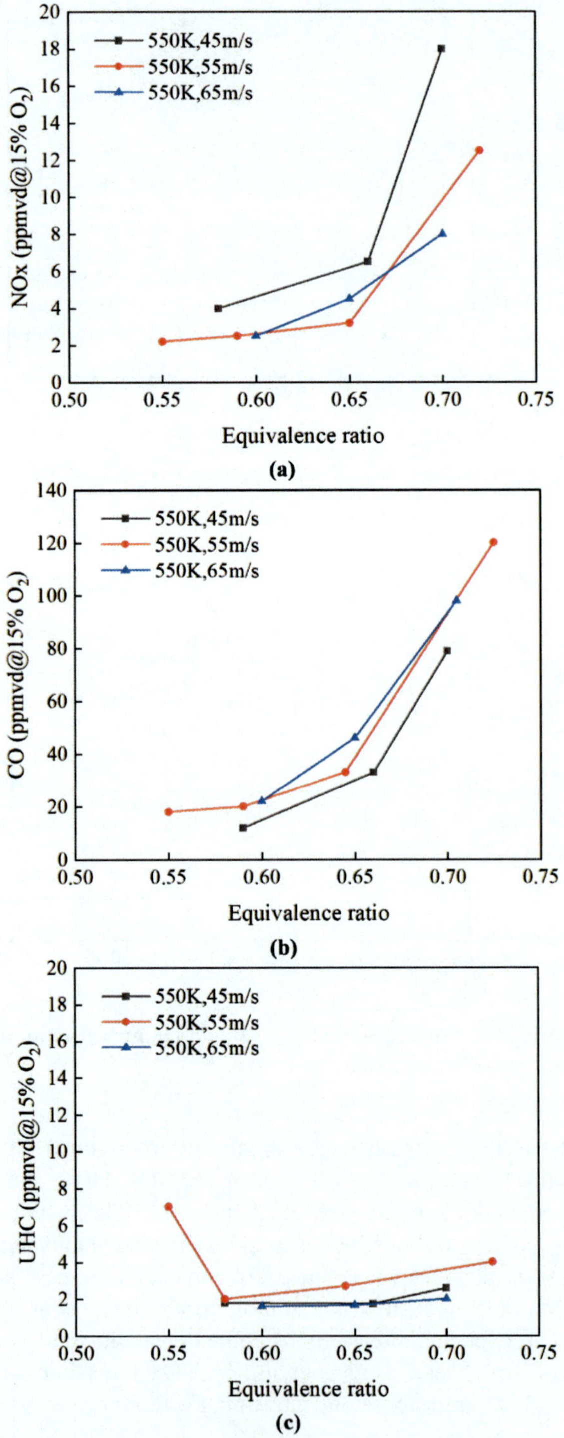

Fig. 8. Relationship between pollutant emission and equivalent ratio under atmospheric pressure (a) NO_X (b) CO (c) UHC

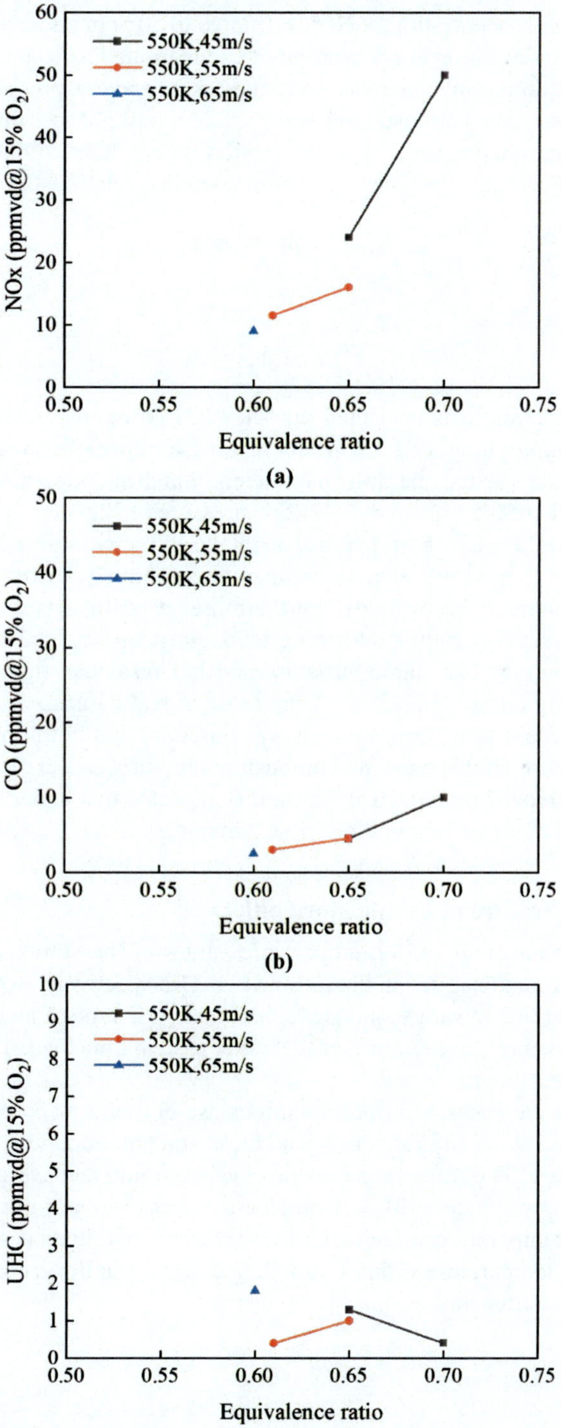

Fig. 9. Relationship between pollutant emission and equivalent ratio under 5 atmospheric pressure (a) NO_X (b) CO (c) UHC

maximum emission concentration of NOx is limited to 50 ppm under all test conditions, while the maximum emission concentration of CO is limited to 10 ppm.

Comparing the emission data under normal pressure and low pressure,, it is observed that the NOx emissions at 5 atmospheres are approximately 24 times higher than the heat removal value at normal pressure. This observation is consistent with the relationship of approximately 0.5 power of pressure. The corresponding relationship is shown in (1).

$$NO_x \text{ (in ppm)} \propto P^n \tag{1}$$

when

$$n = 0.5 \tag{2}$$

Based on estimation, it is projected that the NOx emissions measured at 22.5 atm will increase by approximately 1.1 times compared to those observed at 5 atm. When evaluating the nozzle's actual operating parameters, which include a flow rate of 53.4 m/s, an equivalence ratio of 0.534, and atmospheric pressure of 22.5 atm, it is speculated that NOx emissions would reach about 6ppm at an intake air temperature of 550 K. However, it should be noted that as the inlet air temperature continues to rise, NOx emissions are expected to increase accordingly. Furthermore, it is important to recognize that the quantitative analysis results concerning NOx emission are based solely on nozzle performance screening. The flame structure and hot residence time in single nozzle performance testing differ significantly from those of real flames.

Regarding CO and UHC emissions, it was observed that both pollutants exhibited a significant decrease after increasing combustion pressure, especially UHC emissions, which remained below 2 ppm even at 5 atm. It is expected that under actual component conditions, the UHC emissions will not pose a major issue.

3.2.3 The Temperature of Combustor Outlet

The maximum and average outlet temperature values of the combustor under normal pressure and 5atm are illustrated in Fig. 10 and Fig. 11 respectively. Both figures indicate that under the condition of same equivalent ratio, there is a trend of increasing maximum temperature and average temperature of the outlet gas with the load. This is because the equivalent ratio remains the same.

As the airflow increases, a proportional increase in fuel flow is required in the premixing channel, resulting in more energy addition and consequently, an increase in the exit gas temperature. The difference between the maximum and average outlet temperature values is approximately 40 K. Compared to test results at normal pressure, the maximum outlet temperature value under low-pressure conditions is almost identical to the average outlet temperature value. This indicates a more uniform combustion reaction temperature and a better fuel-air premixing degree.

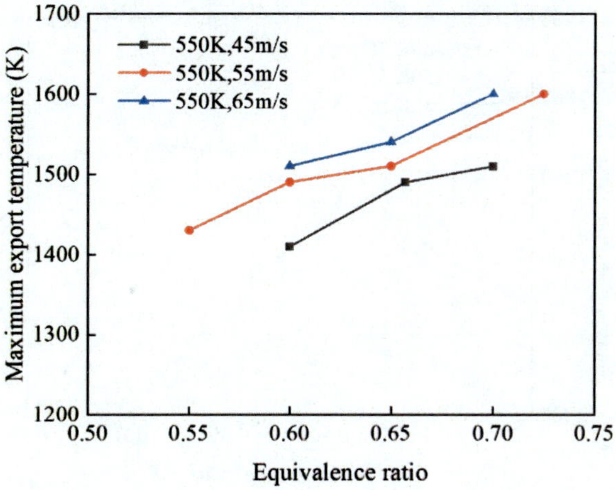

(a)

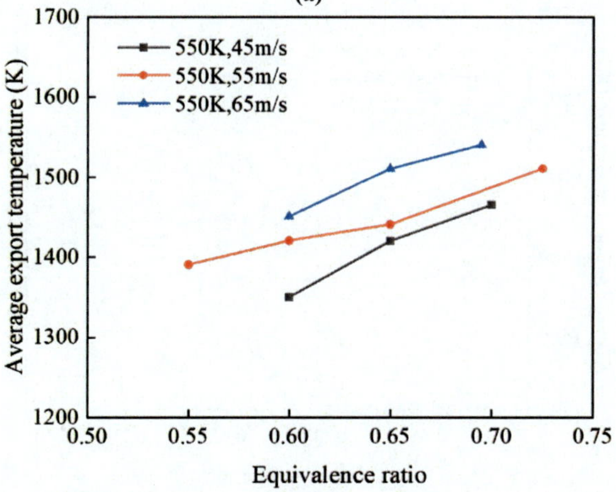

(b)

Fig. 10. Relationship between temperature of combustion chamber outlet and equivalent Ratio under atmospheric pressure (a) Maximum export temperature (b) Average export temperature

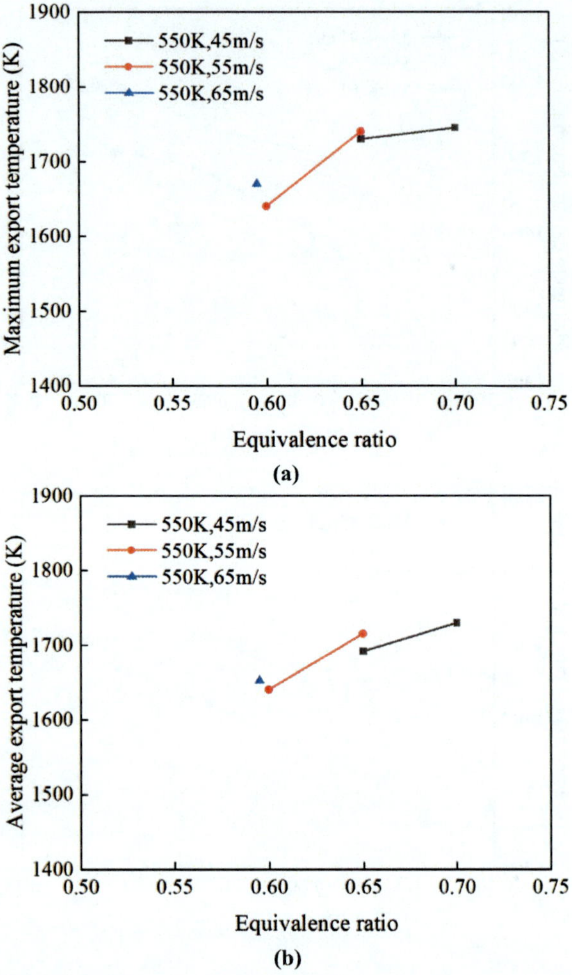

Fig. 11. Relationship between temperature of combustion chamber outlet and equivalent Ratio under 5 atmospheric pressure (a) Maximum export temperature (b) Average export temperature

4 Conclusions

The fuel nozzle is the crucial component of the combustion chamber. The experimental study conducted in this paper aimed to investigate the performance of different fuel nozzles in terms of combustion and emissions at partial loads. The results show that all tested nozzles achieved high combustion efficiency, exceeding 99.8%, with the majority surpassing 99.9%. At 5 atmospheres, all tested nozzles meet the standard requirements with combustion efficiency above 99.99%. Under the condition of partial load, with the same equivalent ratio, both of emissions such as NOx, CO, etc., and exit gas temperature increase with the increase in load, which has a significantly contrasting effect on nozzle selection. The peripheral nozzle end 055(2) was found to outperform the reference end scheme 055. However, it is important to note that NOx emissions are expected to increase as intake air temperatures rise.

References

1. Yang, A., Tian, X., Feng, Y., Fen, Z.: A combustor fuel nozzle orifice flow characteristic research. Dongfang Turbine **04**, 007 (2021)
2. Ji, G.: Effect of burner outlet length on combustion performance in F-class gas turbine. Therm. Turbine **02**, 003 (2021)
3. Ji, Q., Zhao, C., Li, M., Zhang, B.: Experimental study on pressure atomizing injector performance of a heavy-duty gas turbine. China Acad. J. Electron. Publ. House **37**, 002 (2011)
4. Zhao, W., Yu, Z., He, H., Ai, Y., Wang, Y.: Experiment and numerical study on emission characteristics of premixing fuel nozzle in gas turbine combustor. J. Propuls. Technol. **41**, 002 (2020)
5. Ramraj, H., Sundararaj, T., Sekar, C., Arora, R., Kushari, A.: Effect of nozzle exit area on the performance of a turbojet engine. Aerosp. Sci. Technol. **116,** 106844 (2021)
6. Hyun, Y., Senoo, Y., Yamaguchi, M., Hayami, H.: The influences of tip clearance on the performance of nozzle blades of radial turbines (experiment and performance prediction at three nozzle angles). JSME Int. J. **31**(2), 258–262 (1988)
7. Wang, B., Shao, F., Shen, X., Zheng, G., Wu, X.: Inner flow field characteristics simulation of the dual fuel nozzle of gas turbine. Sci. Technol. Eng. **20**(27), 11125–11130 (2020)
8. Wang, W., Li, M., Mou, Y., Xin, J.: Study on wear, ablation and charcoal accumulation of components and nozzles of QD128 gas turbine turbine. Gas Turbine Technol. **32**(3), (2019)
9. Sun, B., Li, S., Su, W., Li, J., Wang, N.: Effects of gas temperature on nozzle damping experiments on cold-flow rocket motors. Acta Astronautica **126**, 18–26 (2016)
10. He, Z., Zhong, W., Wang, Q., Jiang, Z., Shao, Z.: Effect of nozzle geometrical and dynamic factors on cavitating and turbulent flow in a diesel multi-hole injector nozzle. Int. J. Therm. Sci. **70**, 132–143 (2013)
11. Costa, G., Anania, M.P., Biagioni, A., Bisesto, F.G., et al.: Characterization of supersonic gas jets for different nozzle geometries for laser-plasma acceleration experiments at SPARC_LAB. J. Instrum. **17** (2022)
12. Tang, Y., Lou, D., Wang, C., Tan, P., Hu, Z., et al.: Study of visualization experiment on the influence of injector nozzle diameter on diesel engine spray ignition and combustion characteristics. Energies **13**, 5337 (2020)

Check for updates

Analysis of the Influence of Total Pressure Error on Air Data Calibration

Chao Xi(✉)

Chinese Flight Test Establishment, Xi'an, China
xczzz@163.com

Abstract. The installation error of the air data system is comprised of static pressure error and total pressure error. Two basic methods are used in flight test to determine the installation error of the air data system, namely pressure-based method and speed-based method. Total pressure error is usually small enough so that it can be neglected and won't cause significant errors. Sometimes, due to the position error of the pitot tube or the accuracy of the sensor, total pressure measured is different from the real total pressure, this difference can severely influence to the air data calibration result. This paper quantitative analysis the influence of total pressure error on the result of air data calibration, and put forward a solution to the air data calibration method in flight test.

Keywords: Total pressure error · Static pressure error · Air data calibration

1 Introduction

Accurate air data is crucial to a safety flight. As modern aircraft evolve towards a more worry-free operating concept, flight control system, FADEC and other systems are increasingly relying on more accurate and more reliable air data. However, the measured air data always has a small error compared to the actual air data. Generally, the error of the air data consists of principle error, instrument error and installation error. Installation error is caused by variations in air-flow information at different positions of the fuselage, which can vary depending on various configurations (flaps, slats, L.G), Mach number, and angle of attack. In order to estimate the installation error, CFD method, wind tunnel method and flight test method are often used.

2 Air Data Measurement and Calibration Method

So far, civil transport aircraft still uses pitot-static tube to measure the total pressure, static pressure to obtain the airspeed and pressure altitude. The difference between total pressure and static pressure is called impact pressure (Eq. 1). The Mach number can be obtained using The Bernoulli equation or Rayleigh supersonic pitot-static equation (Eq. 2, Eq. 3), and pressure altitude can be obtained using standard atmosphere model.

$$q_c = P_t - P_s \tag{1}$$

Z. Jing and X. Zhan (Eds.): ICASSE 2023, LNEE 1153, pp. 192–199, 2024.
https://doi.org/10.1007/978-981-97-0550-4_15

$$\frac{q_c}{p_s} = \left[1 + \frac{\gamma - 1}{2}M^2\right]^{\frac{\gamma}{\gamma-1}} - 1 \cdots\cdots (M < 1) \tag{2}$$

$$\frac{q_c}{p_s} = \left[\frac{\gamma + 1}{2}M^2\right]^{\frac{\gamma}{\gamma-1}} \left[\frac{\gamma + 1}{1 - \gamma + 2\gamma M^2}\right]^{\frac{1}{\gamma-1}} - 1 \cdots (M \geq 1) \tag{3}$$

$$H_p = f(p_s) \tag{4}$$

$$M = \frac{V}{a} \tag{5}$$

The number γ in Eq. (2) is gas constant and a value of 1.4 can cover most situation in normal flight. Replace the static pressure and speed of sound as sea level parameters of standard atmosphere model in Eq. (2), (3) and (5), we can get indicated airspeed (IAS). If the impact pressure is calibrated, the indicated airspeed is called calibrated airspeed (CAS) as shown in Eq. (6) and (7). When at sea level of the standard atmosphere model, the calibrated airspeed (V_C) is equal to the true airspeed (V_T) and ground speed when there is no wind.

$$V_c = a_{sl}\sqrt{5\left[\left(\frac{q_c}{p_{s-sl}} + 1\right)^{\frac{1}{3.5}} - 1\right]} \cdots\cdots (M < 1) \tag{6}$$

$$V_c = 0.881285 a_{sl}\sqrt{\left(\frac{q_c}{p_{s-sl}} + 1\right)\left[1 - \left(\frac{1}{7\left(\frac{V_c}{a_{sl}}\right)^2}\right)\right]^{2.5}} \cdots\cdots (M \geq 1) \tag{7}$$

From Eq. (2), (3), (4), (6), (7), we can see the accurate IAS, Mach number and pressure altitude rely on accurate total pressure and static pressure. We already know these information are different when measured in different position of the fuselage, in order to calibrate the measured air data, flight test method is used to determine and verify this error.

There are many ways to determine air data error in flight test, for example: tower fly-by method, steady level flight using GPS [1], trailing cone method (Fig. 1), nose boom method (Fig. 2). These methods can be categorized into two types: speed-based method and pressure-based method. Speed-based method tries to obtain the real airspeed, using the real airspeed to obtain the real static pressure, and then the real IAS and pressure altitude can be obtained. The schematic diagram is shown in Fig. 3. Pressure based method tries to obtain the real static pressure, and the pressure altitude can be obtained. Assuming that there is no error in total pressure, we can get real IAS using total pressure and real static pressure. The schematic diagram is shown in Fig. 4. Nowadays, most transport aircraft use retractable traílling cone for air data calibration. For small aircraft, retractable traílling cone system cannot be installed into the aircraft due to the limitations in cabin and sometimes even a primitive trailing cone is hard to be installed, so speed-based method (level flight with GPS) is more commonly used in air data system calibration.

Fig. 1. Trailing cone method

Fig. 2. Nose boom method

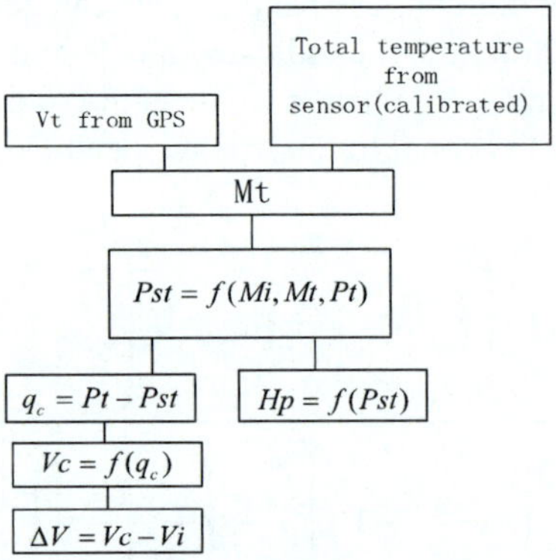

Fig. 3. The schematic diagram of speed-based method

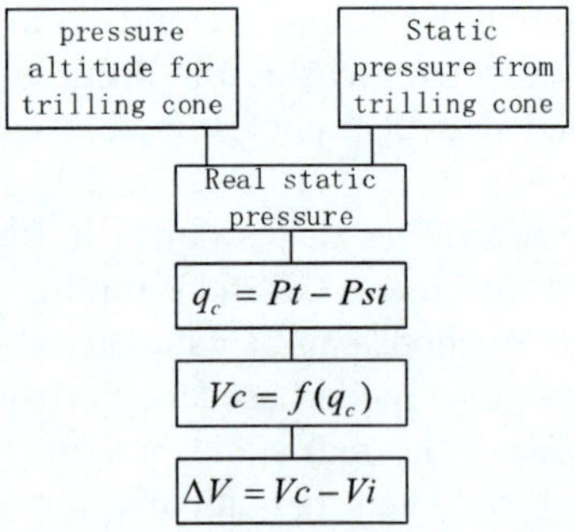

Fig. 4. The schematic diagram of pressure-based method

3 Total Pressure Error and Its Influence on Air Data System Calibration

A large number of wind tunnel tests and flight tests show that total pressure measured can be accurate [2, 3] in a large range of alpha angle and slide angle. However, sometimes the measured total pressure error can be quite significant, particularly during crosswind flight. Figure 5 shows the time history during the taxing of a transport airplane. It shows clearly the installation error of total pressure, the IAS difference from ADC1 and ADC2 during taxing was about 20 km/h to 30 km/h, but the static pressure difference between ADC1 and ADC2 was less than 10 Pa, the total pressure difference was about 140 Pa. Obviously, the IAS difference was caused by total pressure difference. After carefully checking, the total pressure of ADC1 was sourced from the left pitot tube and the total pressure of ADC2 was sourced from the right pitot tube. The schematic diagram of the air data system is shown in Fig. 6. However, almost all transport aircraft is designed this way, surely this is not the real reason that causes the difference on total pressure from two ADCs. Finally, we found that near the front of the two pitot tube was a small bulge of the porthole, the initial grid model for CFD simulation and scaled model for wind tunnel test were covered with smooth surface, so the installation error of total pressure was left until flight test. So, we increased the height of the foundation bed for pitot tube as shown in Fig. 7.

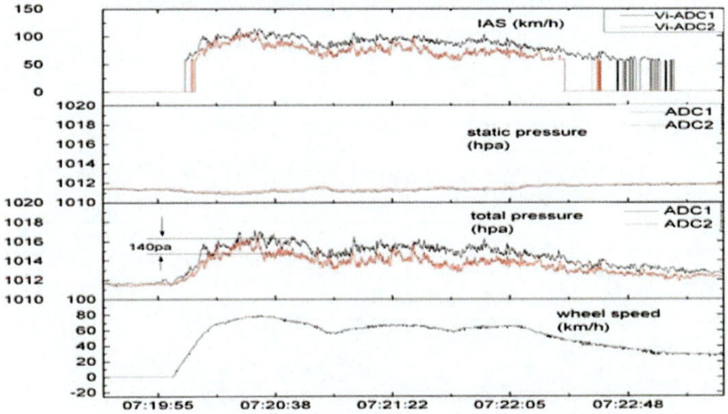

Fig. 5. Time history during taxing

Another set of data is shown in Fig. 8. The IAS difference from the two ADCs was about 10 km/h when flown at 6200 ft height. The static pressure difference between the two ADCs was about 24 Pa (average) and the total pressure difference was about 130 Pa (average). In this case, the aircraft was a small normal type aircraft, one air data system was mounted ahead of the nose and the other air data system was mounted at the wing tip of the aircraft.

As can be concluded from the above examples, if the air data system is not designed properly, the total pressure is possible to be distorted. In these cases, traditional air data

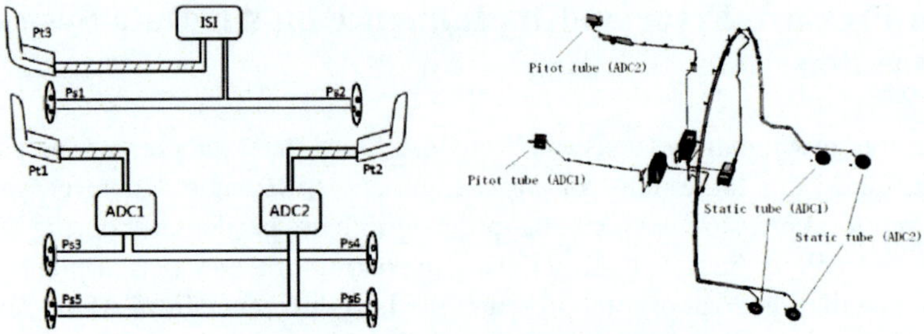

Fig. 6. The schematic diagram of the air data system

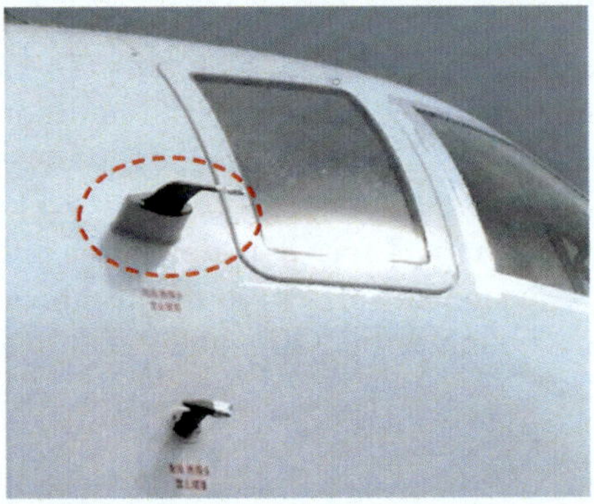

Fig. 7. Pitot-tube on foundation bed

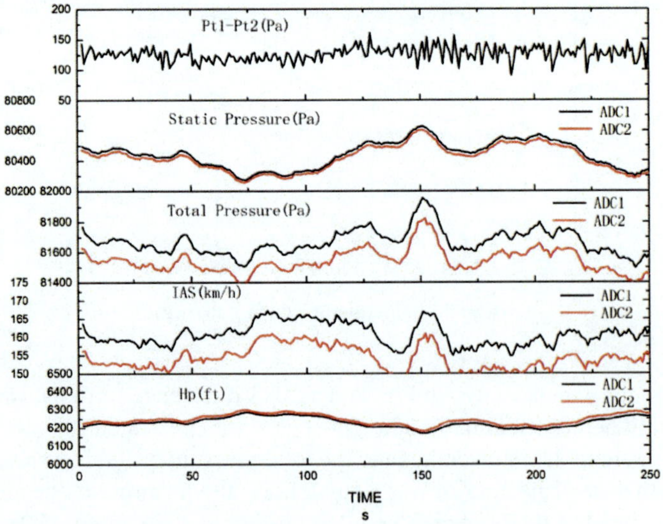

Fig. 8. Time history of the small aircraft

system calibration methods may lead to obvious mistakes. To be specific, using speed-based method could get accurate CAS but wrong pressure altitude; using pressure-based method could get accurate pressure altitude but wrong CAS.

4 Example Analysis

During previous chapters, total pressure error dose exists. Let's analyze air data calibration in two cases. In the first case, if the static pressure from ADC was accurate, the IAS error was caused by total pressure error. Assume that the difference between the real air speed and IAS from ADC was just meet the upper limit or lower limit that FAR25.1323 (c) required: "the airspeed error of the installation, excluding the airspeed indicator instrument calibration error, may not exceed three percent of five knots, whichever is greater, throughout the speed range, from V_{MO} to $1.23V_{SR}$, with flaps retracted, and $1.23V_{SR0}$ to V_{FE} with flaps in the landing position". Using speed-based method for air data calibration, we could get the reference pressure altitude and it may be wrong from the real pressure altitude. Table 1 shows the static pressure error and pressure altitude error using traditional speed-based method. Figure 9 shows the D value of pressure altitude along with different air speed. We can see that the deviation of reference pressure altitude could far more exceed the limit that FAR25.1325 requires for static pressure system since very low air speed. The FAR25.1325 (e) required: "Each system must be designed and installed so that the error in indicated pressure altitude, at sea level, with a standard atmosphere, excluding instrument calibration error, does not result in an error of more than ±30 ft per 100 knots speed for the appropriate configuration in the speed range between $1.23V_{SR0}$ with flaps extended and 1.7 V_{SR1} with flaps retracted, however, the error need not be less than ±30 ft".

Table 1. Pressure altitude error caused by total pressure error

IAS (km/h)	Ps (Pa)	Pt (Pa)	IAS limit (km/h)	Ps error (Pa)	Hp error (m)	IAS limit (km/h)	Ps error (Pa)	Hp error (m)
100	101325	101798.7	109	89.27	−7.43	91	−81.53	6.79
150	101325	102393	159	132.56	−11.03	141	−124.72	10.39
200	101325	103229.1	209	176.5	−14.68	191	−168.53	14.04
300	101325	105627.1	309	285.27	−23.72	291	−240.91	20.07
400	101325	109094.9	412	486.51	−40.42	388	−470.66	39.25
500	101325	113647.7	515	783.38	−65.01	485	−756.56	63.17
600	101325	119394.6	618	1169.73	−96.92	582	−1127.38	94.27
700	101325	126450.5	721	1660.74	−137.33	679	−1596.9	133.78
800	101325	134956.4	824	2275.37	−187.71	776	−2182.23	183.26

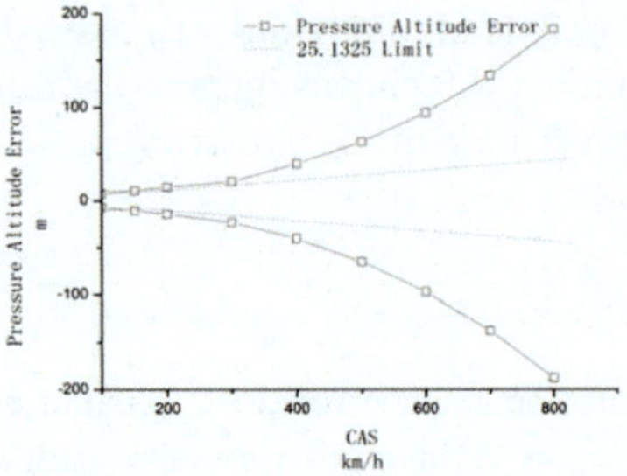

Fig. 9. The error of pressure altitude along with different air speed

In the second case, if the total pressure error was the same as the error shown in Fig. 8 (130 Pa), using pressure-based method for air data calibration, we can get accurate pressure altitude for the static pressure was accurate, the reference IAS was different from the real IAS data. Table 2. Shows the error of IAS using pressure-based method for air data calibration when the total pressure measured is not correct. The black solid line in Fig. 10 shows the D-value of reference CAS and real CAS, the red dashed line shows the limit required by FAR25.1323 for airspeed indicating system, and the red solid line shows the calibration result for CAS using tradition pressure-based method. We can see the error in CAS exceeded the requirement of FAR25.1323 during almost the whole subsonic range, and the error gets larger when the airspeed gets lower.

Table 2. IAS error caused by total pressure error

IAS (km/h)	q_c (Pa)	Errored q_c (Pa)	Reference IAS (km/h)	D-value of IAS (km/h)	Errored q_c (Pa)	Reference IAS (km/h)	D-value of IAS (km/h)
100	473.66	603.66	112.896	−12.896	343.66	85.2203	14.7797
150	1067.96	1197.96	158.873	−8.873	937.96	140.643	9.357
200	1904.14	2034.14	206.722	−6.722	1774.14	193.146	6.854
300	4302.09	4432.09	303.886	−3.886	4172.09	294.968	5.032
400	7769.91	7899.91	403.351	−3.351	7639.91	396.83	3.17
500	12322.69	12452.69	502.657	−2.657	12192.69	497.59	2.41
600	18069.57	18199.57	602.191	−2.191	17939.57	598.115	1.885
700	25125.51	25255.51	701.858	−1.858	24995.51	698.504	1.496
800	33631.43	33761.43	800.21	−0.21	33501.43	798.808	1.192

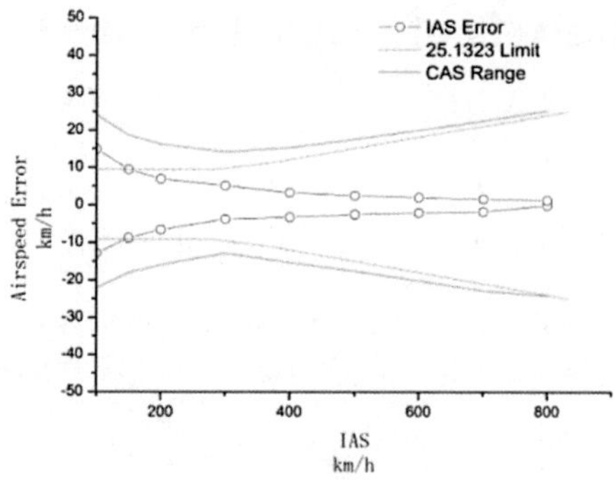

Fig. 10. The reference CAS range along with real CAS

5 Conclusion

We could see the total pressure error dose exists when using flight test method, in such situation, using traditional pressure-based air data calibration method (ignore the total pressure error) could get accurate pressure altitude, but the reference CAS could be deviated from the real CAS, and the smaller the airspeed, the larger the deviation. Using traditional speed-based air data calibration method (ignore the total pressure error) could get accurate CAS, but the reference pressure altitude could be errored considerably. Based on the analysis above, pressure-based air data calibration method (i.e. trailling cone method) was preferred if the system had a more accurate requirement on pressure altitude. In addition, in order to avoid the total pressure installation error, the position of pitot tube must be carefully selected. Additional pitot tubes also can be used to estimate the accuracy of total pressure.

References

1. Liu, H.Y., Liu, L., Cao, F.H.: Reference air data boom method for air data system calibrations. J. Exp. Fluid Mech. **27**(2), 91–94 (2013)
2. Zhang, D.Y., Chen, P., Lu, X.R.: The calibration method for installation error of the air data system using GPS. Flight Dyn. **18**(2), 62–66 (2000)
3. Xi, C., Cheng, T., Liu, J.: Research of a GPS-based airspeed calibration method under crosswind. Aeronaut. Sci. Technol. **26**(05), 53–56 (2015)

Analysis of Impact Properties of Nanocomposites at Micron Scale

Biao Yan(✉) and Zhongcheng Mu

Shanghai Jiao Tong University, Shanghai, China
{20000521yanbiao,muzhongcheng}@sjtu.edu.cn

Abstract. With the increasing number of human activities in space, spacecraft are facing harsh environmental conditions. Space debris is one of the most serious problems that can cause severe structural damage to spacecraft and create new debris during impact. Researchers have found that nanomaterials have high strength, high modulus, and excellent performance. Therefore, applying nanomaterials to spacecraft protective structures is a feasible method. This paper adopts the SPH numerical simulation method to explore the impact resistance characteristics of spacecraft protective materials under high-speed impact by space debris at the micron scale. The impact characteristics during impact are considertableed, and the deformation and damage of spacecraft during impact are analyzed. The simulation results show that the specific impact energy decreases with the increase of target thickness under the condition of the same initial velocity and radius. The reason why the specific impact energy is abnormal at the microscopic scale is that the smaller the thickness of the target is, the smaller the fragments will be, which will lead to an increase in the total fracture surface area of the target fragments and a higher energy absorbed by the target per unit mass.

Keywords: nanomaterials · space debris · AUTODYN simulation · specific impact energy · Micro scale

1 Introduction

With more and more activities of people exploring space, the space environment is getting worse. Spacecraft running in space suffer from ultra-low temperature, strong radiation, atomic oxygen corrosion and space debris impact. Among them, space debris is a significant problem that affects the normal operation of spacecraft in orbit. The research shows that the impact of space debris with the size of 1 μm on the spacecraft can cause permanent damage to the exposed charge coupled devices. When the size of space debris is 10 μm, the impact will cause obvious crater damage, and the space debris with the size of 100 μm can penetrate the thick bulkhead structure of 300 μm to 500 μm [1].

In response to the potential hazards posed by space debris, researchers have carried out extensive research with a view to obtaining materials with high protection and impact resistance properties. Nanomaterials are an emerging material with great potential and are one of the most preferred options for spacecraft protection structures.

Z. Jing and X. Zhan (Eds.): ICASSE 2023, LNEE 1153, pp. 200–210, 2024.
https://doi.org/10.1007/978-981-97-0550-4_16

In the impact test, the projectile hits the target at a certain speed, and the target can reduce the speed of the projectile, that is, reduce the energy of the projectile. The stronger the impact resistance of the target, the greater the energy loss of the projectile during impact. The specific impact energy is used to measure the impact resistance of the target, which is defined as follows

$$E_p^* = \frac{E_P}{\rho_f \pi R^2 h} \tag{1}$$

where Ep is the impact energy, that is, the kinetic energy lost by the projectile during its high-speed impact on the target, ρ_f is the density of the target material, R is the radius of the projectile, and h is the thickness of the target.

Generally speaking, the greater the thickness of the target film, the greater the impact energy of the projectile during impact. This is consistent with the general view that thicker film layers of specific materials will be more difficult to penetrate. Jinho Hyton et al. [2]. Studied the high strain rate impact behavior of polystyrene and compared the energy absorption capacity of polystyrene at impact velocities of 75–550 nm and 350–800 m/s. He found that at the same impact velocity, the smaller the thickness of the polystyrene film, the greater the specific impact energy. Yuwen Zhu [3] found that the specific impact energy was negatively correlated with the film thickness in the study of nano-scale polymer films.

Although researchers have carried out a number of impact tests at the nanoscale and studies of particle impact properties at the macroscale, impact studies at the micron scale, which lies between the two, have rarely been addressed. This paper therefore focuses on investigating the impact properties of nanomaterials by conducting numerical simulation experiments, modelling the SPH of nanomaterials using AUTODYN software, investigating the relationship between specific impact energy and target thickness at the micron scale, and theoretically explaining the mechanism by which this phenomenon occurs.

2 Nano Impact Simulation

2.1 SPH Simulation Geometry Model

The geometric model shown in Fig. 1 was created using AUTODYN software, where the blue part is the circular projectile, the green part is the target, the radius of the projectile is 2.5 um, the thickness of the target is 1 um and the width of the target is 15 um. To save computational resources, the spacing between the projectile and the target is 5 nm.

Particle impact experiments at conventional sizes have been carried out by many people and some progress has been made. This paper focuses on particle impacts at microscopic sizes. The size of space debris is mostly at the micron level, so the order of magnitude of the projectile and target selected for the numerical simulation experiments is at the micron level. We focus on the deformation of the target before and after the impact as well as the change in performance, and the shape of the projectile is chosen to be circular in order to facilitate the study. The velocity of the space debris is of the order of km/s, so the velocity of the projectile is chosen to be of this order. The SPH method

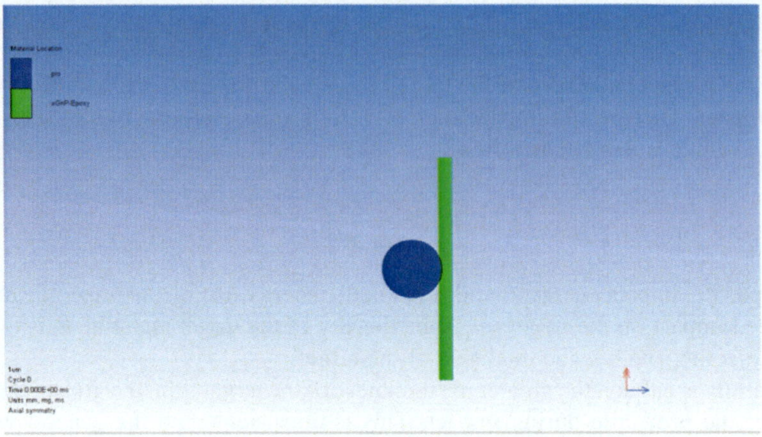

Fig. 1. Schematic diagram of the geometric model in the AUTODYN software simulation

has a huge advantage over other finite element methods in numerical simulations, and both the projectile and the target are modelled using the SPH method.

In order to investigate the relationship between specific impact energy and target thickness at the micron scale, three sets of numerical simulation experiments will be conducted. The radius of the projectile is chosen to be 2.5 um for all three sets of experiments, and the variable for each set is the initial velocity of the projectile, which is chosen to be 1000, 3000 and 5000 m/s. There are six simulated impact experiments in each set. The thicknesses of the targets for the six simulations were set to 0.2, 0.5, 1, 2, 5 and 10 um respectively.

The SPH simulation particle sizes for the three sets of experiments are given in Table 1. Through experimental simulations, the above selection of SPH particle sizes allows the AUTODYN simulation to converge.

Table 1. SPH particle size of three groups of tests

Initial speed m/s	Target thickness/um					
	0.2	0.5	1	2	5	10
1000	0.00001	0.00001	0.00001	0.00002	0.00003	0.0005
3000	0.00001	0.00001	0.00001	0.00002	0.00003	0.0005
5000	0.00001	0.00001	0.00001	0.00002	0.00003	0.0005

2.2 Simulation of Material Models

The material model consists of an equation of state, a strength model and a failure criterion. The equation of state characterises the basic parameters of the material such as density, bulk modulus and specific heat of the material; the strength model characterises the change in material form and contains a range of parameters such as yield stress and shear modulus; the failure criterion gives the scale at which stresses and strains will cause the material to fail.

The equation of state of the projectile is linear. The strength equation is the von Mises strength equation and the failure model is the Hydro (Pmin) failure model to avoid deformation of the projectile during a high speed impact. The material parameters of the projectile are given in Table 2. The density of the projectile is 3.89 g/m3, the bulk modulus is 231 GPa, the reference temperature is 293 K, the shear modulus in the strength model is 152 GPa and the yield stress is 90 GPa. The high modulus of the projectile ensures that no significant deformation of the projectile occurs during the high speed impact.

Table 2. Material model parameters of projectile

Material parameters	projectile
Equation of state	**Linear**
Reference density	3.89000E + 00 (g/cm3)
Bulk modulus	2.31000E + 08 (kPa)
Reference temperature	2.93000E + 02 (K)
Specific heat	0.00000E + 00 (J/kgK)
Thermal conductivity	0.00000E + 00 (J/mKs)
Strength models	**von Mises**
Shear modulus	1.52000E + 08 (kPa)
Yield stress	9.00000E + 07 (kPa)
Failure models	**Hydro (Pmin)**
Hydraulic pull-up limit	−1.01000E + 20 (kPa)

The response of thin film materials to high speed impact conditions involves consideration of the effects of strain, strain rate and temperature. The Johnson-Cook strength model model is generally used to describe the strength limits of materials at large strains, high strain rates, high temperature environments and the failure process. In the Johnson-Cook strength model, the yield stress is determined by strain, strain rate and temperature [4]. Therefore, for the target, we chose the linear equation of state, the Piecewise Johnson-Cook strength model and the Plastic Strain failure model. In the Johnson-Cook model, the yield stress σ_y is related to the equivalent plastic strain ε_p, the equivalent plastic strain rate $\dot{\varepsilon}_p$ and the temperature T as follows.

$$\sigma_y = \left(A + B\varepsilon_p^n\right)\left[1 + C\ln\left(\frac{\dot{\varepsilon}_p}{\dot{\varepsilon}_0}\right)\right]\left(1 - T^{*m}\right) \tag{2}$$

where reference strain rate $\dot{\varepsilon}_0 = 1\text{s}^{-1}$; A is the yield strength of the material at quasi-static conditions; B and n are the effects of strain change; C is the strain rate sensitivity index; and m is the temperature softening factor. The Piecewise Johnson-Cook strength model is the same as the Johnson-Cook model for the strain hardening and thermal softening sections, but the sections are replaced by a segmented line of yield stress against equivalent plastic strain.

The material model parameters for the target are shown in Tables 3 [5].

Table 3. Material Model Parameters of Target

Material parameters	Target
Equation of state	**Linear**
Reference density	1.00000E + 00 (g/cm3)
Bulk modulus	4.17000E + 06 (kPa)
Reference temperature	3.00000E + 02 (K)
Specific heat	0.00000E + 00 (J/kgK)
Thermal conductivity	0.00000E + 00 (J/mKs)
Strength models	**Piecewise JC**
Shear modulus	8.94000E + 05 (kPa)
Yield stress (without plastic strain)	1.00000E-06 (kPa)
Plastic strain#1	2.50000E-02 (none)
Plastic strain#2	5.00000E-02 (none)
Plastic strain#3	1.00000E-01 (none)
Plastic strain#4	3.00000E-01 (none)
Plastic strain#5	4.30000E-01 (none)
Plastic strain#6	0.00000E + 00 (none)
Plastic strain#7	0.00000E + 00 (none)
Plastic strain#8	0.00000E + 00 (none)
Plastic strain#9	0.00000E + 00 (none)
Plastic strain#10	0.00000E + 00 (none)
Yield stress#1	1.00000E + 05 (kPa)
Yield stress#2	1.60000E + 05 (kPa)
Yield stress#3	1.70000E + 05 (kPa)
Yield stress#4	1.40000E + 05 (kPa)
Yield stress#5	1.25000E + 05 (kPa)
Yield stress#6	0.00000E + 00 (kPa)
Yield stress#7	0.00000E + 00 (kPa)

(*continued*)

Table 3. (*continued*)

Material parameters	Target
Yield stress#8	0.00000E + 00 (kPa)
Yield stress#9	0.00000E + 00 (kPa)
Yield stress #10	0.00000E + 00 (kPa)
Strain rate constants	2.40000E-02 (none)
Thermal softening index	1.56000E + 00 (none)
Melting point	5.97000E + 02 (K)
Reference strain rate	1.79700E + 03 (none)
Failure models	**Plastic Strain**
Plastic strain	4.30000E-01 (none)
Crack softening	No

3 Analysis of Numerical Simulation Results

In an impact test, the projectile impacts the target at a certain velocity, which allows the velocity of the projectile to be reduced, i.e. allows the energy of the projectile to be reduced. The velocity of the projectile after it has passed through the target is the residual velocity. The stronger the impact resistance of the target, the greater the energy lost during the impact of the projectile. However, the thickness of the target body is also an important factor affecting the impact of the projectile. Therefore, the specific impact energy is used to measure the size of the impact resistance of the target.

The three sets of simulation models in 2.1 were run to obtain the residual velocity of the projectile after impact with the target, and the corresponding specific impact energy was calculated to obtain the data results shown in Table 4.

Table 4. Variation of residual velocity and specific impact energy with target thickness obtained from three sets of experiments

Target thickness /um	Initial speed m/s	Residual speed m/s	Ep* J/kg
0.2	1000	985.3	946045.08
0.5	1000	967.7	824118.67
1	1000	960.4	503314.09
2	1000	936.4	399227.59
5	1000	820.6	422660.17
10	1000	692.8	337151.59

(*continued*)

Table 4. (*continued*)

Target thickness /um	Initial speed m/s	Residual speed m/s	Ep* J/kg
0.2	3000	2958.9	7939191.44
0.5	3000	2911.5	6783741.82
1	3000	2838.9	6098526.69
2	3000	2712.1	5330964.88
5	3000	2428.6	4022132.98
10	3000	2011.1	3212800.76
0.2	5000	4924.9	24162086.34
0.5	5000	4830.5	21605964.43
1	5000	4695.7	19128436.46
2	5000	4436.1	17248962.76
5	5000	3892.9	12766110.71
10	5000	3098.3	9984681.56

As shown in Table 4, at the same initial velocity of the projectile, the residual velocity of the projectile decreases as the thickness of the target increases, because the thicker the target, the more impact energy of the projectile can be absorbed during the impact. This is because the thicker the target, the more impact energy it can absorb during the impact. However, the opposite result occurs with the specific impact energy, which decreases as the thickness of the target increases.

Plotting the results of all three sets of tests in Fig. 2, it can be seen more graphically that the specific impact energy decreases with increasing target thickness, regardless of the initial velocity of the projectile.

4 Analysis of Results and Theoretical Analysis

4.1 Analysis of Results

Figure 3 shows the target deformation for target thicknesses of 0.2, 1 and 5um. The thinner the target thickness, the more pronounced the overall deformation of the target during the impact process. The deformation during impact is greatest for a 0.2um target, followed by a 1um target, and the smallest for a 5um target. The smaller the thickness of the target, the more small fragments are formed by the impact of the projectile, the more deformation of the target, the more the total fracture surface area of the target fragments increases, the greater the effect of the target per unit mass on the reduction of the velocity of the projectile, resulting in more energy absorption, corresponding to a greater specific impact energy.

Five small fragments were selected separately from the impacted target and the velocities of the small fragments of the target were obtained using the AUTODYN software, as shown in Table 5.

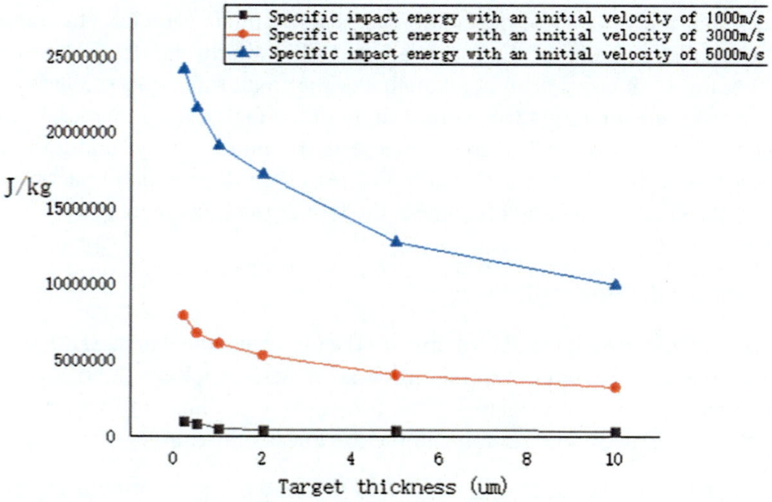

Fig. 2. Changes of specific impact energy with target thickness at different speeds

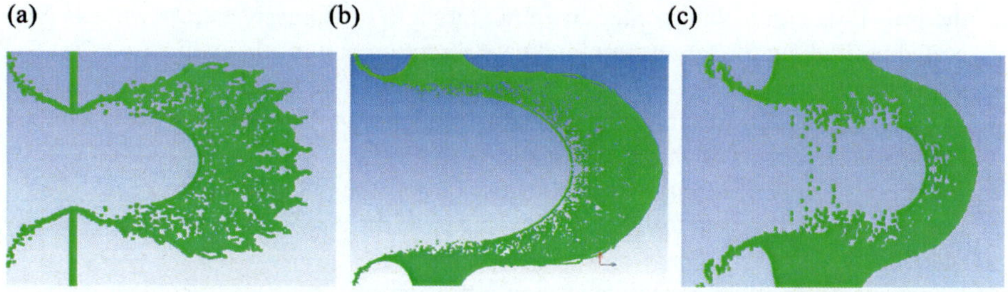

Fig. 3. (a) Deformation of the target with a target thickness of 0.2 um. (b) Deformation diagram of the target with a target thickness of 1 um. (c) Deformation diagram of target body with target body thickness of 5 um.

Table 5. Velocities at observation points for target fragments

Target thickness/um	Speed of fragment 1 m/s	Speed of fragment 2 m/s	Speed of fragment 3m/s	Speed of fragment 4 m/s	Speed of fragment 5 m/s
0.2	4641.9	4835.6	5048.4	4813.2	4936.2
1	3294.8	3740.5	4035.5	3864.9	4212.8
5	2884.1	3110.6	3010	3271.6	2628.9

The average velocity of the five small fragments of the 0.2 um target was 4855.06 m/s, the average velocity of the five small fragments of the 1 um target was 3829.7 m/s and the average velocity of the five small fragments of the 5 um target was 2981.04 m/s. It can be seen that the velocity of the small fragments of the 0.2 um target is higher than that of the small fragments of the 1 um target, and similarly the velocity of the small fragments of the 1 um target is higher than that of the small fragments of the 5 um target. The velocity of the 1 um target fragment is higher than that of the 5 um target fragment. This means that the smaller the target thickness, the more energy is absorbed by the small fragments formed by the deformation of the target during the impact. This is the reason why the smaller the target thickness the higher the impact energy.

4.2 Analysis of the Impact Process

In order to fully understand the deformation of the target and the damage situation during the impact, an example from the simulation was chosen to analyze the six stages of the impact process. The initial velocity of the projectile was chosen to be 5000 m/s, the radius of the projectile R was 2.5 um and the thickness of the target h was 0.2 um.

(a) The projectile has not started to move and the material state of both the target and the projectile are in elasticity. The spacing between the projectile and the target is 5 nm.
(b) The projectile starts to move and gradually comes into contact with the target, when the projectile comes into contact with the target, a compressive stress wave is generated both along the target film thickness direction and in-plane direction [5]. The stress wave in the plane of the target body can be divided into radial stress as well as tangential stress. At this point the state of the part of the target body impacted by the projectile begins to change from elastic to plastic.
(c) As the projectile continues to impact the target, the state of the impacted target becomes increasingly deformed and the bottom of the target becomes almost entirely plastic.
(d) The projectile continues to impact and a portion of the target in contact with the projectile begins to fail and deform.
(e) The continued impact of the projectile caused most of the failure deformation in the target, while target cracks appeared in the center of the target. Around the center of impact, the principal stresses along the radial and tangential directions are almost equal, resulting in an extremely high crack density [3].
(f) At this point the central part of the film has all failed and deformed, and many small fragments have formed. At the same time a small part of the unimpacted middle part of the target becomes plastic. At these high strain rates, the crack formation in the middle part of the target leads to heating and thermal softening of the adiabatic film, further reducing the stress required for the crack to expand. At the same time, the outer radial and tangential cracks intersect to form islands of undeformed film (Fig. 4).

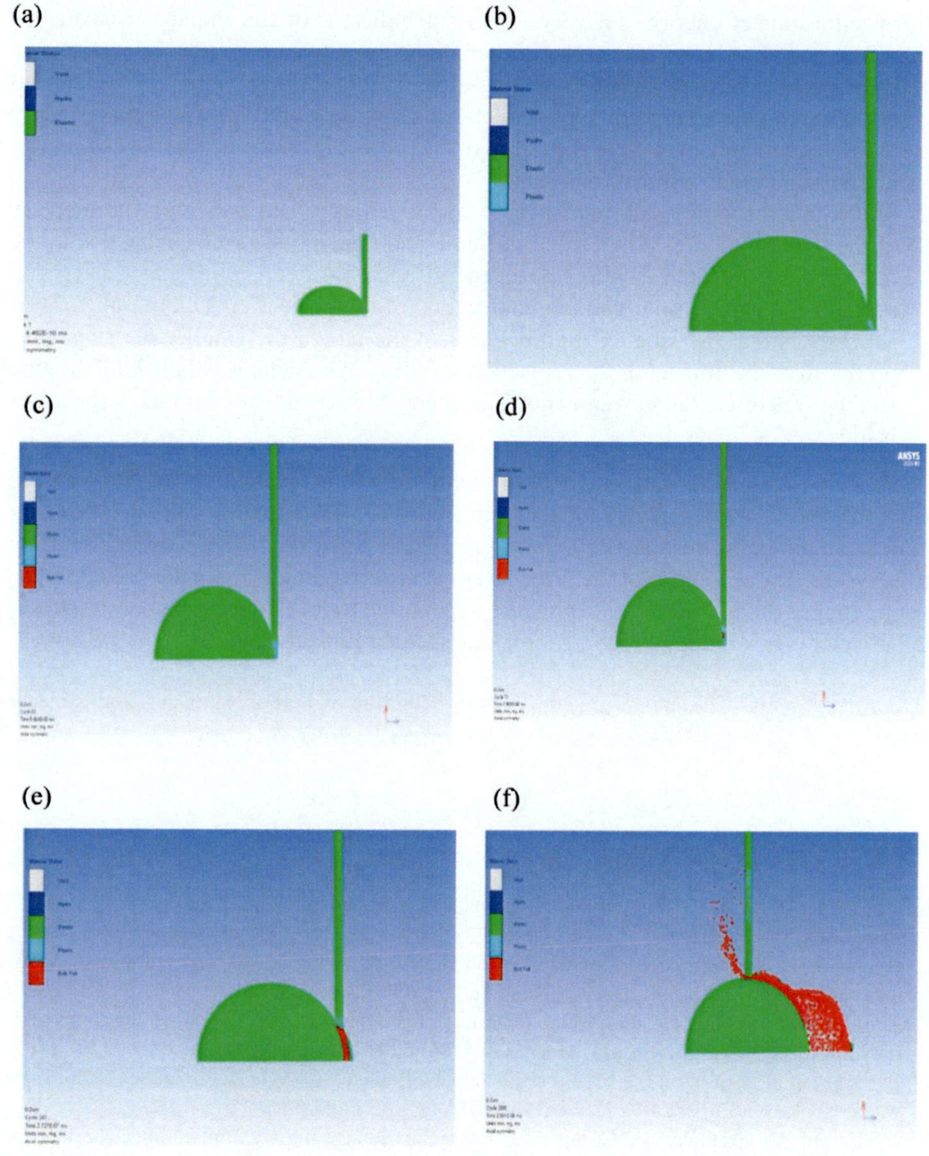

Fig. 4. Schematic diagram of the process of a projectile passing through a target

5 Conclusion

In this paper, SPH simulation modelling was carried out by using AUTODYN software and the following conclusions were obtained.

(1) Specific impact energy is a very important indicator of the impact resistance of the target. In order to investigate the relationship between specific impact energy and target thickness, three sets of numerical simulation experiments were carried out. Through the simulation experiments, it is found that the specific impact energy increases with the increase of the radius of the projectile under the high-speed impact of the projectile in the macron scale.
(2) At the macroscopic scale, the specific impact energy increases with the increase of the target thickness. However, our numerical simulation results show that at the microscopic scale, the specific impact energy decreases with increasing target thickness for the same initial velocity and radius of the projectile. The reason for this anomaly is that the smaller the thickness of the target, the smaller the fragments produced by the impact of the projectile, resulting in an increase in the total fracture surface area of the target fragments. The energy absorbed per unit mass of the target is higher.

However, this study still has the shortcoming of not examining the relationship between specific impact energy and target thickness at smaller scales. The specific quantitative equations for the specific impact energy and target thickness need to be further investigated. The velocities of the projectiles we have simulated are in the kilometer per second range, and it would be worthwhile to continue to investigate whether impacts at higher velocities would also have anomalous specific impact energies.

Acknowledgments. The authors gratefully acknowledge the support of National Natural Science Foundation of China (11972234).

References

1. Drolshagen, G.: Effects of hypervelocity impacts from meteoroids and space debris. ESA: TEC-EES/2005.302/GD (2005)
2. Hyon, J., et al.: Projectile impact shock-induced deformation of one-component polymer nanocomposite thin films. ACS Nano **15**(2), 2439–2446 (2021)
3. Zhu, Y., et al.: Scaling for the inverse thickness dependence of specific penetration energy in polymer thin film impact tests. J. Mech. Phys. Solids **161**, 104808 (2022)
4. Zhao, C., et al.: Study on the blast resistance performance of single steel plate concrete shear walls. Explosion Impact **40**(12), 24–36 (2020)
5. Manasrah, A.: GraphenNanoComposites for hypervelocity impact applications.The University of Mississippi. ProQuest Dissertations Publishing 10262932 (2017)
6. Hyon, J., et al.: Extreme energy absorption in glassy polymer thin films by supersonic micro-projectile impact. Mater. Today **21**(8), 817–824 (2018)

Simulation Research on Air Distribution Optimization of Civil Aircraft Cabin Ventilation System

Rongjia Jin(✉)

Shanghai Aircraft Design and Research Institute, Shanghai 201210, China
283912047@qq.com

Abstract. The cabin ventilation system of civil aircraft is mainly composed of distribution ducting, which convey the mixed air from the mixing unit and the trim air to various areas of the cabin through the cabin distribution ducting to provide ventilation for the cabin. The design of the system directly affects the comfort of the passengers. This paper takes the cabin ventilation system of a civil aircraft as the research object, and investigates the air distribution of the left side of the rear cabin ventilation system. Flowmaster software is used to model the cabin ventilation system, and the air distribution of the cabin ventilation system is simulated, which meets the air distribution requirements of the cabin ventilation system, and provides a reference for the optimization design of the cabin ventilation system of civil aircraft.

Keywords: cabin ventilation system · Air distribution · Simulation calculation

1 Introduction

The cabin ventilation system is an important component of the air conditioning system of civil aircraft. Its function is to distribute air supply to the cabin with appropriate temperature, pressure, and humidity to ensure that passengers have a comfortable cabin temperature, pressure environment, and sufficient fresh air [1]. With the rapid development of the civil aviation transportation industry, the comfort of the aircraft cabin environment has been widely concerned, and the design of the ventilation system closely related to the cabin comfort is crucial. At home and abroad, there are many researches on the simulation and test of aircraft cabin flow field environment [2–4] and cabin air quality simulation and test [5, 6], but there are few researches on the simulation of air distribution ducting flow.

The cabin ventilation system of civil aircraft will mix the air from the mixing unit with the trim air, and then transmit it to each area of the cabin through the cabin distribution ducting to distribute air supply for the cabin with appropriate temperature, pressure and humidity, so as to ensure that the passengers have a comfortable cabin temperature, pressure environment and sufficient fresh air.

Z. Jing and X. Zhan (Eds.): ICASSE 2023, LNEE 1153, pp. 211–219, 2024.
https://doi.org/10.1007/978-981-97-0550-4_17

During the ground test of the original configuration of the cabin ventilation system of a certain civil aircraft, it was found that the test value of the cabin air distribution flow was inconsistent with the design value. The test results of the ventilation system on the left side of the rear cabin are taken as an example. Compared with the design values, the test values of the air supply flow in the emergency door area located in the front section of the rear cabin are larger than the design values, and the test values of the air supply flow in the cabin riser ducting area, lavatory area, galley and door area in the rear section of the rear cabin are smaller than the design values. In hot weather conditions on the ground, the lower temperature in the front section of the rear cabin and the higher temperature in the rear section of the rear cabin affect the comfort of passengers in the cabin. Therefore, it is necessary to optimize the air distribution of the cabin ventilation system in order to improve the air distribution uniformity and comfort of the cabin ventilation system.

2 Air Distribution Simulation Model of Cabin Ventilation System

2.1 Simulation Model of Non-standard Air Resistance Components

In this paper, the aerospace module of Flowmaster is used to model the cabin ventilation system, and the physical model is converted into a simulation model. For the simulation of the electronic equipment cabin ventilation system, the incompressible steady-state analysis model is chosen.

In order to improve the accuracy of the simulation model, it is necessary to conduct CFD simulation of the air resistance characteristics of several non-standard ducting components and interior components in the cabin ventilation system, so as to obtain the air resistance parameters of these components.

As an example of a combination of a buffer cavity and a grille for the cabin interior, due to the complexity of its geometric structure, this component cannot directly calculate its air resistance characteristics using the model library in Flowmaster. The geometric structure is shown in Fig. 1. After CFD simulation, the air resistance curve of this component can be obtained, as shown in Fig. 2. The air resistance characteristics are added to the simulation model can make the calculation results of air distribution more accurate.

The comparison table between the physical model and the simulation model of the cabin ventilation system is shown in Table 1.

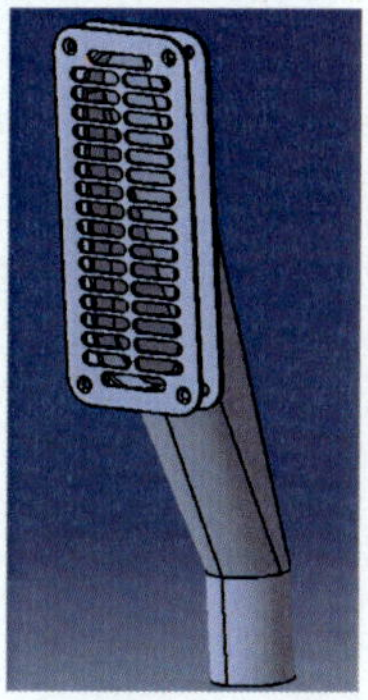

Fig. 1. Examples of geometric models for non-standard ducting component

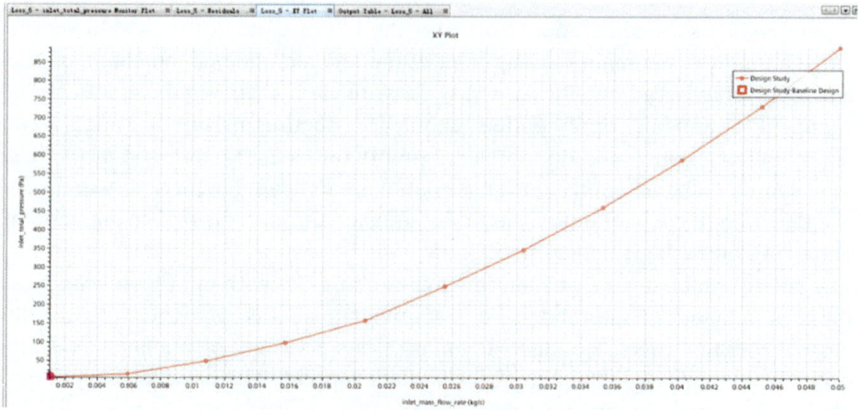

Fig. 2. Calculation results of air resistance characteristic curve

Table 1. The comparison between the physical model and the simulation mode

Boundary Name	Boundary Type
Cabin ventilation system inlet	Flow source
Air distribution ducting	Pipe, Bend, Junction, Transition, Loss
Restrictor ring	Orifice
Cabin ventilation system outlet	Pressure source

2.2 Build Simulation Models

According to the modeling principles, the modeling results of the rear left side of the cabin ventilation system for a certain type of civil aircraft as shown in Fig. 3.

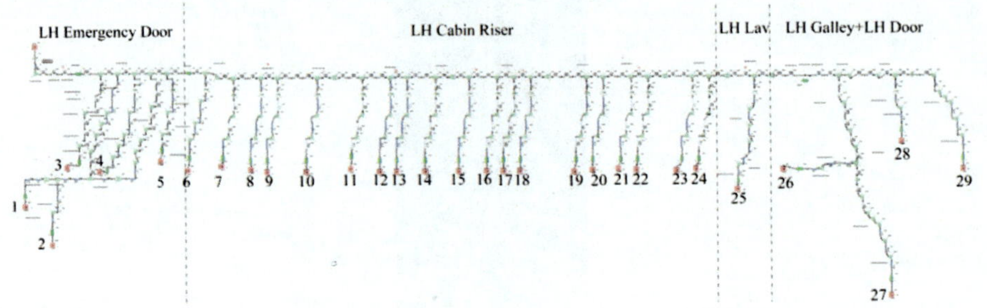

Fig. 3. Simulation model of the left side of the rear cabin ventilation system of a Certain Civil Aircraft

Figure 3 shows the simulation model for the left rear side of the cabin ventilation system. The inlet is the main ducting inlet at the left rear side of the cabin ventilation system, represented as a flow source. The outlets are various air supply outlets, including emergency door area air supply outlets, cabin riser ducting air supply outlets, lavatory air supply outlets, galley and door area air supply outlets, all expressed in the form of pressure sources. The positions of each air outlet are indicated numerically in the Fig. 3, and the flow rate of each air outlet should be adjusted through the orifice diameter of the restrictor ring on each air outlet ducting.

The key to establish air distribution simulation model with the Flowmaster is to set the input parameters of the model. For components such as pipes, bends, transitions, junctions, restrictor rings, etc., the Flowmaster has a variety of mathematical models to choose from, and has certain engineering data support. The parameter types of each simulation element are shown in Table 2.

Table 2. Air simulation parameter table of cabin ventilation system

Component Types	Parameter 1	Parameter 2	Parameter 3	Parameter 4
Pipe	Diameter	Length	Roughness	Friction coefficient
Bend	Diameter	Turning radius	Roughness	
Transition	Main diameter	Branch diameter	Branch angle	
Loss	Resistance loss coefficient	Flow area		
Restrictor ring	Pipe diameter	Orifice diameter		
Flow source	Flow	Temperature		
Pressure source	Pressure			

For the cabin ventilation system simulated in this paper, the parameter settings of each component are designed according to the system parameters of a certain type of civil aircraft.

2.3 Model Validation

In order to verify the simulation model of the left side of the rear cabin ventilation system in this paper, the simulation value and ground test value were used for comparison and verification. The verification results are shown in Fig. 4.

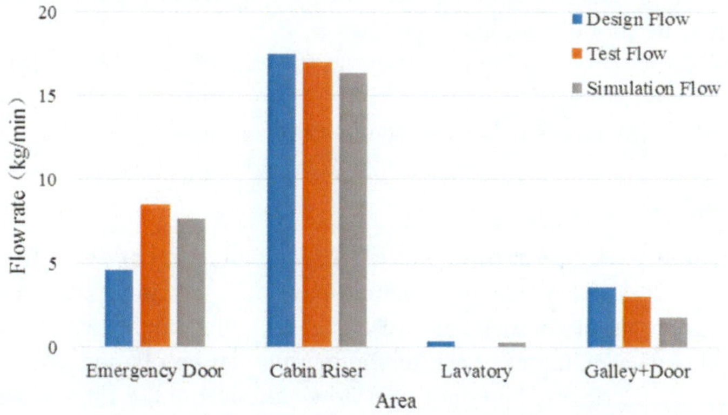

Fig. 4. Verification results of air distribution simulation model on the left side of the rear cabin

According to the verification results, the simulation value on the left side of the rear cabin also has the problem of insufficient rear flow compared to the design value. The test value in the lavatory area is 0, which is considered as a test error and is not used as a reference for validation. The simulation and test error for the galley and door areas is relatively large, with a value of - 40%. It is considered that the reference flow value for the galley and door areas is small, resulting in an amplification of the calculation error. In summary, the simulation model results in this article are consistent with the trend distribution of the test results, and can be used for optimal design.

3 Simulation and Results Analysis

3.1 Simulation Procedure

The simulation calculation steps for air distribution optimization of the cabin ventilation system in this paper including the matching calculation of the orifice diameter of the flow restrictor ring and the verification of the air distribution results. The specific steps are as follow:

a) Classify each component according to the module type in Table 2, and input the parameters into the calculation model;

b) Input the flow characteristic parameters of each non-standard ducting component into the loss module in the calculation model;
c) Input the target flow values of the inlet and outlet of the model into the model, and calculate the size of each flow restrictor ring through the flow balance model to obtain the optimized restrictor ring parameters on each branch;
d) Add the optimized restrictor ring parameters to the current restrictor module;
e) Add the total flow at the inlet flow source model, and add the outlet pressure of each branch to the outlet pressure source model;
f) Determining the total inlet flow according to the input conditions and add it to the flow source module of the inlet;
g) Determine the pressure of each air outlet according to the input conditions and add it to the pressure module at the outlet;
h) Adding the target flow value of each outlet to the pressure source module of each outlet;
i) The outlet flow values are calculated using an incompressible model.

3.2 Results Analysis

There are a total of 29 restrictor rings on each air outlet ducting of the ventilation system on the left side of the rear passenger compartment of a certain type of civil aircraft to ensure that each air supply branch has a restrictor ring that can be used to regulate flow. The optimized orifice diameter of the restrictor ring can be obtained by matching and calculating the restrictor ring and the target flow value using the Flowmaster, as shown in Table 3.

Table 3. Verification results of air distribution calculation model on the left side of the rear cabin

Area	Number of restrictor ring	Optimized orifice diameter of restrictor ring (m)
Emergency door area	1	0.0168
	2	0.0168
	3	0.0168
	4	0.0175
	5	0.0175
Cabin riser ducting area	6	0.0175
	7	0.0175
	8	0.0175
	9	0.0190
	10	0.0190

(continued)

Table 3. (*continued*)

Area	Number of restrictor ring	Optimized orifice diameter of restrictor ring (m)
	11	0.0190
	12	0.0190
	13	0.0192
	14	0.0192
	15	0.0192
	16	0.0192
	17	0.0206
	18	0.0206
	19	0.0206
	20	0.0205
	21	0.0208
	22	0.0207
	23	0.0140
	24	0.0280
Lavatory area	25	0.0262
Galley and door area	26	0.0249
	27	0.0250
	28	0.0170
	29	0.0170

The comparison of air distribution optimization results on the left side of the rear cabin are shown in Fig. 5 and Fig. 6, including the comparison of flow design values, flow before optimization, and flow after optimization. From the comparison results, it can be seen that, with the constant total inlet flow, the flow value and design value of each outlet further converge through optimization. The average deviation between the optimized supply air flow and the design value in each region is 2%, and the maximum deviation is 6.4%, meeting the optimization goal.

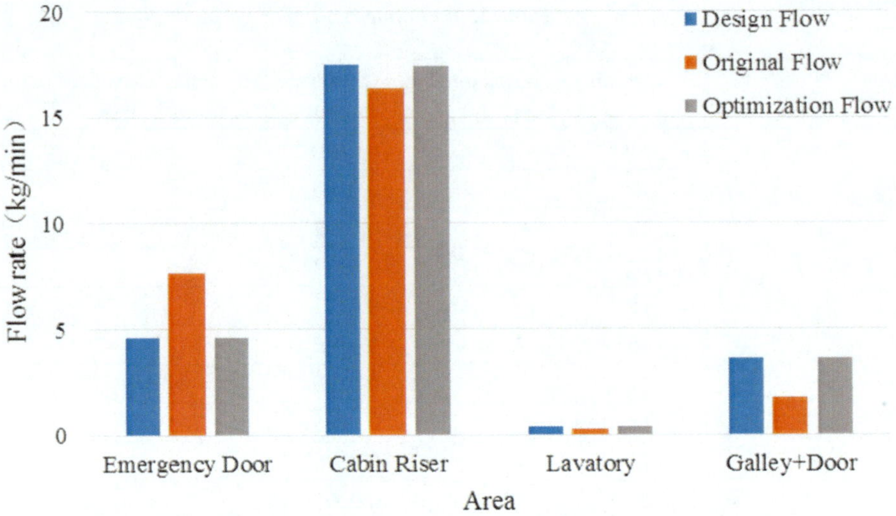

Fig. 5. Optimization results of air distribution on the left side of the rear cabin

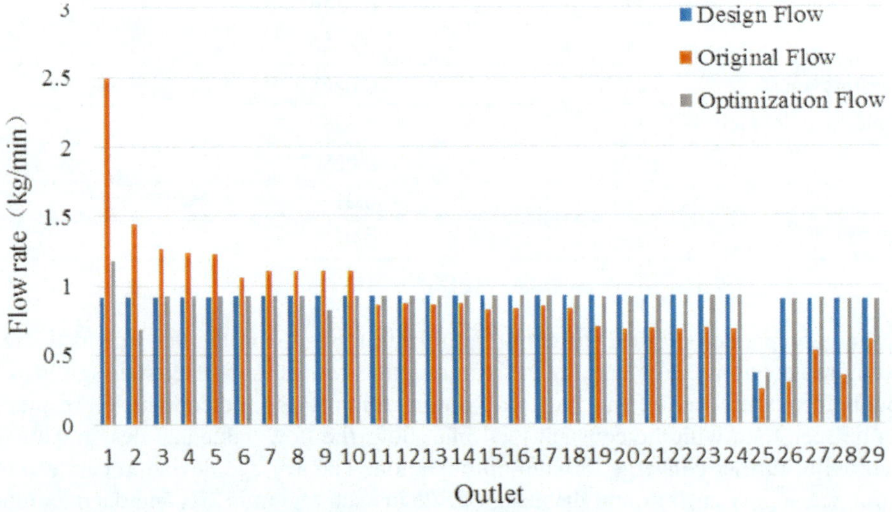

Fig. 6. Optimization results of air distribution at each air outlet on the left side of the rear cabin

4 Conclusion

In this paper, the cabin ventilation system of a certain type of civil aircraft as the research object, and investigates the air distribution of the left side air distribution system in the rear cabin.The left side air distribution system in the rear cabin was simulated using the Flowmaster, and the air distribution flow was optimized.

The results show that the air distribution simulation by using the restrictor ring size calculated by the flow balancing model can meet the optimization design objectives of the cabin air distribution system.

References

1. Bruno, L.J., Hipsky, H.W., et al.: Aircraft Environmental Control System. US20190225343A1 (2019)
2. Zhang, T.T., Li, P., Wang, S.: A personal air distribution system with air terminals embedded in chair armrests on commercial airplanes. Build. Environ. **47**, 89–99 (2012)
3. Zítek, P., Vyhlídal, T., Simeunovic, G., et al.: Novel personalized and humidified air supply for airliner passengers. Build. Environ. **45**(11), 2345–2353 (2010)
4. Mo, H., Hosni, M.H., Jones, B.W.: Application of particle image velocimetry for the measurement of the airflow characteristics in an aircraft cabin. ASHRAE Trans. **109**, 101–110 (2003)
5. Zhang, T., Chen, Q.Y., Lin, C.H.: Optimal sensor placement for airborne contaminant detection in an aircraft cabin. Hvac R Research **13**(5), 683–696 (2007)
6. Nagda, N.L., Hodgson, M.: Low relative humidity and aircraft cabin air quality. Indoor Air **11**(3), 200–214 (2001)

Check for updates

Investigation of the 2-D Distribution Form of Bearing Stress in a Single-Bolt Single-Shear Metal-Composite Hybrid Joint

Yuxiang Lin and Xiaojing Zhang(✉)

Shanghai Jiao Tong University, Shanghai, China
{Linyushit,zhangxj76}@sjtu.edu.cn

Abstract. The force analysis of bolts in single-bolt, single-shear mechanical joints is a nonlinear problem that affects the calculation of flexibility and forms the basis of the calculation of the load-distribution ratio in multi-bolt joints. This paper simplifies the force analysis of bolts based on beam theory and establishes a method for determining the distribution of bolt bearing stress in a two-dimensional plane through the first derivative of the shear force of the bolt shank. Based on the good strain correlation between the finite element model and experimental results, three types of composite material layering, three total thickness variations, six metal plate thicknesses ratios, and eight levels of fastener clamping force were considered. The fitting of data indicates that linear distribution can effectively describe the majority of joint conditions. However, when the composite layer is too thick (such as S3) or the thickness of the metal plate exceeds 2.4 times the thickness of the composite plate (such as M6), uniform and linear distributions cannot be used to describe the situation.

Keywords: composite material · mechanical joints · bearing stress distribution

1 Introduction

In mechanical joint structures, the force analysis of bolts is a complex issue. Bolt deformation includes bending, shearing, and compression. Therefore, a rational design of bolt diameter, composite material thickness, and layup sequences is crucial in structural design. To solve these kinds of problems, many scholars (Tate, 1946; Tang, 2012; Lian et al., 2014; Sinthusiri and Nassar, 2020) simplified bolt shanks to beams subjected to bending moments and bearing forces. This allows for the further calculation of fastener flexibility. Tate and Rosenfeld (Tate, 1946) assumed that the bearing stress was uniformly distributed over the cross-sections of plates and bolts, then shearing and bending deformations were calculated. Tang (2012) also assumed that the bearing stress on the bolt shank was uniformly distributed along the thickness. Lian et al. (2014) assumed that the bearing stress was linearly distributed along the bolt shank, and provided the parameters k and q for the linear distribution. Then the parameters were corrected by finite element models. Meanwhile, through extracting the contact force of the finite element

Z. Jing and X. Zhan (Eds.): ICASSE 2023, LNEE 1153, pp. 220–230, 2024.
https://doi.org/10.1007/978-981-97-0550-4_18

model Sinthusiri and Nassar (2020) pointed out that when the plate thickness ratio is 1:3, the bearing stress on the thick plate is non-uniform along the thickness direction and is closer to a quadratic function distribution. Kou calculated the distribution of bearing stress using the elastic foundation model (Kou and Xu, 2018), but as a function of bolt deflection. It is easy to calculate the flexibility of a beam using the uniform distribution of bearing stress and linear distribution of bearing stress, although higher-order polynomials, exponential functions and other functional forms can cause more difficulties. There has been no validation of the distribution form of bearing stress on bolt shanks in the above studies. It is worth noting that many factors can also affect the form of bearing stress distribution. The thickness of the plate directly affects the length of the shank. In practical engineering, the ratio of bolt diameter to single plate thickness is often taken to be around 1.5 (Tang, 2012), making the bolt a short and thick beam that is primarily subjected to shear deformation to prevent excessive bending deformation. Since the contact relationship between the bolt and the plate is changed, clearance fit is one of the factors that mostly influence the distribution of bearing stress (McCarthy and McCarthy, 2003). Starting with the analysis of the bolt shank unit cell, this paper presented a method for obtaining the distribution of bearing stress through theoretical deduction. After the finite element models were validated by static testing of single-bolt, single-shear hybrid joints, the correctness and applicable range of uniform distribution (Uni) and linear distribution (Lin) was further studied.

2 Theoretical Basis of the Analysis

2.1 Derivation of the Bearing Stress Distribution

Viewing the bolt as a two-dimensional beam with a rectangular cross-section, a unit cell of a bolt shank is analyzed in Fig. 1(a). Due to the bearing force applied, shear forces $F_x(z_1)$ and $F_x(z_2)$ are respectively exerted on the upper and lower surfaces, according to the Timoshenko Beam Theory (Temusinco and Gail, 1978). To ensure the balance of the unit cell, the bending moments M_1 and M_2 generated during the bending process of the bolt should also be taken into account on the upper and lower surfaces. The distribution form of the bearing stress on the bolt is unknown. The force equilibrium equation in the x direction is shown in Eq. 1.

$$F_x(z_1) - F_x(z_2) = \int_{z_1}^{z_2} q(z) \tag{1}$$

When the length of the unit cell is extremely small, the bearing stress can be assumed to be uniformly distributed on the unit cell, as q(z) in Fig. 1(b). This process can also be viewed as taking the derivative of both sides of Eq. 1, or to be written as Eq. 2.

$$\lim_{z_1 - z_2 \to 0} \frac{F_x(z_1) - F_x(z_2)}{z_1 - z_2} = q(z_1) \tag{2}$$

Therefore, the linear density of the bearing stress in a certain coordinate z is equal to the first-order derivative of shear force at this place, with a unit of N/mm. If the confidence of the linear distribution of shear force in the thickness direction is high, then the distribution of the bearing stress tends to be uniformly distributed (Uni). Also, if shear force is quadratically distributed, then bearing stress is linearly distributed (Lin). In the following text, CS and MS is the abbreviation of composite-shank contact area and metal-shank contact area, respectively, as shown in Fig. 2.

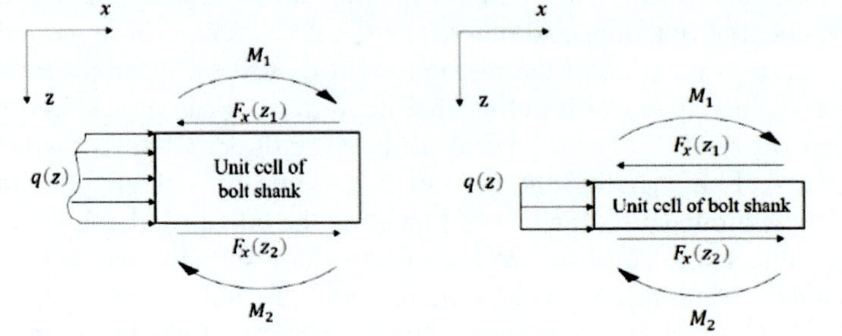

(a) Force analysis of an unit cell (b) force analysis of a very short unit cell

Fig. 1. Force analysis of bolt shank unit cells

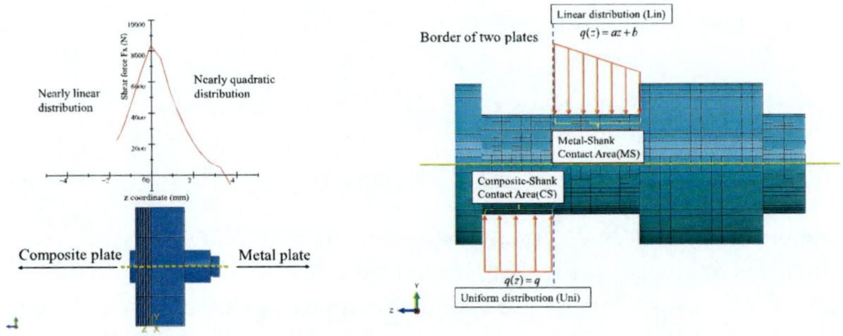

(a) Shear force of bolt shank (b) Abbreviations of contact areas and distribution forms

Fig. 2. The distribution forms and abbreviations

2.2 Statical Method Used

In statistics, the coefficient of determination R^2 is used measuring confidence in the predictive power of a certain distribution. It can be calculated by Eq. 3:

$$R^2 = 1 - \frac{\sum_{i=1} (\hat{y}_i - y_i)^2}{\sum_{i=1} (y_i - \bar{y})^2} \tag{3}$$

The range of R^2 is between 0 and 1. In this research, an R^2 value greater than 0.95 is considered to indicate a high correlation between the predicted model and the real situation.

3 Modeling and Testing

3.1 Test Set-Ups

Tensile testing of metal-composite hybrid joints in single-bolt, single-shear configuration was conducted on the MTS Landmark 250 kN testing machine. Six specimens were tested, and the final result was taken as the average of each specimen in each group. The metal plate material was aluminum alloy with an elastic modulus of 72.4 GPa and a thickness of 0.4 times that of the composite plate. The composite plate was made of 24 layers of balanced and symmetrically laid composite, with a total thickness of 4.48 mm. The out-of-plane bending of the specimens was constrained using bending limit fixtures according to standard (ASTM International, 2017). The fastener used was a high-locking bolt with a diameter of 4.76 mm made of Inconel 718 nickel-chromium alloy with an elastic modulus of 199 GPa. The set-up of the experiment was shown in Fig. 3. Eight strain gauges were arranged on the surface of each test piece to measure the local strain level during the tensioning process. The geometric parameters of the test piece and the position of the strain gauges was shown in Fig. 4. After that, the test pieces were repeatedly loaded to 5kN four times in the linear section, and the strain results of the four peak loads were averaged.

Fig. 3. Test set-up

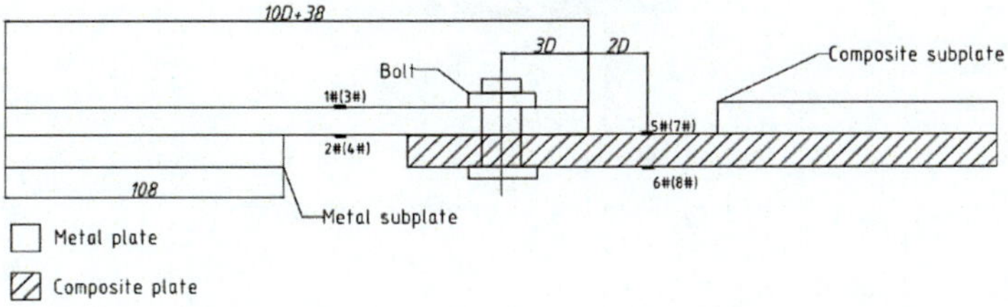

Fig. 4. Specimen and geometrical parameters

3.2 Finite Element Modeling

A finite element model was developed using Abaqus 6.14–2 software. The entire model was divided by structured mesh using C3D8R elements, and the composite layers were established using the solid method. The tangential contact relationship between the bolt shank and the hole was set as frictionless, and the tangential contact between the plates was set with a friction coefficient of 0.15 to simulate the load shared by the frictional force in the test. All degrees of freedom at the end face of the composite plate were constrained, and a 5kN tensile force was applied at the end face of the metal plate, with freedom in the tensile direction released, as shown in Fig. 5. A simplified fixture was assembled and the out-of-plane direction was constrained to ensure that the structure did not experience excessive out-of-plane bending. After the analysis was completed, the tensile component of the free body cut was extracted at intervals of 0.01mm in the post-processing module as the shear force acting on the bolt shank, as shown in Fig. 6. The data was extracted and the shear force diagrams of Fig. 2(a) were plotted.

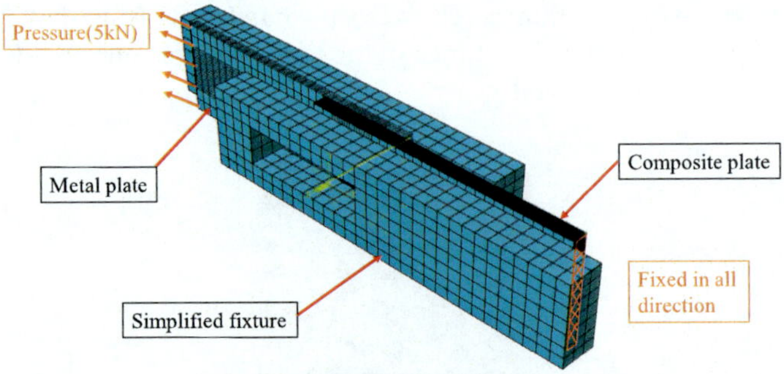

Fig. 5. Boundary conditions

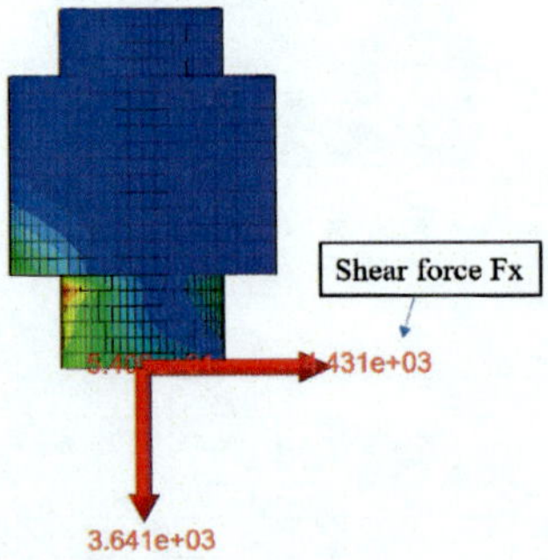

Fig. 6. Free body cut of the bolt shank

3.3 Model Validation

For mechanical joints, it is difficult to directly obtain the local stresses of the bolt or hole through experiments. By placing strain gauges on the surface of the two plates, the load-carrying situation of the bolt and plate can be effectively reflected. This paper verified the validity of the finite element model by comparing the strain gauge data of the two plates. Comparison between experimental and finite element strains of strain gauges 1# ~ 8# are presented in Table 1. It can be seen that the maximum strain error did not exceed 15%, indicating the validity of the finite element model.

Table 1. Strain comparison between test results and FE results

Strain	1#	2#	3#	4#	5#	6#	7#	8#
Test results	837	1007	843	1075	680	288	771	260
FE results	896	1044	896	1043	772	245	755	260
Error	7.05%	3.67%	6.29%	−2.98%	13.53%	−14.93%	−2.08%	0.00%

4 Analysis Results

4.1 Analysis Matrix

Four factors are listed in Table 3: layup sequence, total composite thickness, metal thickness ratio, and clamping force, which are represented by the codes L, S, M, and F, respectively. Three kinds of typical lay-up sequences are chosen according to (He et al., 2016), as shown in Table 2. The total laminate thickness is 3.73mm in group L. On the basis of L1 layup, the total thickness of the laminate is changed by altering the number of times the layup is stacked. Here, s/2s/3s represent 20/40/60 layers respectively. The thickness ratio is defined as the ratio of metal plate thickness to composite thickness. The clamping force is defined as the pressure generated on the bolt section by the tightening effect. The specific parameter variations are shown in Table 2–3 (where bold letters represent the basic level).

Table 2. Typical lay-ups to be analyzed.

Lay-up codes	Lay-up sequences	Angle ratio (0°:90°:45°)	Total number of layers
L1	[45°/0°/−45°/0°/45°/90°/−45°/0°/45°/−45°]s	30:60:10	20
L2	[45°/0°/−45°/0°/90°/0°/45°/0°/−45°/0°]s	50:40:10	
L3	[45°/0°/0°/−45°/0°/0°/0°/90°/0°/0°]s	70:20:10	

Table 3. Parameter variations.

Parameters	code	Lay-ups	Thickness ratio	Clamping force(MPa)
Thickness of composite plate	S1	s (20 layers)	1	200
	S2	2 s (40 layers)		
	S3	3 s (60 layers)		
Thickness of metal plate	M1	s	0.2	200
	M2		0.4	
	M3		0.6	
	M4		1	
	M5		2	
	M6		3	
Clamping force	F1	s	1	0.2
	F2			50
	F3			100
	F4			150
	F5			200
	F6			300
	F7			400
	F8			600

4.2 The Effect of Layup Sequences (L)

Figure 7 illustrates the determination coefficient R^2 obtained by linear and quadratic regression of shear stress using the least squares method for different layup sequences. For the given three layup sequences L1, L2, and L3, the R^2 of the linear distribution (Lin) of bearing stress is high in both the Metal-Shank contact area (MS) and the Composite-Shank contact area (CS), indicating that a linear distribution is desirable. However, for MS, the R^2 of uniform distribution (Uni) is only around 0.9, which is insufficient to describe the distribution of bearing stress. For the L3 layup, the R^2 of uniform distribution is slightly lower than 0.95, while for L1 and L2 it exceeds 0.95. This indicates that the increase in the proportion of 0° layup will also slightly increase the non-uniformity of the distribution of bearing stress. The data also shows that the influence of the three layup proportions on the distribution of bearing stress is extremely small, although the layup proportions can affect the bearing strength of the joint (Chen et al., 2013). Table 4 shows the distribution forms of bearing stress in the MS and CS under the variable levels of L1 to L3.

4.3 The Effect of Layup Thickness (S)

Under the S1 layup, the bearing stress distribution in CS can be approximated by Uni, while the bearing stress distribution in MS can be described by Lin, as shown in Fig. 8.

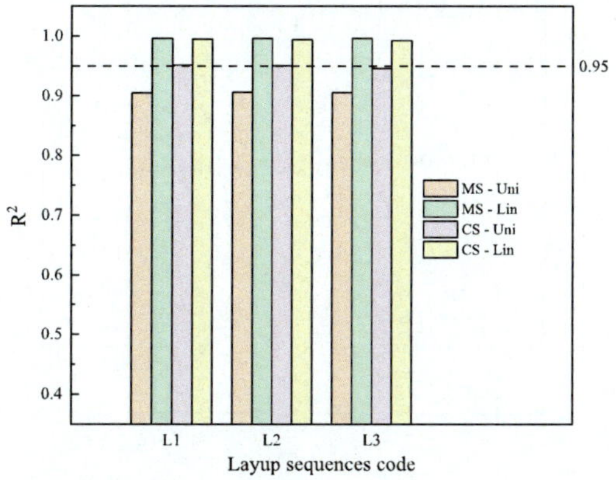

Fig. 7. R^2 in MS and CS under different layup sequences (L)

Table 4. Suitable distribution forms for each contact area.

	MS-Uni	MS-Lin	CS-Uni	CS-Lin
L1		✓	✓	✓
L2		✓	✓	✓
L3		✓		✓

Linear distribution cannot even describe the bearing stress in MS under the S3 layup. As the thickness of the layer increases, the fitting performance of each line decreases accordingly. This is due to the fact that a larger thickness represents a longer bolt length, which alters the local contact relationship, resulting in a more uneven distribution of bearing stress. So it's worth noticing that the total thickness, or to say, bolt length is an important factor to be taken into consideration. In engineering practice, the length/diameter ratio is usually 1–2 (Aircraft Design Manual editorial Board, 1983) (Table 5).

4.4 The Effect of Thickness Ratio (T)

Table X reveals that in the majority of cases, linear distribution can effectively fit the bearing stress distribution on both the MS and CS. When the thickness of the metal plate exceeds 2.4 times of the composite thickness, the distribution is too uneven that none of the two types can describe it well. In this situation, the bolt length/diameter is around 2.7. Additionally, it can be observed that the relatively thicker plate in this structure exhibits a more uneven distribution of bearing stress. When the thickness of the composite plate is the same as that of the metal plate (M4), the bearing stress distribution on the metal plate displays a stronger degree of non-uniformity, as shown in Fig. 9.

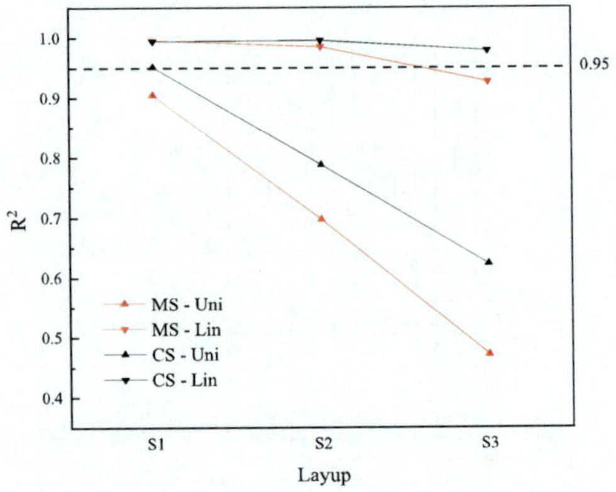

Fig. 8. R^2 in MS and CS under different layup thickness (S)

Table 5. Suitable distribution forms for each contact area.

	MS-Uni	MS-Lin	CS-Uni	CS-Lin
S1		✓	✓	✓
S2		✓		✓
S3				✓

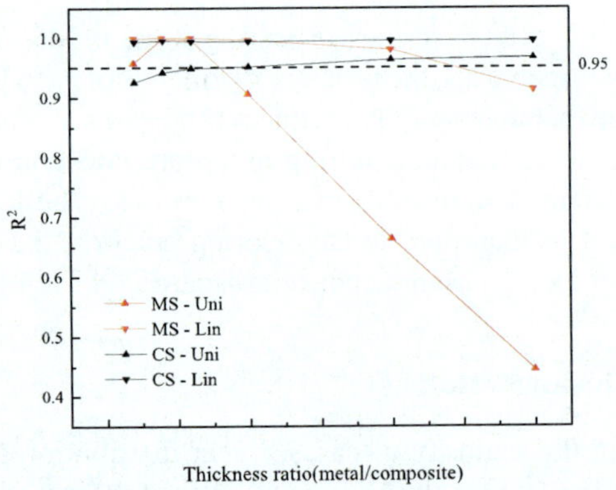

Fig. 9. R^2 in MS and CS under different thickness ratio (M)

4.5 The Effect of Clamping Force (F)

The trend depicted in Fig. 10 indicates that as the clamping force increases, the value of R^2 also increases correspondingly. This is because the load transmitted through the bolt-hole interaction is smaller, resulting in less deformation of the bolt shank, which to some extent improves the non-uniformity of the bearing stress. It's worth noting that according to [M], the bearing stress of high-strength bolts is generally optimal at around

Table 6. Suitable distribution forms for each contact area.

	MS-Uni	MS-Lin	CS-Uni	CS-Lin
M1	✓	✓		✓
M2	✓	✓		✓
M3	✓	✓		✓
M4		✓	✓	✓
M5		✓	✓	✓
M6			✓	✓

Table 7. Suitable distribution forms for each contact area.

	MS-Uni	MS-Lin	CS-Uni	CS-Lin
F1		✓		✓
F2		✓		✓
F3		✓		✓
F4		✓		✓
F5		✓	✓	✓
F6		✓	✓	✓
F7		✓	✓	✓
F8		✓	✓	✓

20% of the bolt material's ultimate strength (about 1300MPa for Inconel 718), and may not necessarily reach the 600MPa of the test plan group. In this paper, the value above 400MPa is only a model exploration.

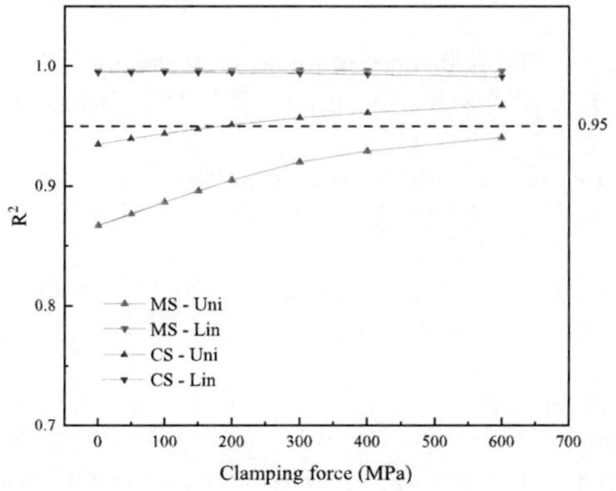

Fig. 10. R^2 in MS and CS under different thickness ratio (F)

5 Conclusion

1) The 2-D distribution of the bearing stress on the bolt shank can be derivated from the first-order derivation of the shear force distribution. The linearly distributed shear force indicates uniformly distributed bearing stress, and the quadratically distributed shear force indicates linearly distributed bearing stress. It is relatively convenient to calculate the flexibility of fasteners using these two distribution forms.
2) The layup sequences barely affect the bearing stress, but as the thickness of the composite layers increases, the distribution of compression force becomes significantly more uneven due to the longer length of the bolt shank.
3) The bearing stress distribution in most of the joint situations can be solved by linear distribution. However, when the length/diameter ratio of bolt shanks exceeds 2.7, the R^2 of linear distribution is lower than 0.95, which means it does not adequately describe the distribution at this point.
4) Larger clamping force obviously relieves the unevenness of the bearing stress. Along with the increase in clamping force, the R^2 of uniform distribution and linear distribution shows an upper trend.

References

Aircraft Design Manual editorial Board: Aircraft design, manual National Defense Industry Press, Beijing (1983)

ASTM International (2017): Standard test method for bearing response of polymer matrix composite laminates. https://www.astm.org/d5961_d5961m-17.html. Accessed 26 Mar 2023

Chen, K., Liu, L., Wang, H.: Combined effects and mechanism of interference fit and clamping force on composite mechanical joints. Acta Mater. Compos. Sin. **30**(06), 243–251 (2013). https://doi.org/10.13801/j.cnki.fhclxb.2013.06.035

He, B., Ge, D., Mo, Y., Du, X.: Tensile tests and strength estimation for double-lap single-bolt joints in T800 carbon fiber reinforced composites. Acta Mater. Compos. Sin. **33**(07), 1540–1552 (2016). https://doi.org/10.13801/j.cnki.fhclxb.20151020.002

Kou, J., Xu, F.: Three-dimensional bearing stress distribution at bolt-hole in quasi-isotropic laminates. Acta Mater. Compos. Sin. **35**(12), 3360–3367 (2018). https://doi.org/10.13801/j.cnki.fhclxb.20180315.001

Lian, T., Huang, Q., Yin, Z., Su, X.: The modification of the formula for the aircraft structural fastener's flexibility. J. Mech. Strength **36**(04), 555–559 (2014). https://doi.org/10.16579/j.issn.1001.9669.2014.04.019

McCarthy, M.A., McCarthy, C.T.: Finite element analysis of effects of clearance on single shear composite bolted joints. Plast. Rubber Compos. **32**(2), 65–70 (2003). https://doi.org/10.1179/146580103225001390

Sinthusiri, C., Nassar, S.A.: Load distributions in bolted single lap joints under non-central tensile shear loading. ASME Int. Mech. Eng. Congress Expos. **84485**, V02AT02A005 (2020). https://doi.org/10.1115/IMECE2020-24652

Tang, Z.: Investigation of fastener flexibility for joint structure. Sci. Tech. Eng. **12**(09), 2178–2182+2192 (2012). https://doi.org/10.3969/j.issn.1671-1815.2012.09.044

Tate, M.B.: Preliminary investigation of the loads carried by individual bolts in bolted joints (1946). https://ntrs.nasa.gov/citations/19930081668

Temusinco, S., Gail, J.: Mechanics of Materials. Science Press, Beijing (1978)

The Impact of Bleeding Slot Angles on the Performance of a Compressor

Jiang Junjian, Qiang Xiaoqing(✉), and Zhao Ziheng

Shanghai Jiao Tong University, Shanghai, China
Qiangxiaoqing@sjtu.edu.cn

Abstract. Several stages of a multi-stage axial compressor have been investigated to study the impact of interstage bleeding on compressor total performance and local flow field. Three different angles of bleeding slot (30°, 45°, 60°) were presented to make the numerical simulation of the flow field. As the result shows, the slot has a significant effect on the performance of the compressor. At a slot angle of 45°, the main flow loss is low and the efficiency is the highest. At an angle of 60°, the efficiency drops, the main flow loss is excessive and there is a large separation.

Keywords: bleeding air · compressor · slot angle · numerical simulation

1 Introduction

Bleeding air systems are critical to the safe and reliable operation of aircraft engines. Compressor bleed air at the proper pressure and temperature can be used for cabin air conditioning, engine intake anti-icing and cooling of hot components. In short, the bleed air system, which is different from the mainstream, is an airflow network that connects various parts of the engine. Therefore, the bleed air system is closely related to the overall performance of the engine and the performance of each component. With the development of aero-engine technology and the increase of turbine intake temperature, the flow rate of the bleed air system is also increasing.

In recent years, scholars at home and abroad have conducted various studies on the influence of the bleeding air system structure on the performance of the compressor. Siggeirsson [1, 2] made an insightful study of the effect of air diversion on stability. Leishman [3, 5] studied circular holes, slot and slot with ramps, and then compared multiple groups of conditions under different bleeding air rate. Peltier et al. [6]conducted a numerical study on the aerodynamic characteristics of two slot structures. Mare [7] investigated the effect of changes in the mean flow field due to different bleeding air rates on the aerodynamic flexibility of compressor components. Grimshaw et al. [9] have studied the losses when the bleeding air rates vary under the condition that the radius of the corner of the slot is determined. Maadi [8]attempted at supersonic inlets for bleeding air. Sivagnanasundaram [10, 11] has conducted profound research for centrifugal presses Cong-Truong Dinh [12] studies inside the bleeding slot by changing the inlet and outlet

Z. Jing and X. Zhan (Eds.): ICASSE 2023, LNEE 1153, pp. 231–239, 2024.
https://doi.org/10.1007/978-981-97-0550-4_19

angle of the slot and so on. Wellborn [?] designed a direct bleeding air slot at the shroud of the blade and placed it at the leading edge of the stator for suction.

Liang Dong et al. [13] used Rotor35 as an arithmetic example for numerical simulations, and used the structure without the bleeding air slot as a reference for the calculation of the round hole and slot bleeding air, respectively. Zhou Wenxiao [14] of Shanghai Jiao Tong University studied the intermediate stage of a multi-stage axial flow compressor and designed three different structures of rectangular, circular and heterogeneous bleeding air slots according to the required maximum bleeding air volume.

From the above analysis, it can be learned that domestic and foreign scholars have paid more attention to the influence of the overall structural changes of the induced gas tank on the performance of the compressor, and most of them are single-stage compressors or single-row cascade. On this basis, this paper focuses on the effect of the variation of the angle between the bleeding air slot and the main flow channel on the performance characteristics of the compressor. The effect of the bleeding air on the overall performance as well as the local characteristics of the multi-stage compressor is further investigated.

2 Numerical Methods

This paper uses NUMECA for CFD operations. The compressor model simulated in this paper selects the intermediate stage of a compressor and its two adjacent stages of blades for study. The location of the intermediate bleeding air is located at the outlet of the static blade. The three different slots structures are shown in Fig. 1.

The width of the slots are all 33 mm and the height of the slots are all 66mm. But the angles between the slots and the main flow channel are 30°, 45° and 60°, respectively, and the bleeding air flow rate is 1% of the main flow rate. Numeca Fine/Turbo software's pre-processing module AutoGrid5 is used to generate the mesh of the main flow channel. The blade model adopts O4H topology, and the mesh near the wall is partially encrypted.

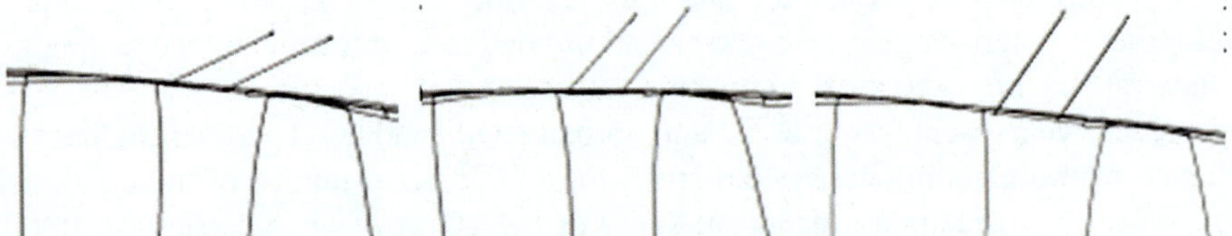

Fig. 1. Structures of the Different Angle slots

2.1 Mesh Independence

In order to find the grid configuration that can reduce the computational effort and ensure the accuracy of the computational results, the mesh independence was studied and five grids were made for the computation of single channel 0.9M, 1.8M, 3.6M, 7.2M, and 14.4M(unit:million), respectively. And absolute total pressure and isentropic efficiency have plotted against each other in Figs. 2 and 3.

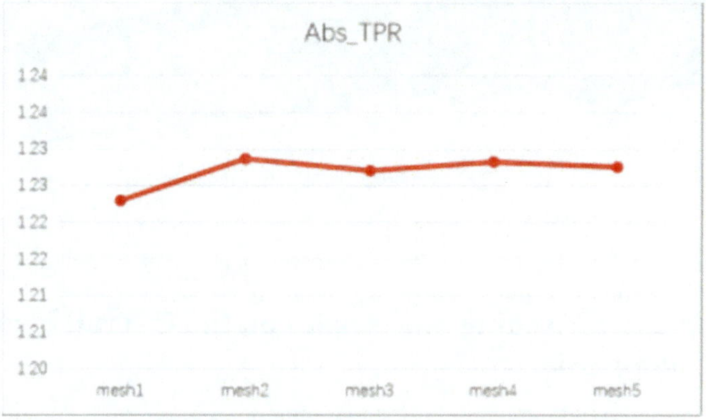

Fig. 2. Absolute Total Pressure Ratio in Different Meshs

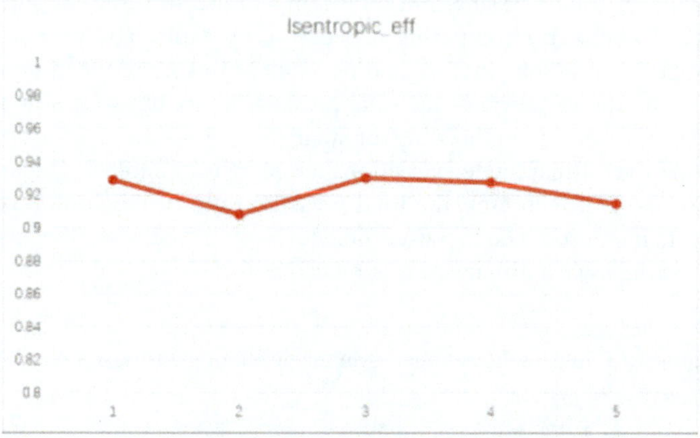

Fig. 3. Isentropic efficiency in Different Meshs

By comparison, a single channel 3.6M mesh is selected for numerical calculations. The grid is shown in Fig.4. The numerical solution is based on perfect Gas, the modelling of turbulence is based on the Spalart-Allmaras equation, the mathematical model is turbulent Navier-Stokes equation. The time step is based on the fourth-order Runge-Kutta method, and the hybrid plane method is used for the stator and rotor intersection interface. The distribution of total pressure, total temperature and airflow angle along span of the blade and then these data can be the boundary condition at the inlet. The outlet of the channel and the slot are given a static pressure at the mean radius in a radial equilibrium equation, respectively.

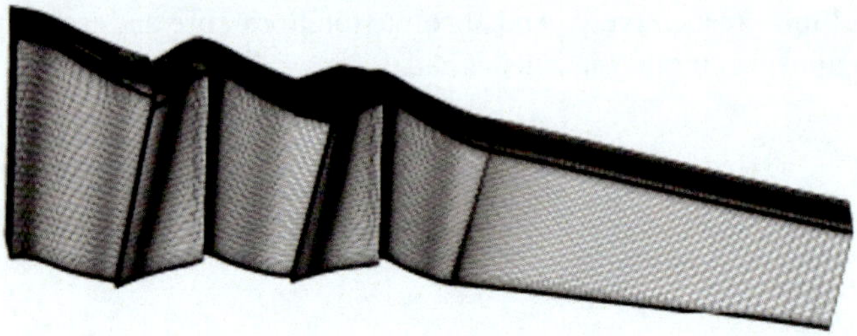

Fig. 4. Mesh of the Channel

3 The Impact of Bleeding Slot Angles on the Overall Performance of the Compressor

Firstly, analysis the impact of bleeding air on the overall performace of the compressor. The solution of the original channel without slot will be shown to compare with the solution with slot.

The contours of total pressure and total temperature in different angles slot have shown in Figs. 5 and 6. By comparing the total temperature and pressure contours it can be seen that the total temperature and pressure varies more when the angle of slot is 45°and 60°, and the variation of the total pressure along the span slows down at the outlet of the static blade of the bleeding air stage.

Some of the overall parameter variations are shown in Table 1. It can be seen that as the angle of the slot increases, the total pressure ratio of the compressor decreases and the flow rate increases. The highest efficiency is achieved when the angle of the slot reaches 45° and the lowest efficiency is achieved at 60°.

Table 1. Compare the parameters in the different plans of slot bleeding

Parameters	ori	30°	45°	60°
mass flow(kg/s)	605.85	626.66	636.13	648.90
TPR	1.4473	1.4088	1.3838	1.3177
efficiency	0.90352	0.90750	0.91256	0.85641

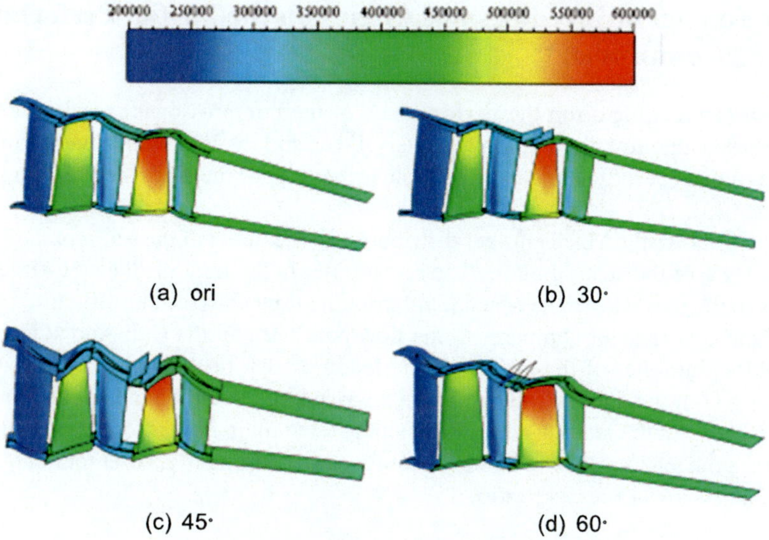

Fig. 5. Contour of total pressure (Pa) in different angles slot

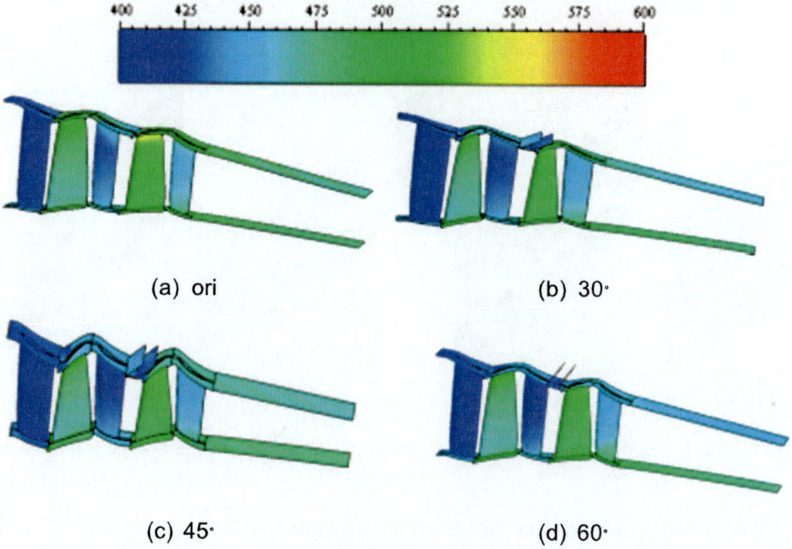

Fig. 6. Contour of total temperature (K) in different angles slot

4 The Impact of Bleeding Slot Angles on the Partial Performance of the Compressor

The effect of bleeding airon the performance of the compressor is mainly reflected in the changes in the flow field at the boundary layer and the end wall. Firstly look at the variation of each aerodynamic parameter near the inlet of the slot when the angle of the slot is changed.

Figure 7 shows the Mach number distribution at the outlet of the stator of the bleeding air stage, 99% of the span of blade. The rectangular in the figure is the slot's inlet which connects to the main flow. The Mach number of the flow with the slot structure increases in the place near the slot compares to the flow path without the slot. And at an angle of 60, the Mach number of the airflow at the leading edge of the slot is even close to 1. Combining with the data in Table 1, it can be assumed that when the angle is increased the angle between the leading edge of the slot and the main flow is too large, causing the airflow past the leading edge to be sucked out faster, resulting in a lower total mainstream pressure and a lower pressure ratio.

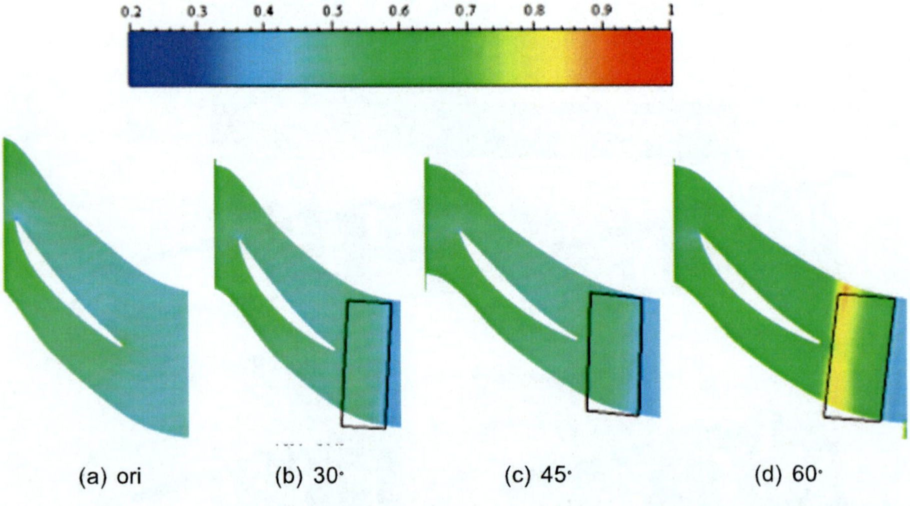

Fig. 7. Contour of mach number at the outlet of stator of bleeding air stage (span = 99%)

Figure 8(a) and 8(b) show the variation of the angular deflection angle of the airflow at the inlet and outlet of the stator of the bleeding air stage along the blade span, respectively. Where Vr and Vm are the circumferential and meridional velocities respectively. It can be clearly observed that the effect of the bleeding air on the airflow extends to the full blade height and affects the flow field before the bleeding air slot. At the inlet of the stator, the deflection angle of the flow path with the bleeding air structure increases by approximately 2 below 0.8 span.

Above 0.8 span, the deflection angle of the flow with the bleeding air structure is reduced by approximately 2. At the outlet, however, the deflection angle increases across the entire span, with a maximum increase of almost 20 degrees.

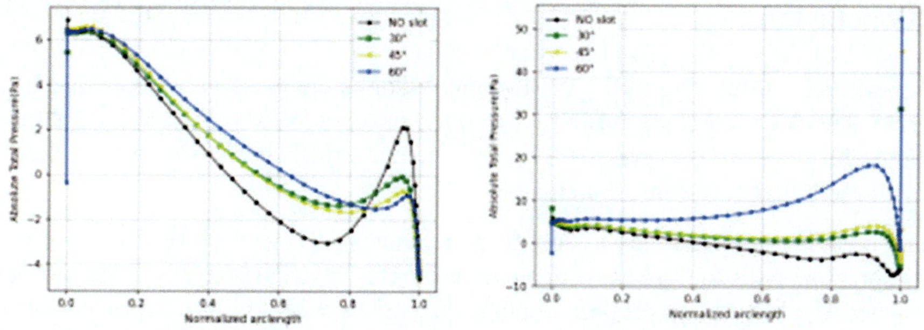

(a) inlet (b) outlet

Fig. 8. Variation of airflow angle along span of blade at the stator of bleeding air stage

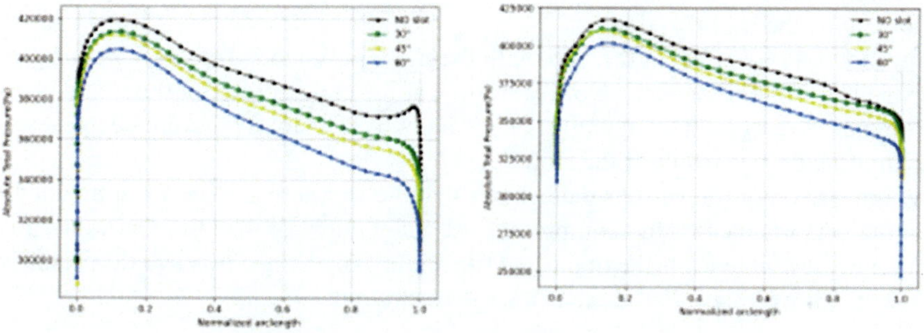

(a) inlet (b) outlet

Fig. 9. Variation of total pressure along span of blade at the stator of bleeding air stage

It is worth noting that the increase is greatest at both the inlet and the outlet with an bleeding air slot angle of 60. Combined with the Mach number distribution in Fig.7, it is assumed that the large angle of the bleeding air slot causes the airflow at the leading edge of the bleeding airslot to separate, resulting in a sharp change in airflow deflection angle at locations greater than 0.8 span. When the angle of the slot is 30° and 45°, the airflow is less separated, resulting in a more stable deflection angle.

Figure 9(a) and 9(b) show the distribution of the total pressure along the span at the inlet and outlet of the stator respectively. It can be found that the total pressure decreases along the span at both the inlet and outlet. As the angle of the slot increases, the total pressure decreases more and more. Above 0.8 span, the total pressure decreases the most.

5 Conclusion

This paper takes three stages of a multi-compressor as the research objects. Air is took off from stator of second stage in order to find out the impact of the bleeding structure on the performance of compressor. Through the numerical simulation of flow field and some research, here are some conclusions:

1. The bleeding structure has a significant impact on the overall performance of the compressor, with the greatest increase in efficiency occurring when the slot angle reaches 45°, but a sharp drop in efficiency occurring when the slot angle reaches 60°. The total pressure coefficient decreases and the flow rate increases as the angle of the slot increases.
2. The slot has a more pronounced effect on the flow field at the end wall. At 0.99 span, the main flow mach number increases compared to the flow channel without the slot structure, and the mach number of airflow increases more at the leading edge of the slot than at the trailing edge.
3. The slot has less effect at the hub, with the more pronounced changes starting at 0.8 span. The angle of deflection of the airflow at the end wall is smoother and changes from negative to positive at angles of 30° and 45°. The total pressure decreases in all span of blade as a result of the slot.
4. A comparison of the various data shows that the slot angles have a significant effect on the performance of the compressor. At a slot angle of 45°, the main flow loss is low and the efficiency is the highest. At an angle of 60°, the efficiency drops, the main flow loss is excessive and there is a large separation.

Acknowledgements. This work was supported by the National Science and Technology Major Project (J2019-II-0017–0038) and the Fundamental Research Funds for the Central Uni- versities.

References

1. Siggeirsson, E. M. V., Andersson, N., Wallin, F.: [ASME ASME Turbo Expo 2018: Turbomachinery Technical Conference and Exposition - Oslo, Norway (Mon- day 11 June 2018)] Volume 2B: Turbomachinery - Numerical and Experimental Study on Bleed Impact on Intermediate Compressor Duct Performance," 2018
2. Siggeirsson, E. M. V. Andersson, N. Burak Olander, M.: Numerical and Ex- perimental Aerodynamic Investigation of an S-Shaped Intermediate Compressor Duct With Bleed," in ASME Turbo Expo 2020: Turbomachinery Technical Confer- ence and Exposition, 2020, vol. Volume 2D: Turbomachinery, V02DT36A017, doi: https://doi.org/10.1115/gt2020-15336. [Online]. Available: https://doi.org/10.1115/GT2020-15336
3. B. et al.: Effects of inlet ramp surfaces on the aerodynamic behavior of bleed hole and bleed slot off-take configurations, J. Turbomachinery, 2007
4. Leishman et al.: effects of bleed rate and endwall location on the aerodynamic behavior of a circular hole bleed Off-Take, J Turbomach, 2007
5. Leishman, B.A., Cumpsty, N.A.: Mechanism of the Interaction of a Ramped Bleed Slot With the Primary Flow. J. Turbomach.Turbomach. **129**(4), 669–678 (2007). https://doi.org/10.1115/1.2752193

6. Peltier, V. Dullenkopf, K. Bauer, H.-J.: Numerical investigation of the aerody- namic behaviour of a compressor Bleed-Air system, In: ASME Turbo Expo 2014: Turbine Technical Conference and Exposition, 2014, vol. Volume 2A: Turbomachin- ery, V02AT37A021, doi:https://doi.org/10.1115/gt2014-25822
7. Mare, L. D., Simpson, G., Mueck, B., Sayma, A. I.: Effect of bleed flows on flutter and forced response of core compressors, In: Asme Turbo Expo: Power for Land, Sea, & Air, 2006
8. Grimshaw, S. D. Brind, J. Pullan, G. Seki, R.: Loss in axial compressor bleed systems, In: ASME Turbo Expo 2019: Turbomachinery Technical Conference and Exposition, 2019
9. Maadi, S. R. Sepahi-Younsi, J.: Effects of bleed type on the performance of a supersonic intake, 2021
10. Sivagnanasundaram, S., Spence, S., Early, J., Nikpour, B.: An impact of various shroud bleed slot configurations and cavity vanes on compressor map width and the inducer flow field, In: ASME Turbo Expo: Turbine Technical Conference Exposition, 2016
11. Sivagnanasundaram, S., Spence, S., Early, J., Nikpour, B.: An investigation of compressor map width enhancement and the inducer flow field using various configurations of shroud bleed slot, 2010, pp. 1701–1710
12. Dinh, C.-T., Vu, D.-Q., Kim, K.-Y.: Effects of Rotor-Bleeding Airflow on Aerodynamic and Structural Performances of a Single-Stage Transonic Axial Com- pressor. Int. J. Aeronaut. Space Sci **21**(3), 599–611 (2019). https://doi.org/10.1007/s42405-019-00239-5
13. Dong, LI., Zhiqiang, G., Canping, C., Shan, G.: Numerical simula- tion on flow field of axial compressor after bleeding, J. Aerospace Power, vol. 32, no. 2, p. 9, 2017
14. W. X. Zhou, X. Q. Qiang, J. F. Teng and Y. U. Wen-Sheng, "Numerical Research of Inter-stage Bleeding in a Multistage Axial Compressor," Energy Conservation Technology, vol. 29, no. 6, pp. 490-494, 2011https://doi.org/10.3969/j.issn.1002-6339.2011.06.003

Numerical Investigation of Diffuser Curvilinear Meridional Shape on Centrifugal Compressor Stage Performance

M. O. Goryukhin, V. V. Eremenko(✉), A. E. Mikhailov, A. B. Mikhailova, and D. G. Krasnoperov

Federal State Budgetary Educational Institution of Higher Education, Ufa University of Science and Technology, Bashkortostan, Russia
eremenko.vladislav@net.ugatu.su, vlad.eremenko@yandex.ru

Abstract. Centrifugal compressors are widely used in a variety of applications due to their advantages. Centrifugal compressors have a larger frontal area for a given mass flow rate compared with axial compressors, which limits their application in aircraft propulsion systems. Centrifugal compressors for aircraft/rotorcraft applications are commonly made up of a radial impeller with a radial vaned diffuser and a radial-to-axial bend with an exit guide vane. The main requirements for a centrifugal compressor are high efficiency, surge margin and dimensional limitations. This is a study of a diffuser design concept with curvilinear meridional shape aiming to reduce the overall radial dimensions of the centrifugal compressor stage without any sacrifice of compressor performance.

Keywords: Centrifugal Compressors · Vaned Diffuser · "Radial-axial" Diffuser

1 Introduction

Centrifugal compressors are applied to turbochargers, auxiliary power units, and gas turbine engines because of their high-pressure ratio, relatively wide operating range, simple design, and cost benefits [1].

Centrifugal compressors have a larger frontal area for a given mass flow rate compared with axial compressors, which limits their application in aircraft propulsion systems. Increased frontal area causes the propulsion system drag to increase and complicates the centrifugal compressor-turbine matching. Centrifugal compressors for aircraft/rotorcraft applications are commonly made up of a radial impeller with a radial vaned diffuser and a radial-to-axial bend with an exit guide vane [2].

In research to date, the majority of research and design has focused on extending the operating range or improving the compressor efficiency. However, reducing the compressor dimensions and frontal area is of great importance for aircraft/rotorcraft applications.

High pressure compressor exit corrected mass flow rate is one of the fundamental limitations for the design of turbofan engines for regional aircraft or business jets. Axial

Z. Jing and X. Zhan (Eds.): ICASSE 2023, LNEE 1153, pp. 240–249, 2024.
https://doi.org/10.1007/978-981-97-0550-4_20

high-pressure compressor with rear centrifugal compressor stage is a concept that allows to overcome the aforementioned limitation, increase the engine overall pressure ratio, and design bypass ratio. Radial dimensions of axial-radial high-pressure compressor rear centrifugal compressor stage have a major effect on overall engine architecture and components matching.

This is a study of a diffuser design concept with curvilinear meridional shape aiming to reduce the overall radial dimensions of the centrifugal compressor stage without any sacrifice of compressor performance.

2 Baseline Compressor

This study was carried out on an open-source state-of-the-art centrifugal compressor stage – NASA High Efficiency Centrifugal Compressor for Rotorcraft Applications (HECC) (Fig. 1) [3]. Experimental centrifugal compressor stage NASA HECC design is based on the famous CFD validation case NASA CC3 compressor stage [4]. NASA HECC stage overall performance data is shown in Table 1.

Table 1. Design-intent performance ($N_C = 100\%$, 0.012 inch tip clearance, suppressed inlet $p_{0,0} = 11$ psia)

Parameters	Value
Rotational speed	21789 RPM
Pol. Eff	>0.88
Total to total pressure	4.85
Inlet flow rate	10 kg/s
Stability Margin	12

The main parts of the HECC geometry are a set of three blade rows figuring out the radial impeller, radial vaned diffuser, and the axial exit guide vane, Fig. 1. In this configuration, the impeller is composed of fifteen pairs of main and splitter blades, the diffuser row is composed of twenty pairs of main and splitter blades and EGV has sixty cascade-type blades. The impeller blades have elliptical leading edges. Trailing edges are also elliptical which differs from the typically blunter trailing edges of trimmed impellers. Impeller blades are backswept and have a 0.3 mm tip clearance.

The measurements of the NASA HECC used in the present study for validation are taken from the experiments which have been conducted in NASA Small Engine Components Test Facility. The Small Engine Components Test Facility and compressor measurements stations are shown in Fig. 2.

In these experiments, HECC impeller was operated at five different speeds in the range between surge and choke lines. The geometry of the HECC, the data of the measurements, and the measurement instrumentations used in the test rig are available in the NASA contractor report [3].

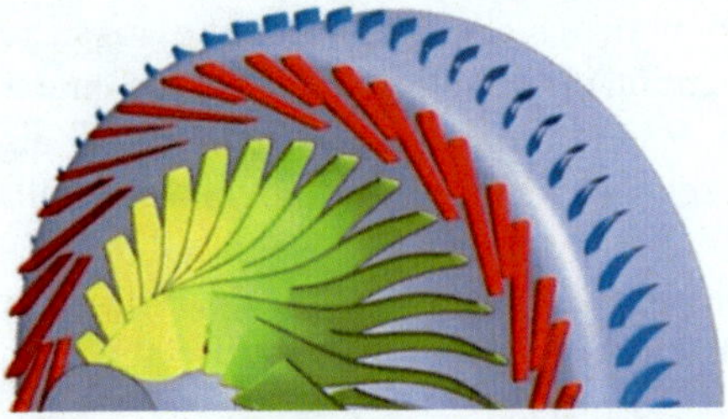

Fig. 1. Experimental centrifugal compressor stage NASA HECC

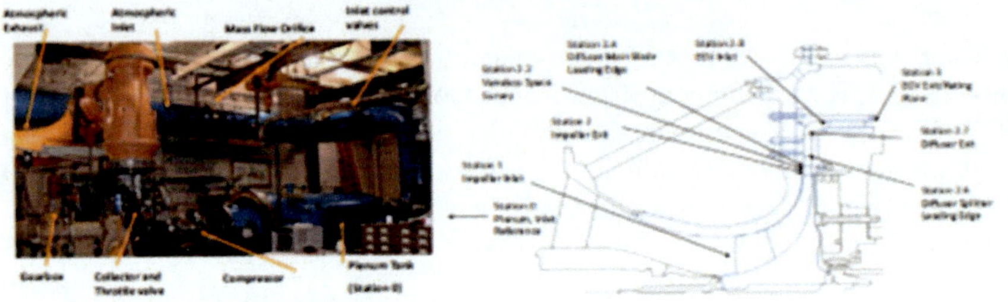

a b

Fig. 2. a – Small Engine Components Test Facility, b – Compressor Measurement Stations

3 Centrifugal Compressor Numerical Modelling Validation

3.1 Numerical Investigation Set-Up

The commercial CFD code Ansys CFX was used to conduct numerical simulations for this investigation, and ANSYS Turbogrid was used for the meshing of all fluid domains. The computational domain consists of the rotating impeller, stationary diffuser, and EGV. In the CFD analysis for both the impeller and the diffuser one main and one splitter blade are selected and the EGV is represented by only one blade.

At the inlet of the centrifugal compressor, the total pressure of 101325 Pa and the total temperature of 288,15K are defined for airflow. The design rotational speed of the impeller is 21789 rpm, in the present study 90%, 95%, 100%, and 105% of this design speed are used. There is an elongated inlet domain aimed to form a realistic velocity field with a developed boundary layer at the impeller inlet Mass flow exit boundary conditions were implemented near design loading while average static pressure exit boundary conditions were applied near the choke operating condition [5]. The change in the type of boundary conditions is justified due to the high sensitivity of the calculation to the static pressure in the region of minimum values [6]. One blade from each blade set is used to decrease computational costs and save computational time. As a result of this path selection from inlet to outlet in the computational domain, the angles for the impeller, diffuser, and the EGV are 24°, 18°, and 6°, respectively. Thus, the pitch ratio between these three domains is not equal which requires the usage of the mixing-plane interface between the domains. This treatment allows for flow in both directions across the interface and averages the flow leaving the impeller circumferentially in such a way as to conserve momentum, mass, and energy [7]. The NASA HECC stage computational domain is shown in Fig. 3. As suggested in the literature a BSL-EARSM turbulence model is selected [8, 9].

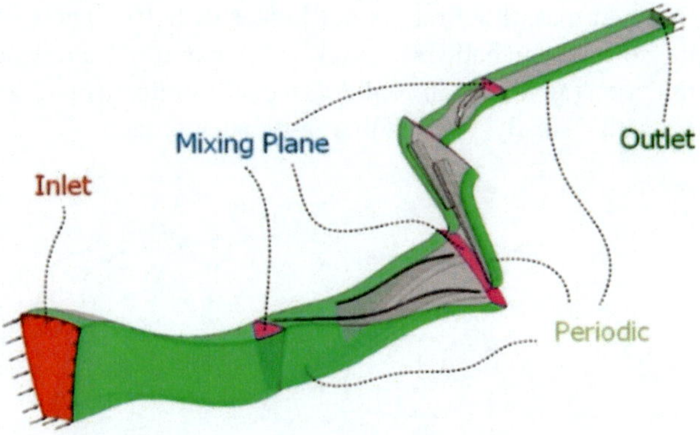

Fig. 3. NASA HECC stage computational domain

In the CFD simulation, governing equations were discretized using the three-dimensional finite-volume method; a high-resolution advection scheme is used to solve the equations more accurately in this study. Furthermore, convergence criteria were determined by ensuring the root-mean-square(RMS) values of mass and momentum were under $1 \times 10 - 5$ and imbalances in mass, momentum, and energy were below $1 \times 10 - 2$ [10] (Fig. 4).

Fig. 4. y^+ in the computational domain

The maximum values of y^+ occur at the interface between the impeller and the vaned diffuser, as well as on the impeller blades, due to the highest magnitudes of absolute velocities in these areas [5]. Therefore, y^+ of the wall-adjacent nodes will be averaged and evaluated over these areas.

3.2 NASA HECC Model Validation

CFD results computed for NASA HECC compressor are compared to data in Fig. 5. The overall trends of the CFD predictions show good agreement with rig data, considering that the CFD model uses the same running tip clearance for all operating speeds. The difference between experimental and predicted data considering the total-to-total pressure ratio and corrected mass flow rate is not higher than 1%. The CFD model tends to overpredict total-to-total adiabatic efficiency, the maximum discrepancy is about 5%. Overall, these simulations gave us high confidence in our ability to predict the compressor stage performance with a steady, RANS/mixing-plane approach.

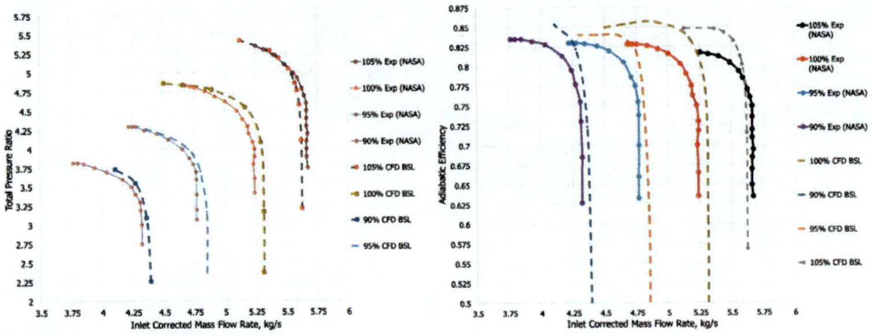

Fig. 5. Performance map of NASA HECC stage measurements (solid) versus steady-state CFD results (dashed) at design clearance

4 Vaned Diffuser and Exit Guide Vane Optimization

The meridional Mach number distribution is characterized by hub flow separation in the radial-to-axial bend channel, Fig. 6, which has a negative effect on stage efficiency and operational range. The radial vaned diffuser in the NASA HECC stage has reduced radial dimensions compared with the baseline NASA CC3 stage. Flow separation in the radial-to-axial bend channel is caused by increased radial diffuser exit velocity in the NASA HECC stage.

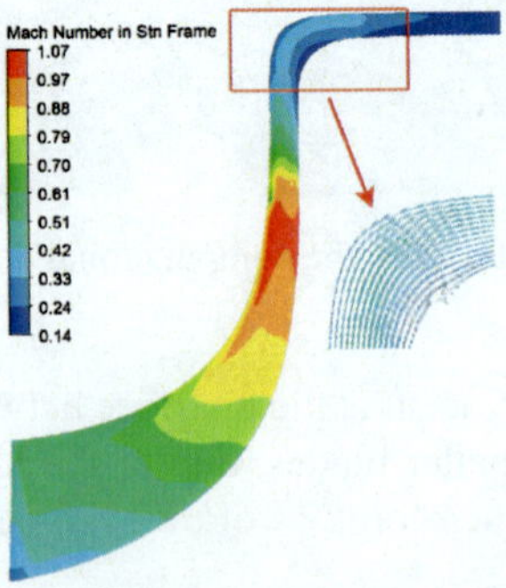

Fig. 6. NASA HECC Mach number meridional distribution

An objective of this study was to achieve a 5–10% radial reduction of the diffuser, bend channel, and exit guide vane flow path with no overall performance penalty. Six different diffuser-bend channel-exit guide vane system architectures are considered, Fig. 7.

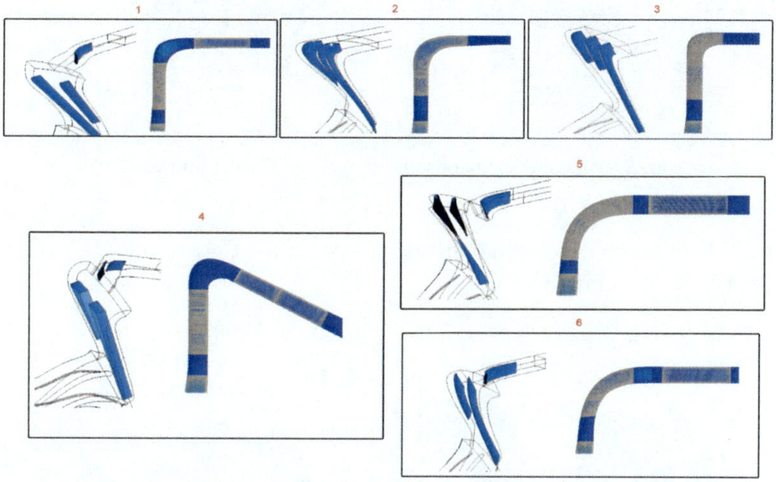

Fig. 7. Centrifugal compressor stage diffusion systems designs

Description of centrifugal stage diffusion systems designs on Fig. 7:

Case 1 – baseline diffusion system;

Case 2 – Radial vaned diffusion system with integral service routings with one main and one splitter blade [11, 12];

Case 3 – Radial vaned diffusion system with integral service routings with one main and two splitter blades [11, 12];

Case 4 – circumferentially-split diffusion system with inboard and outboard airfoil segments [13];

Case 5 – curvilinear in meridional shape "radial-axial" diffuser with axial exit guide vane;

Case 6 – optimized curvilinear in meridional shape "radial-axial" diffuser axial with exit guide vane.

Based on a series of CFD calculations of various diffusion system designs, it was identified that Case 6 provides reduced diffusion system total pressure loss and reduced exit guide vane radial dimensions. In addition, due to the improved flow structure in the diffusion systems blade rows, the stage operating range has been increased.

A comparison of the radial dimensions and lengths of Case 1 and Case 6 is shown in Fig. 8.

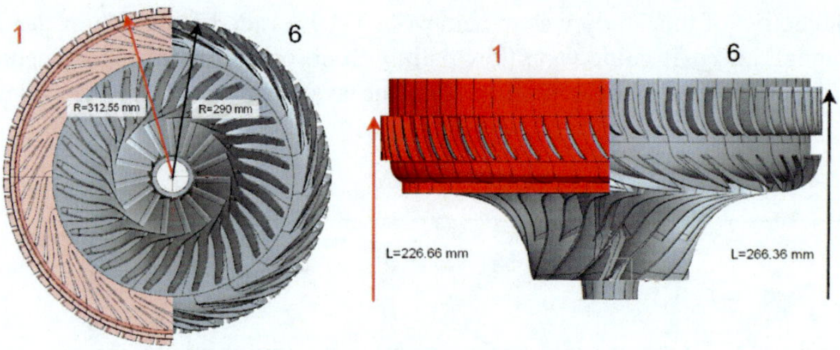

Fig. 8. Radial dimensions and lengths of Case 1 and Case 6

Flow fields in compressor stage diffusion system for Case 1 and Case 6 are shown on Figs. 9, 10, 11, 12, and 13.

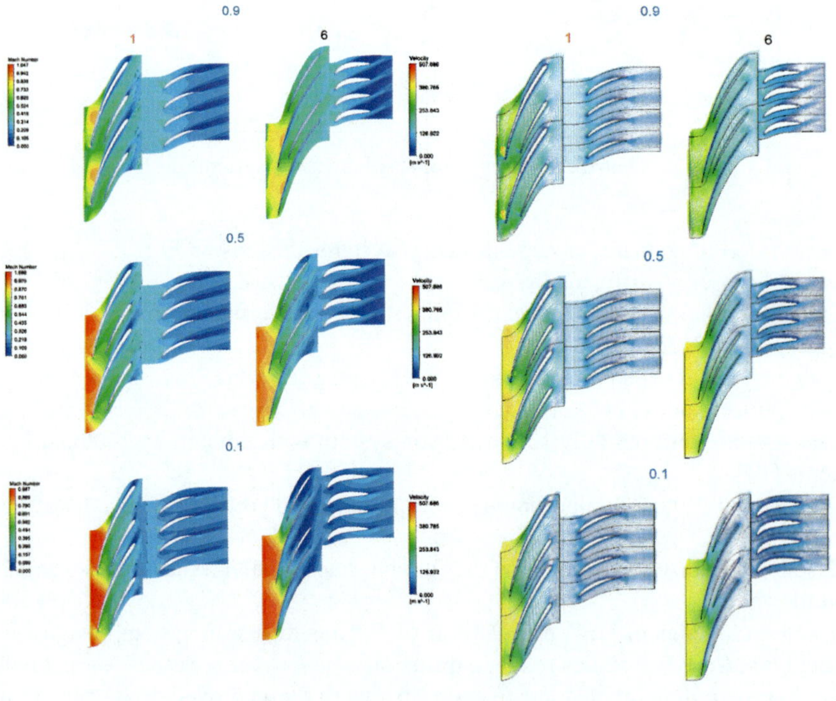

Fig. 9. Blade-to-Blade Mach number distribution and velocity vector distribution for Case 1 and Case 6 at 10% span, 50% span and 90% span ($P_{aveout} = 485\text{kPa}$)

Comparison of blade passage flow field shows a Mach number reduction in Case 6 with a curvilinear "radial-axial" diffuser, reduction of flow separation at span 50% and span 90%. This leads to an increase in the operating range by 3% and stage total-to-total pressure ratio by 0.48%.

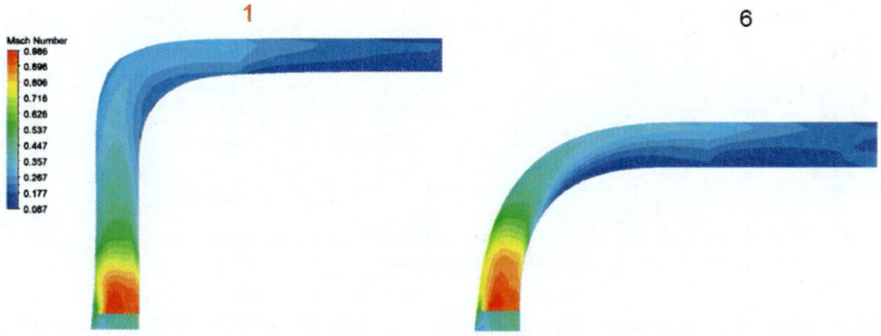

Fig. 10. Meridional Mach Number distribution for Case 1 and Case 6 ($P_{aveout} = 485$kPa)

At the same time, the flow field in the meridional plane shows that diffusion system diameter reduction does not significantly increase the flow separation area.

Fig. 11. Meridional Turbulence Kinetic Energy distribution for Case 1 and Case 6 ($P_{aveout} = 485$kPa)

As shown in Fig. 11, energy loss and entropy rise in the diffusion system is mainly influenced by an impeller tip clearance flow field. The meridional shape of the "radial-axial" diffuser and diffuser blade inlet and outlet angles are the variables used to optimize the diffusion system flow field.

In Fig. 12, you can see a smoother decrease in velocity in the diffusing system with a curvilinear "radial-axial" diffuser, which has a favorable effect on the total pressure loss. Figure 13 shows the overall performance for the baseline NASA HECC compressor stage and proposed compressor stage with a curvilinear "radial-axial" diffuser.

A proposed compressor stage with a curvilinear "radial-axial" diffuser has reduced by 7.2% radial dimensions compared with the baseline NASA HECC stage. As shown in

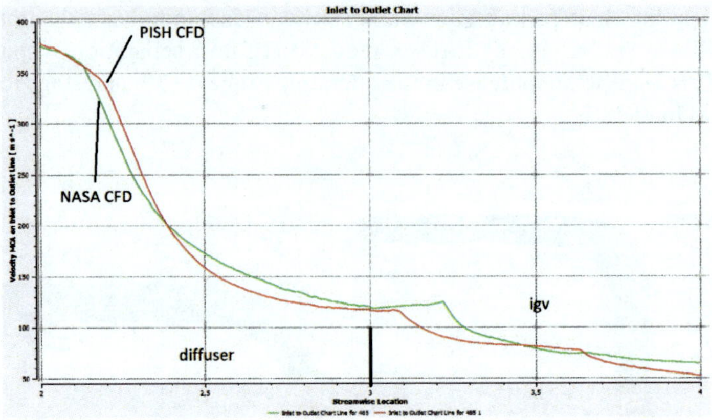

Fig. 12. Comparison of mass-flow averaged velocity distribution in diffusion system for Case 1 and Case 6 (P_{aveout} = 485kPa)

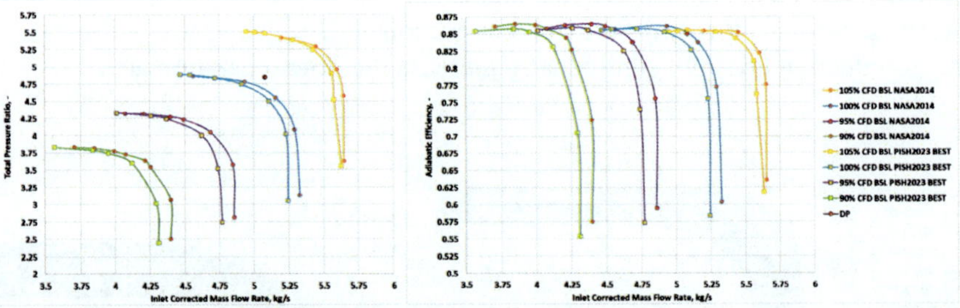

Fig. 13. Overall performance of baseline NASA HECC compressor stage and proposed compressor stage with curvilinear "radial-axial" diffuser

Fig. 13, the reduction of stage radial dimensions has no negative effect on the compressor stage's overall performance (Table 2).

Design point surge margin is defined as:

$$S_M = \left(\frac{PR_S/\dot{m}_S}{PR_D/\dot{m}_D} - 1 \right) \bullet 100\%.$$

Table 2. Surge margin performance for design point

S_M, %	
NASA CFD 10,07	PISH CFD **13,47**

A proposed compressor stage with a curvilinear "radial-axial" diffuser has an increased design-point stall margin by 3% compared with the baseline NASA HECC stage.

5 Conclusion

In this study, systematic CFD analysis has been carried out for a radial compressor stage NASA HECC to reduce radial dimensions with no overall performance penalty. The main conclusions are:

1) Radial vaned diffusion system with integral service routings with one main and one/two splitter blades with no exit guide vane is not an effective diffusion system with reduced radial dimensions considering a big overall performance penalty. The same results are obtained for circumferentially-split diffusion systems with inboard and outboard airfoil segments.
2) A diffusion system with a curvilinear meridional shape "radial-axial" diffuser and axial exit guide vane provides an operational range increase with no adiabatic efficiency and pressure ratio penalty.
3) Additive manufacturing process could be utilized effectively to manufacture the centrifugal compressor stage diffusing system with a curvilinear "radial-axial" diffuser and axial exit guide vane.

The results obtained will help in understanding the physics of the processes occurring in the vaned diffusers and exit guide vanes of centrifugal compressors.

References

1. Casey, M., Robinson, C.: Radial Flow Turbocompressors. Cambridge University Press (2021)
2. Dave, S., Shukla, S., Jain, S.: Design and analysis of centrifugal compressor. In: Technology Drivers: Engine for Growth, pp. 283–288. CRC Press (2018)
3. Medic, G., et al.: High efficiency centrifugal compressor for rotorcraft applications (No. NASA/CR-2014-218114) (2014)
4. McKain, T.F., Holbrook, G.J.: Coordinates for a high performance 4: 1 pressure ratio centrifugal compressor (No. E-10833) (1997)
5. Gooding, W.J., Meier, M.A., Key, N.L.: The impact of various modeling decisions on flow field predictions in a centrifugal compressor. J. Turbomach. **143**(10) (2021)
6. Zhang, L., Kritioti, L., Wang, P., Zhang, J., Zangeneh, M.: A detailed loss analysis methodology for centrifugal compressors. J. Turbomach. **144**(5), 051013 (2022)
7. Denton, J.D.: The calculation of three-dimensional viscous flow through multistage turbomachines (1992)
8. Menter, F.R.: Best Practice: Scale-Resolving Simulations in ANSYS CFD. ANSYS Germany GmbH, 1 (2012)
9. Menter, F.R.: Turbulence Modeling for Engineering Flows. Ansys, Inc. (2011)
10. Atac, O.F., Yun, J.E., Noh, T.: Aerodynamic design optimization of a micro radial compressor of a turbocharger. Energies **11**(7), 1827 (2018)
11. Barton, M.T., Kington, H.L., Palmer, D.L.: Radial vaned diffusion system with integral service routings US7717672B2 Patent (2006)
12. Choi, J.H., Kim, S.M.: Diffuser for Compressor US20190226493A1 Patent (2018)
13. Nasir, S., Nolcheff, N.: Cristopher frost diffuser assemblies for compression systems US10989219B2 Patent (2019)

Modeling Method of Specimen Repair Techniques from Polymer Composite Material

Artem Shubin(✉) and Konstantin Shramko

Moscow, Russia
arttemshubin@gmail.com

Abstract. Paper describes three main repair techniques of aircraft polymer composite material structures: adhesive bonded external repair, mechanically fastened external doubler, stepped (overlapping) repair. Modeling method of this techniques is then proposed, justified and performed in Siemens Simcenter 3D. Results of each technique are analyzed and compared to each other. Main conclusions are: suggested modelling method is proved to be informative and reliable; bonded external repair and mechanically fastened external doubler technics are fast, but not firm repair methods, while stepped repair is more preferable for loaded structures utilized in long-term life cycles, required smooth skin finish.

Keywords: PCM · repair techniques · FEA

1 Introduction

Polymer composite materials are used very widely in different spheres of industry, while this paper is mostly concerns to aircraft industry. The reason is that, for instance, in space industry there are structural elements made of PCM (polymer composite material), but usually spacecraft are non-reusable and these structural elements doesn't subject to long term cyclic loads, thus rarely need to be repaired.

Nowadays aircraft industry began to use PCM very widely and almost all main aircraft's structural elements are made of PCM (Fig. 1), therefore there is need to repair its structures made of composite material during aircraft utilization. And the problem is that there are quite a lot of repair methods without approved static strength, moreover there are still open questions: how is better to do element modelling, how to model and what type of interaction use for glued connection between repair and parent parts.

This paper includes: description of repair methods, their finite element modelling, analysis of acquired results, comparing of different methods and recommendations for future researches. It is very important to analysis construction bearing strength of undamaged and repaired specimen. The reason is: even though structure is repaired, there is a possibility of stress concentration appearance, which may cause failure in the most stressed structural elements.

Z. Jing and X. Zhan (Eds.): ICASSE 2023, LNEE 1153, pp. 250–260, 2024.
https://doi.org/10.1007/978-981-97-0550-4_21

Fig. 1. Composite structure parts used in generations of civil aircraft (bottom – the newest)

2 Damage Types and Repair Methods

Aircraft subjected to a various loads and environmental conditions. Main damages for PCM structure in aircraft are due to fatigue, impacts and manufacture errors. There are such types of damage as [1]:

- crack
- delamination
- fiber or matrix breakage
- honeycomb cells damage
- moisture saturation
- overheat
- manufacture thickness and density variation

There are exist different methods of repair procedures of PCM. Main methods which will be discussed in this paper are: adhesive bonded external repair, mechanically fastened external doubler, stepped (overlapping) repair.

2.1 Adhesive Bonded External Repair

This repair process is quite simple, as it is just binding PCM or metallic splice (Fig. 2, a) over damaged zone using a paste or film-based adhesive (Fig. 2, b). Before that it's important to clean the surface for adhesion reasons. This can be done by using a vacuum or cloth to take out dust, and degrease surface with special solution.

Adhesive bonded doubler is quite efficient method in structure strength restoration, but there are several disadvantages. It's not applicable in critical airflow areas, for example, at forward fuselage zone, where there are different air-flow dependent sensors, such as altimeter, pitot and others. This type of repair is forbidden at such zones, because it greatly changes airflow and therefore indications of sensors could be faulty. Other disadvantages are: quite long process time, not weight efficient and increased stiffness in the region.

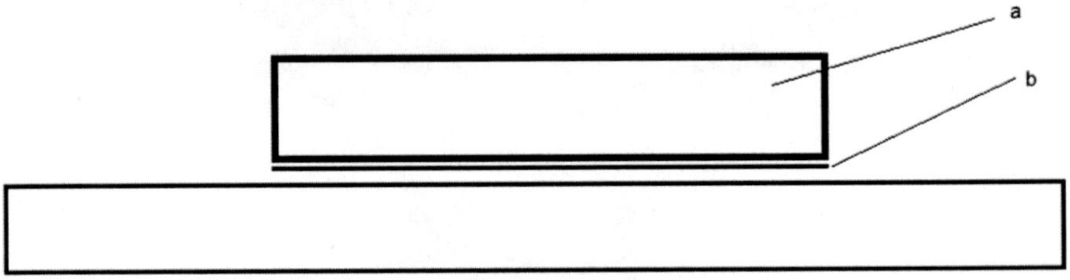

Fig. 2. Scheme of adhesive bonded external repair technique

2.2 Mechanically Fastened External Doubler

This method is similar to adhesive bonded external repair, but instead of binder it uses bolts, rivets or screws to fasten an external composite or metallic plate onto a damaged zone (Fig. 3). Advantages of this technique are: fast in-field repair in case time is matter, what is important in aviation. Disadvantages of this repair method is that it leaves a residual stress at the fastener region which can deteriorate the structural integrity of the PCM especially from fatigue point of view, and also affects air-flow in critical areas as well as has poor weight efficiency and increased stiffness.

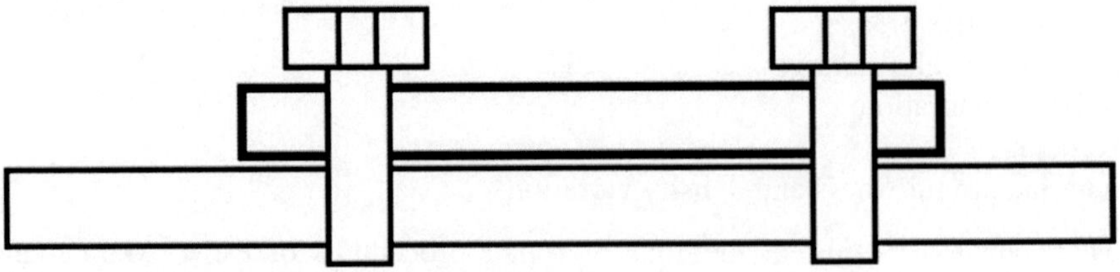

Fig. 3. Scheme of mechanically fastened external doubler repair technique

2.3 Stepped (Overlapping) Repair

This repair method is similar to the scarf method with soft patch but instead of an edge tapering cutout, during stepped repair machine removes ply by ply PCM with a certain constant offset from previous ply. Dry PCM patch (Fig. 4, a) has the same lay-up as parent panel, except plus one same first ply on bottom. Patch, is then bonded with a film of glue (Fig. 4, b) and finished with vacuum infusion method. According to conducted tests and finite element (FE) modelling, this method proved to be reliable, because the weakest link for static tests is regular zone of repaired specimen (Francis Collombet et al. 2019) [2]. Advantages of this method are obvious, it is quite weight efficient, it has smoother outer surface, so might be used in critical air-flow areas, but also that stress distributes more evenly for each ply. On the other hand, due to low angle of stepping, technician have to remove more material, this could be a problem for thick panels. Stepped repair is more complicated, than any others methods before, and requires considerable technician skill.

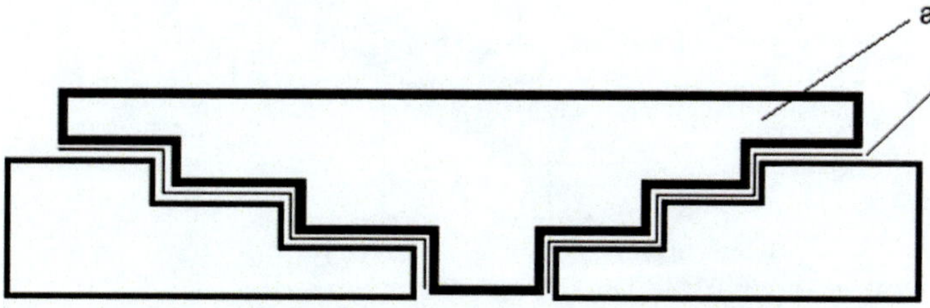

Fig. 4. Scheme of stepped repair technique

3 Finite Element Analysis

For particular case of PCM structure damage there is a problem to validate and certificate repair method, because PCM is more difficult material, in contrast to metallic, so there are still uncertainties of the repair methods. To prove, that in particular zone of aircraft, particular repair method is valid, there exist FE analysis, which can assure airworthiness of that structural element. Here arises a problem of what methodology to use for FE simulating of the structure.

In this paper there will be focus on static tensile simulation of repaired PCM laminate with typical symmetrical balanced lay-up for aircraft from 20 layers and thickness of one ply equal to 0,2 mm. Modelling and simulations are done in Siemens Simcenter 3D.

Main goal of this paper is to perform simulation with solids, but not with 1D, 2D, shell elements, because they are not suitable for stress, strain and displacement analysis of laminate through thickness, as well as analyze stress at boundaries between different elements. Therefore, in this paper will be conducted solid modelling and simulation of three repair methods.

For modelling was taken typical specimen of composite with thickness equal to 25 mm and length selected according to repair method. So, all method specimens have the same section, which area equal to 100 mm^2. At one end applied tensile force is equal to 5 kN, at the other - specimen is fully fixed, specimen is simply supported from upper and lower sides to prevent its bending.

3.1 Finite Element Analysis of Adhesive Bonded External Repair

General view of model is in the Fig. 5. Specimen consist of parent laminate (lower one) and aluminum patch (upper one) from one side only. Laminate is connected to patch by thin film of glue. Patch and glue are meshed with 3D swept elements, laminate is extruded from 2D elements into similar 3D swept and each of solid's mesh is connected to the other accordingly.

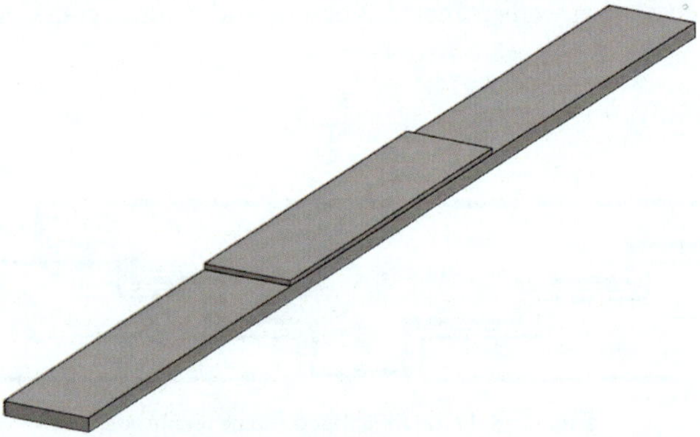

Fig. 5. Adhesive bonded external repair model

After simulation we can see, that glue has great stress concentrations at the begin and the end of patch. Patch has moderate stress concentrations at the edges. Stress concentration in laminate at the edges of patch in upper ply is quite large and reducing to the bottom layers. Stress in laminate under patch reduces to very low number, but in patch stress is large. Element's stress results through the length of specimen are shown in Figs. 6 and 7. All values of stress are in MPa.

In Fig. 6: (a) – laminate stress in XX (longitudinal) of: red – upper ply (closer to the patch), blue – middle ply, green – bottom ply (farther from the patch); (b) – laminate stress in YY (transverse in horizontal plane); (c) – laminate stress in XY; (d) – patch stress in XX; (e) – patch stress in YY; (f) – patch stress in XY.

In Fig. 7: (a) – glue stress in XX; (b) – glue stress in YY; (c) – glue stress in XY;

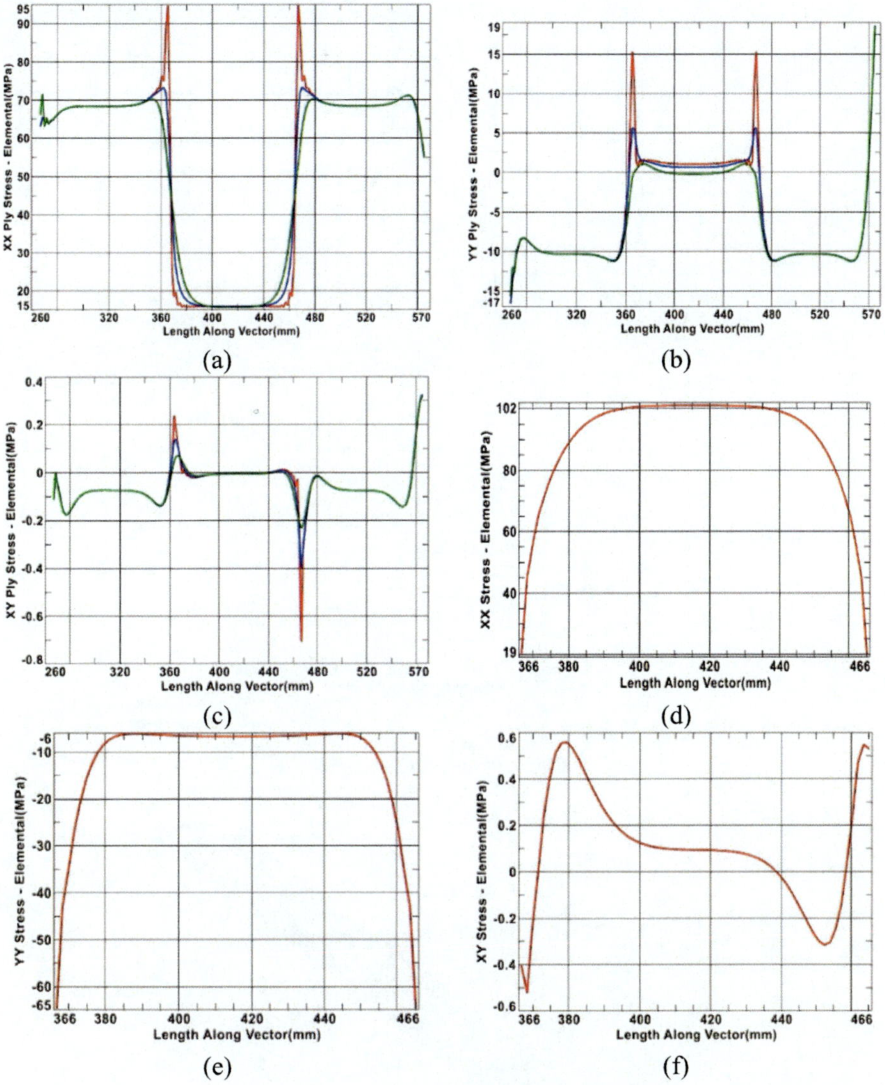

Fig. 6. Laminate and patch stress graphs

3.2 Finite Element Analysis of Mechanically Fastened External Doubler

Model of the fastened external doubler repair method is in Fig. 8. This specimen consists of laminate, external aluminum doubler and five bolt joints. Laminate and doubler have holes for bolts installation, and so they are connected by bolt joints with preload equal to 80 N. Laminate and doubler are meshed the same way as in previous method, bolts and nuts are meshed with 3D tetrahedral elements. Nut is fixed to bolt by rigid glue simulation object, contact simulation objects are set for each according element.

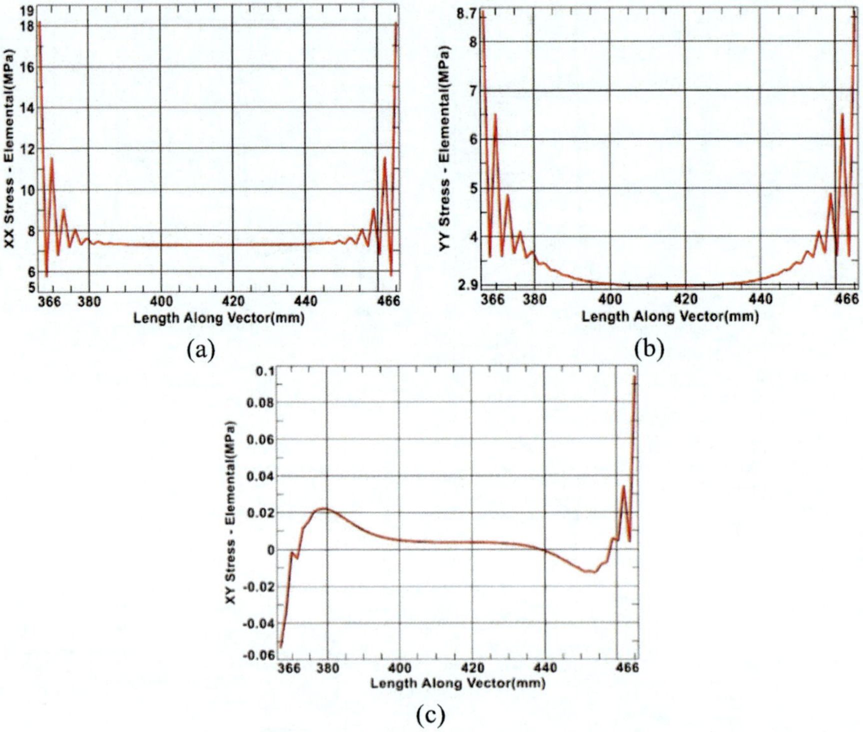

Fig. 7. Glue stress graphs

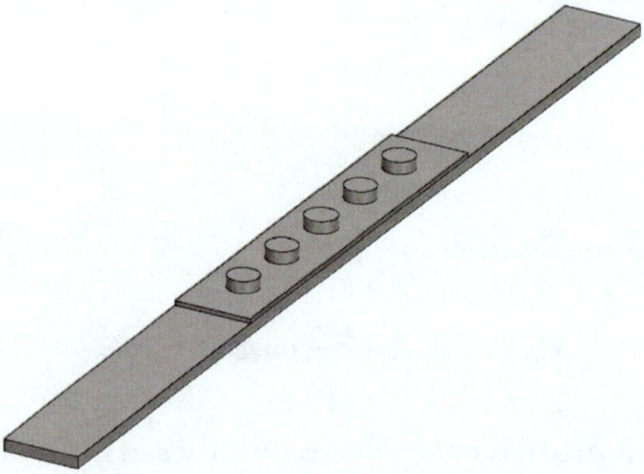

Fig. 8. Mechanically fastened external doubler repair model

Stress distribution between upper and bottom plies under doubler is even. At the transition zone between doubler and laminate stress concentration of laminate is low, at the hole's location stress concentration is very large. Bolted doubler experiences the largest stress concentration also at hole's locations. In Fig. 9 there are stress distribution through length of specimen's elements. Values are taken aside of bolt's holes.

In Fig. 9: (a) – laminate stress in XX of: red – upper ply, blue – middle ply, green – bottom ply; (b) – laminate stress in YY; (c) – laminate stress in XY; (d) – doubler stress in XX; (e) – doubler stress in YY; (f) – doubler stress in XY.

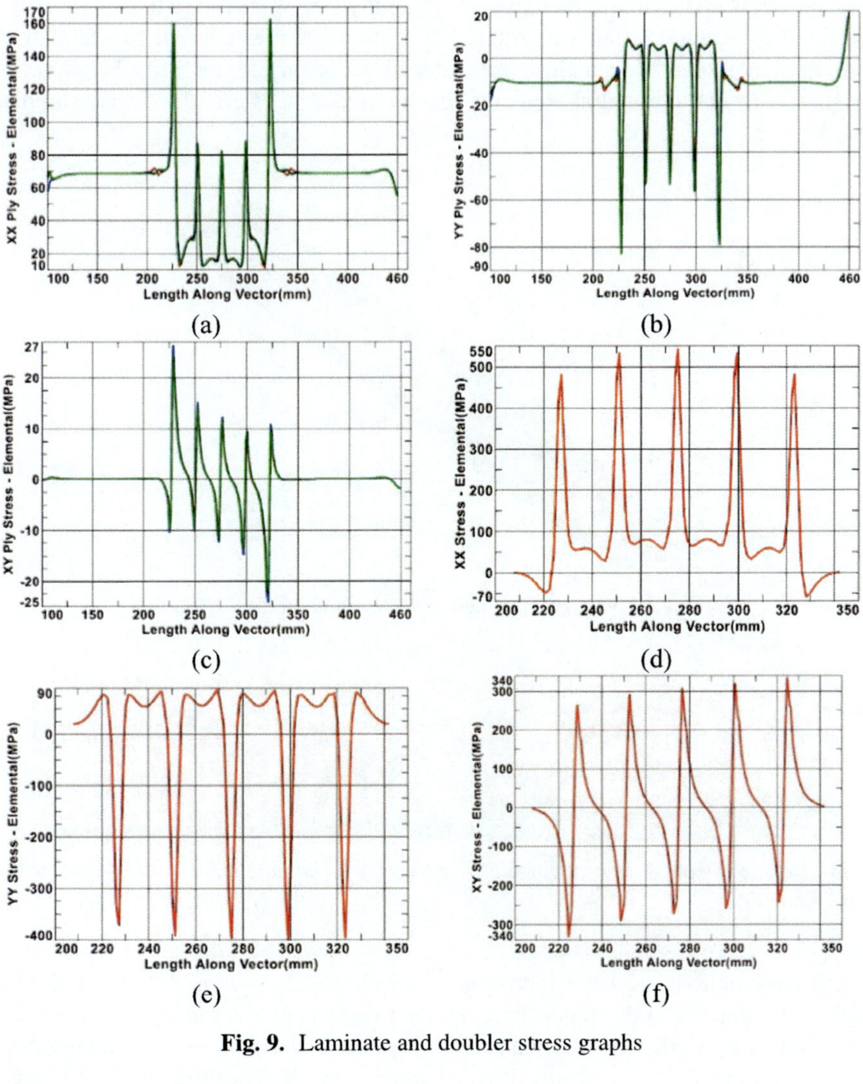

Fig. 9. Laminate and doubler stress graphs

3.3 Finite Element Analysis of Stepped (Overlapping) Repair

This repair method is the most complicated to model. Problem of FE model of such repair technic is in how to model interaction between parent and repair element correctly.

One of the methodologies done in Francis Collombet's paper is to use shell elements for the composite laminate and volume elements for the adhesive: joint rigid beam elements [2]. Another approach, as was mentioned, is to use solid elements to simulate this method.

General view and detailed FE model are in Figs. 10 and 11. Because it is symmetrical problem only one parent laminate was modeled with repair laminate having one extra ply at the top. Between laminates there is solid glue. Ply step was chosen to be 6 mm, which correspond to around 2° tapering angle [3]. Laminates meshed and extruded similar to previous methods, except plies were draped according to each ply location. Glue is meshed with 3D tetrahedral elements. Each contact faces are glued with simulation object feature.

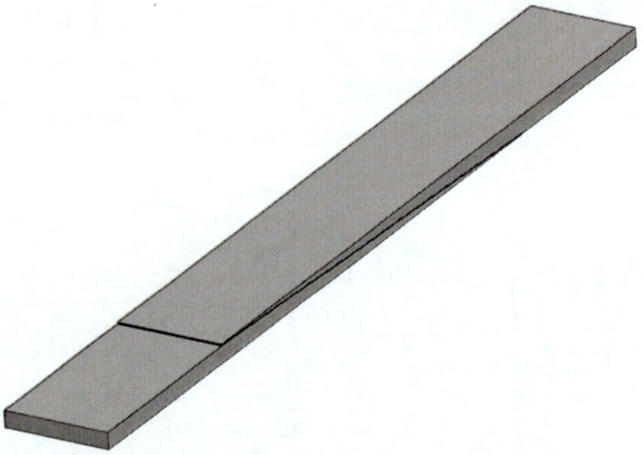

Fig. 10. Stepped repair method model

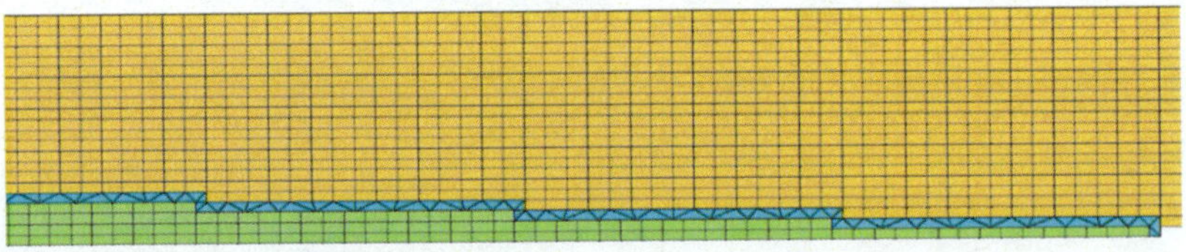

Fig. 11. Detailed stepped repair method FE view (green – parent laminate, cyan – glue, orange – repair laminate)

After calculations we have such results: all plies of both laminates have allowable stress concentrations under repair laminate, but right after at regular zone, stress in the upper ply is a bit higher than regular. Glue experience larger stress concentration, but it is in normal band. Stress results of specimen's elements are shown in Fig. 12, where (a) – parent laminate stress in XX of: red – upper ply, blue – middle ply, green – bottom ply; (b) – parent laminate stress in YY; (c) – parent laminate stress in XY; (d) – glue stress in XX; (e) – glue stress in YY; (f) – glue stress in XY.

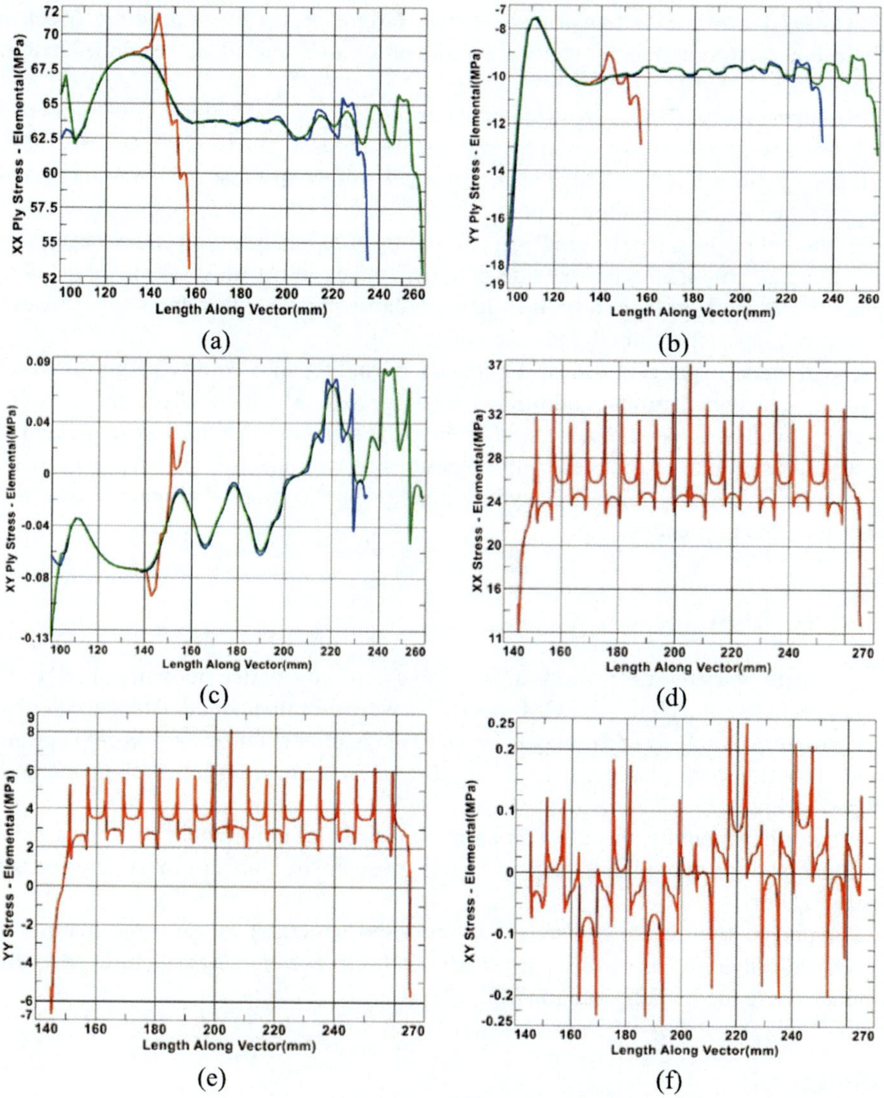

Fig. 12. Laminate and glue stress graphs

4 Analysis and Compare of Results

Adhesive bonded external repair method is quite easy and fast method of repair, depending only on glue hardening time. But huge problem of this method is stress concentration in the upper ply of laminate at the edges of patch, which will certainly affect fatigue properties of this method and possibly cause this ply failure.

Mechanically fastened external doubler method is similar in effectiveness to bonded external repair, stress in general distributed more evenly in laminate, but at holes there are large stress concentrations, that also will reduce fatigue properties of structure. Mechanically fastened external doubler method is more preferable than adhesive bonded external repair.

In both methods can be used different materials of patch and doubler, it will affect how much stress will carry repair patch from laminate, but also change stiffness of structure and possibly cause bending. These methods can be suitable for fast and temporary repairs of PCM in non-critical air-flow zones.

Method of stepped repair is the best comparing stress distribution and concentration. Weak place of this method is outside of repair zone, and it must be also considered, because stress in upper layer is higher, than regular stress. Glue has stress concentrations in zones of plies steps, though they are not critical.

Use of metallic patch in adhesive bonded external repair method changes local stiffens of laminate, which produce additional bending stress. Mechanically fastened external doubler as well as stepped repair methods have the same effect. One way to prevent this effect is to use patch or doubler from both sides of laminate, but for stepped repair in theory solution could be to do steps symmetrically about center of laminate, which might be not possible in practice.

5 Conclusion

Used modelling method is proved to be reliable and informative upon obtained results.

Different repair technics of PCM specimens were described, modelled and calculated in FE program. Analysis and comparison were made after acquiring stress concentration and distribution data of specimens. The most suitable for long-term repair method is stepped technic.

It's worth to mention, that all of repair method produce non-regular stress in structure, that must be considered before repair, or design such structures in most dangerous for damage zones.

For perspective future research it is proposed to model in the same manner and analyze described repair methods for sandwich honeycomb PCM structures, as well as calculate these methods for fatigue life.

References

1. Kasharina, L.A., Makhsidov, V.V., Smirnov, O.I., Ruzakov, I.A.: Identification of defects in polymeric composite materials by fiber bragg grating response (review). Part I (2018). https://doi.org/10.18577/2307-6046-2019-0-2-97-104
2. Collombet, F., et al.: Proof of a composite repair concept for aeronautical structures: a simplified method. Mech. Indust. (2019). https://doi.org/10.1051/meca/2020056
3. Orsatelli, J.-B., Paroissien, E., Lachaud, F., Schwartz, S.: Bonded flush repairs for aerospace composite structures: a review on modelling strategies and application to repairs optimization, reliability and durability. https://doi.org/10.1016/j.compstruct.2022.116338

Process Approach as a Tool of the Knowledge-Intensive Industry Organization Management System

Alena Zastrovskaya, Natalia Khudoliy, and Ekaterina Rybakova(✉)

Moscow, Russia
rybakova3105@yandex.ru

Abstract. This article describes the importance of introducing and subsequent management of the concept of the process approach in organizations in the knowledge-intensive industry. In the modern world, organizations have to work in conditions of continuous change, they need to make important decisions in a timely manner, look for ways to motivate employees, try to take into account the interests of an increasing number of stakeholders and quickly respond to any changes in external and internal factors. That is why knowledge-intensive enterprises more often try to implement a process approach.

Keywords: Process Approach · Process Management · Quality Management System · PDCA

1 Introduction

In modern conditions, for the successful management of an organization, it is necessary to have an effective quality management system (hereinafter referred to as the QMS). To build such a system, it is necessary to be guided by the standards of the ISO 9000 series. In GOST ISO 9001 - 2015, one of the requirements is the construction of a QMS based on a process approach. All businesses use processes to achieve a desired outcome [1].

Process management is the most important element of the management system of a modern organization. Methods of process management are actively developing, new ones appear and existing tools for describing and regulating processes are being improved. Approaches and tools for managing processes based on indicators (metrics) are also actively used [5].

Application of the process approach in the QMS allows:

- understand and consistently comply with the requirements;
- consider processes in terms of adding value to them;
- to achieve effective functioning of processes;
- improve processes based on the evaluation of data and information.

Z. Jing and X. Zhan (Eds.): ICASSE 2023, LNEE 1153, pp. 261–267, 2024.
https://doi.org/10.1007/978-981-97-0550-4_22

The process approach (hereinafter referred to as PA) is of particular importance for the aviation industry, since quality issues are inextricably linked with safety issues, which makes the cost of error in making decisions in the field of quality extremely high [2].

2 Why Process Approach

PA is one of the concepts of organization management, which was formed in the 80s of the XX century. Based on this concept, all activities of the organization imply a system of interrelated processes, which in turn are in a manageable state and measurable.

The main term of this approach is the concept of "process". In accordance with GOST R ISO 9000-2015 (clause 3.4.1), a Process is a set of interrelated and (or) interacting activities that use inputs to obtain the intended result. An important component of the process is the systematic nature of actions, i.e. process actions should be repetitive, not random [1].

Using the process approach will increase:

- efficient use of resources;
- customer satisfaction;
- the effectiveness of the organization;
- trust in the organization;
- efficiency of the management system.

PA was developed and used to create horizontal links in the organization. Such management helps to coordinate the process more efficiently. In the event of a problem, a solution can be found without the involvement of senior management, i.e. at the level of ordinary employees of the departments that participate in this process.

It is worth noting the main advantage of PA, in contrast to functional management, is that PP allows you to concentrate your attention not on a specific unit, but on the result of a process as a whole [3].

PA changes the concept of structure in the organization - the process becomes the main element, the management of the organization is based not on the management of departments, but in the management of processes [4].

When structuring processes, it is necessary to consider the selected process first at the macro level, which is convenient for tracing the relationship with other processes of the system, and then decomposing to the level of the defined task or problem. In Quality Management, it is necessary to decompose each process down to the elementary level. When analyzing and detailing the process, the following criteria should be taken into account:

- the presence of problems in the process, failure, zero performance;
- high risk of errors in the process;
- the need to define the role of participants in the process.

The degree of controllability of the process are indicators such as effectiveness and efficiency [5]. PA enables an organization to manage the organization's Quality Management System processes so that overall performance can be improved. PA involves the systematic definition and management of processes and their interaction in such a way as to achieve the intended results in accordance with the quality policy and the strategic direction of the organization. To improve processes and PA in general, the Plan-Do-Check-Act (hereinafter referred to as PDCA) cycle and risk-based thinking should be applied [6].

Further, we'll consider the sequence of steps for applying the process approach in an organization using the PDCA cycle.

2.1 PLAN

- Define the purpose, scope and policy of the organization.

  The organization must analyze the needs, expectations of interested parties, on their basis to determine the goals, policies of the organization. Monitor and provide feedback to stakeholders to better understand their needs and expectations as they may change.
- Determine the processes of the organization and their sequence.

  The organization shall determine the necessary processes to achieve the planned objectives and their sequence. The description of each process should include inputs and outputs, management and control methods, resources, possible improvements, interaction with other processes.
- Determine process owners.

  The management of the organization should determine those responsible, their powers, responsibilities, subordination, functions, and ensure their competence for the effective functioning of the process and the achievement of intended results.
- Develop the necessary documentation.

  The organization needs to determine the processes that require documentation (in accordance with GOST R ISO 9001-2015), and determine how to document them. The choice of processes depends on the size of the organization, the type of activity, the complexity of the processes [2].
- Determine the interaction of processes, as well as identify the risks of these processes.

  The organization shall determine the interrelated processes, the required process inputs and outputs, and the risks of unintended results. Determine the necessary metrics and controls to convert inputs into planned outputs, and ensure that the organization has accounted for all significant organizational risks.
- Develop a process control system.

  The organization needs to determine how monitoring and measurement will be carried out. Define indicators of efficiency and effectiveness of the processes, and how the results of the audit will be recorded [2].

2.2 DO

- Define resources.

  The organization must determine the necessary resources to carry out the processes (personnel, materials, information, infrastructure, etc.)
- Perform actions.

  The organization must perform actions, measurements, monitoring, use controls to achieve planned results [2].

2.3 CHECK

- Check the results of the process for compliance with the intended goals.

  The organization needs to compare the results obtained and the planned goals in order to establish whether all the requirements are met. For this, monitoring, measurement, and audit data are used [2].

2.4 ACT

- Improve processes by making changes based on the data obtained as a result of process reviews.

  The organization should make changes to the process to improve its effectiveness and efficiency. If problems arise, the organization must identify the problem, analyze the data collected, determine the cause of the problem, correct it, and evaluate the effectiveness of the actions taken. In the absence of problems, the organization should direct its efforts to improve the process, customer satisfaction [2].

3 Classification of Processes

The most common classifications of processes in an organization are:

- **Management processes (development processes).**

  Responsible for managing the organization as a single system. These include goal-setting, planning activities, coordination of activities, coordination of actions of individual elements. This group contains processes that determine the trends and directions of development of the main processes.
- **Basic processes.**

  This group contains the processes by which an organization creates a product or provides services. Such processes are cross-functional within the organization, interacting with both the customer (client) and partners.
- **Supporting processes.**

  This group contains processes that create the infrastructure of the organization [7].

The basic concept of the process architecture is shown in Fig. 1.

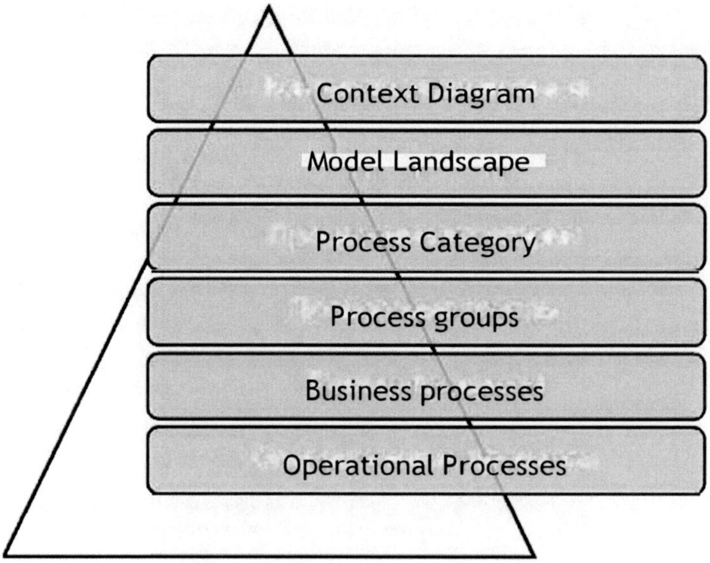

Fig. 1. Architecture of the processes

Context diagram - formed by type of activity (most often according to IDEF0 notation), reflects the general connection with the external environment.

The landscape of the model is a model of the relationship of large categories (management processes/main processes/supporting processes).

Process categories - the composition of process categories (management processes/main processes/supporting processes).

Process groups (optional) - grouping within process categories.

Business processes - regulation of processes with an algorithm.

Operational processes are procedures for employees.

The standard algorithm for such a concept is shown in Fig. 2.

According to this concept, it is possible to build the following hierarchy of internal regulatory documents:

- System-wide (Policies, goals and guidelines).
- Internal corporate (Corporate QMS model, corporate regulatory documents - organization standard, provisions and schemes of standard processes).
- Internal divisional (branches) processes and procedures (Corporate Model, divisional regulatory documents - division standards, division regulations, division instructions, and a block diagram of subprocesses (divisional).

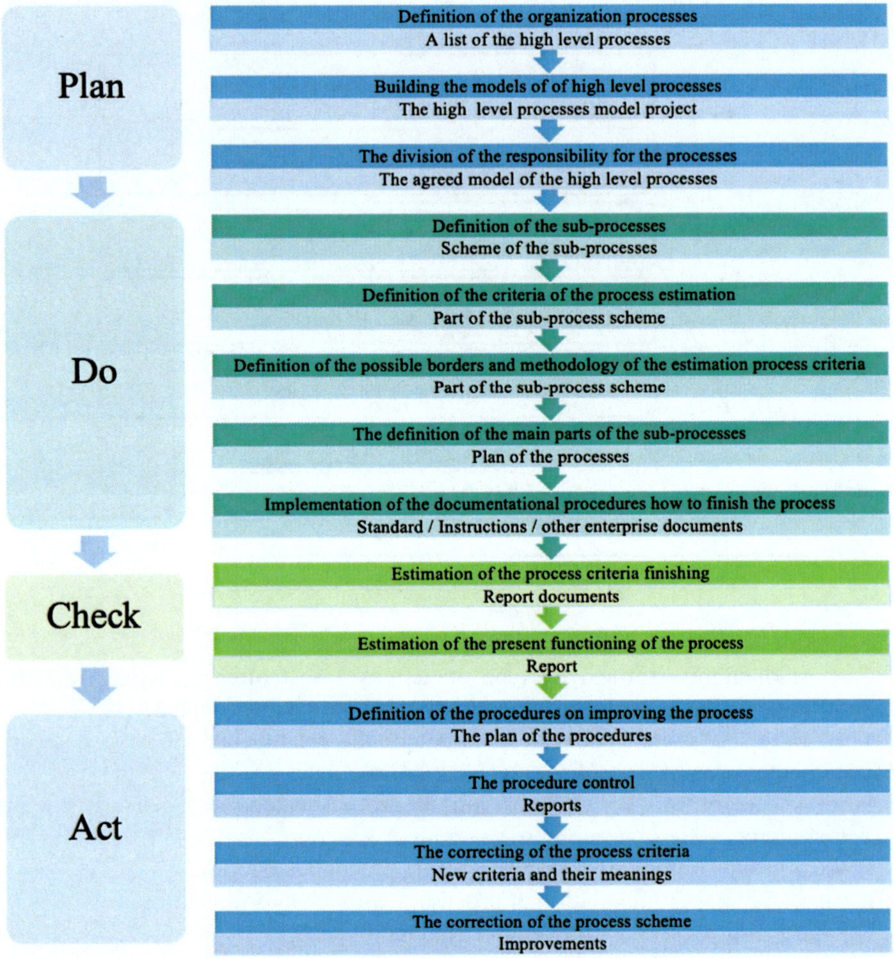

Fig. 2. Architecture of the processes

4 Conclusion

Process management as an approach to managing an organization provides for the need to both improve an individual process in particular and the system (network) of processes as a whole. World practice shows that a management system built on the principles of a process approach is more efficient and effective than a functional system of equal scale.

Acknowledgements. We express our gratitude to our supervisors and professors for valuable advice when planning the study and recommendations on the preparation of the article. The authors also gratefully acknowledge the helpful comments and suggestions of the reviewers, which have improved the presentation and our colleagues from the industry.

References

1. GOST R ISO 9000-2015. Quality Management System = Quality Management Systems. Fundamentals and Vocabulary. – Reissue. Sept 2015 with rev.1. – Instead of GOST R ISO 9000-2011; Introduction 09/28/2015, 49p. Publishing House Standartinform (2015). UDC 658.562.014:006. 354. Group T59
2. GOST R ISO 9001-2015. Quality Management Systems. Requirements = Quality Management Systems - Requirements. – Reissue. Sept. 2015 with rev.1. – Instead of GOST R ISO 9001-2011; Introduction 09/28/15, 32 p. Standartinform Publishing House (2015). UDC 658.562.014:006.354. Group T59
3. Grachev, A., Kastorskaya Yu, L., Voinsova, E.: Practical implementation of the "system of systems" (SoS) approach. Standards and quality. In: Advertising and Information Agency (2020)
4. Lapidus, V., Kastorskaya, L., The system of systems approach and its application to improve efficiency in management and business. Standards and quality. In: Advertising and Information Agency (2021)
5. Repin, V.V.: Business processes. In: Modeling, Implementation, Management. Vladimir Repin. 2nd edn. 512p. Mann, Ivanov and Ferber (2014)
6. Repin, V.V., Eliferov, V.G.: Process approach to management. In: Modeling of Business Processes. Vladimir Repin, Vitaly Eliferov, 544p. Mann, Ivanov and Ferber (2013)
7. Eliferov, V.G., Repin VV Business Processes: Regulation and Management, 319p. Infra-M (2005)
8. Harrington, J.: Perfection of process management. Per. from English. A.L. Raskin; Under Scientific. In: Bragin, V.V. (ed.). RIA "Standards and Quality", 192p. (2007)

Author Index

A
Amoah, Rachael 134

C
Chen, Fang 178
Chen, Yong 13, 32, 52
Chen, Yuhuang 178

D
Dong, Dayong 111
Dong, Qinpeng 111
Dong, Xiaoxu 61

E
Eremenko, V. V. 240

F
Fang, Yang 125

G
Goryukhin, M. O. 240
Gou, Linfeng 104

H
Hu, Lei 178
Hu, Yile 145
Hu, Zhihao 159

J
Jiang, Zongting 104
Jin, Rongjia 211
Junjian, Jiang 231

K
Khudoliy, Natalia 261
Krasnoperov, D. G. 240

L
Lee, Cheng Wei 134
Li, Haomin 145
Li, Huihui 104
Lin, Yuxiang 220
Liu, Hangqi 32
Liu, Xiaohua 159
Luo, Yue 52

M
Ma, Wei 134
Ma, Xianchao 111
Meng, Hua 111
Meng, Yu 178
Mikhailov, A. E. 240
Mikhailova, A. B. 240
Mu, Zhongcheng 200

P
Peng, Ling 75

R
Rybakova, Ekaterina 261

S
Sang, Zhenkun 178
Shi, Zhong 1
Shramko, Konstantin 250
Shubin, Artem 250

T
Tao, Qitian 159

W
Wang, Gechen 75
Wang, Guoqing 52, 89

Z. Jing and X. Zhan (Eds.): ICASSE 2023, LNEE 1153, pp. 269–270, 2024.
https://doi.org/10.1007/978-981-97-0550-4

Wang, Jingshi 89
Wang, Miao 52, 61, 75, 89
Wang, Tianlong 13
Wang, Xin 75
Wang, Yuexing 61
Wu, Dan 1

X
Xi, Chao 125, 192
Xiaoqing, Qiang 231
Xu, Wenjun 75

Y
Yan, Biao 200
Yan, Jiazhi 61
Yan, Xudong 1
Yang, Hui 125
Yang, Zhi 1
Ye, Shuijin 13
Yin, Zhuo 32
Yu, Yin 145

Z
Zastrovskaya, Alena 261
Zhang, Wei 145
Zhang, Xiaojing 220
Zhao, Chunling 111
Zhao, Meng 32
Zhao, Shitong 75
Zhao, Zekai 145
Zhong, Kelin 52
Zhu, Zhengdi 125
Ziheng, Zhao 231
Zou, Zehua 89